Leaving Certificate Geography

Editor
Aidan Culhane

Design & Layout
Paula Byrne, Liz Murphy

Artwork
Niamh Lehane
Michael Phillips
Philip Ryan
Unlimited Design Company, Dublin

Acknowledgements

The author and publishers wish to acknowledge and thank the following:

Gorta; the Indian Embassy, Dublin; Slide File, Dublin for photographs used in this book; Kevin Dwyer for aerial photographs and the Ordnance Survey Office for permission to reproduce maps (based on the Ordnance Survey) by permission of the Government (Permit No. 6508).

LEAVING CERTIFICATE GEOGRAPHY

GUS HEALY

FOLENS

ISBN 0 86 121 5885

© Gus Healy 1997

Produced in Ireland by
Folens Publishers
Broomhill Road,
Tallaght,
Dublin 24.

All rights reserved. No part of this publication may be reproduced or transmitted in any form or by any means electronic, mechanical, photocopying, recording, or otherwise without the prior written permission from the publisher.

This book is sold subject to the conditions that it shall not, by way of trade or otherwise, be lent, re-sold, hired out or otherwise circulated without the Publishers' prior consent in any form or cover other than that in which it is published and without similar conditions including this condition being imposed on the subsequent purchaser.

The publisher reserves the right to change, without notice, at any time the specification of this product, whether by change of materials, colours, bindings, format, text revision or any other characteristic.

While considerable effort has been made to locate all holders of copyright material used in this text, we have failed to contact some of these. Should they wish to contact Folens publishers, we will be glad to come to some arrangement.

While every effort has been made to ensure that the information in this book is accurate, the publisher cannot accept responsibility for any errors.

Contents

CHAPTER 1	*The Planet* *1*	
	Continental Drift – Plate Tectonics	3
	Earthquakes	6
	Volcanoes and Vulcanicity	7
	Types of Rocks – A Classification	14
	Questions	19
CHAPTER 2	*Weathering and Mass Movement* *22*	
	Types of Weathering	23
	Mass Movement	26
	Questions	28
CHAPTER 3	*Underground Water and Limestone Regions* *29*	
	Source of Underground Water	29
	Limestone or Karst Regions	32
	Questions	37
CHAPTER 4	*Rivers and River Basins* *39*	
	The Origin of Rivers	39
	The Development of a River Valley	43
	Human Societies and River Valleys	54
	Questions	56
CHAPTER 5	*Glaciation and Landforms* *58*	
	Types of Glaciers	59
	Features of Glacial Erosion	60
	Glacial Deposition	63
	Causes of Glaciation	71
	Questions	73
CHAPTER 6	*Coastal Landforms – Marine Erosion and Deposition* *75*	
	Coastal Erosion	76
	Features of Marine Deposition	81
	Classification of Coastlines	84
	Questions	89

Leaving Certificate Geography

Contents

CHAPTER 7 *Wind and Desert Landforms* *91*

    Desert Landscapes 92
    Wind and the Irish landscape 99
    Desertification 99
    Questions 100

CHAPTER 8 *Weather and Synoptic Charts* *101*

    Synoptic Charts 105
    Questions 110

CHAPTER 9 *Climate and Natural Vegetation* *112*

    Hot Climates 114
    Temperate Climates 121
    Cold Climates 126
    Questions 129

CHAPTER 10 *Soils* *130*

    Classification of Soils 130
    Questions 139

MAPS

    The World – Physical 140
    The World – Political 142
    Questions 144

CHAPTER 11 *Population* *145*

    Birth Rates, Death Rates and Natural Increase 147
    Changes in Space – Migration 149
    Population Distribution and Density 153
    Population Structure 155
    Population – The Future 157
    Questions 161

CHAPTER 12 *Urbanisation* *164*

    Growth of Towns and Cities 165
    Models Of Urban Structures 171
    Functional Zones Within a City 174
    Problems of Urban Areas 176
    Cities in the Third World 177
    Questions 179

CHAPTER 13	*Agriculture* *181*	
	Factors Controlling Agricultural Production.........	184
	Types of Farming	195
	Agriculture – The Future.....................................	200
	Questions...	202

CHAPTER 14	*Fishing* *203*	
	The Importance of Fishing..................................	203
	Fishing Methods..	204
	Fishing – The Future	207
	A Solution – Aquaculture?..................................	208
	Questions...	215

CHAPTER 15	*Forestry*... *217*	
	Distribution of Forests...	218
	Case Study 1 – Sweden.......................................	223
	Case Study 2 – Ireland	226
	Questions...	228

CHAPTER 16	*Minerals and Energy*........................ *230*	
	Mining Methods and Environmental Effects	231
	Energy	234
	Questions...	244

CHAPTER 17	*Manufacturing Industry* *247*	
	Factors Influencing Manufacturing Industry.........	249
	Environmental Impact of Industry	256
	The World's Major Industries	258
	Questions...	265

CHAPTER 18	*Tourism*... *266*	
	The Growth of Tourism.......................................	267
	Tourist Attractions..	269
	Effects of Mass Tourism	274
	Tourism – The Future..	276
	Case Studies	276
	Questions...	287

Leaving Certificate Geography

Contents

CHAPTER 19 *Transport and Trade* **289**

Factors Influencing the Development of Transport 290
Types of Transport 293
Trade 303
Questions 305

CHAPTER 20 *Worlds Apart – The First and Third Worlds* **307**

Developing Economies – The Search for a Definition 308
The Third World – The Origins of Poverty 309
Causes of The Third World – Exploding The Myths 317
Cities of The Third World 320
Developing Economies – A Solution 322
Questions 328

CHAPTER 21 *Striving for a Balance — Problem Regions* **332**

Types of Problem Regions 333
Case Studies – Problem Regions in France 341
Questions 345

CHAPTER 22 *Ordnance Survey Maps and Aerial Photographs* **348**

Map Reading 348
Map Interpretation 357
Uses of Aerial Photographs 387

CHAPTER 23 *Areas of recent political conflict* **391**

Afghanistan 392
Sudan 392
Rwanda – Burundi – Zaire 392
The Kurds 393
Other Areas of Conflict 393

INDEX **396**

Introduction

Leaving Certificate Geography comprehensively covers Section A and Section B of the syllabus for both Higher and Ordinary Levels, as well as Question 18 on the Ordinary Level Paper. In short it answers three of the four questions necessary for Higher Level and five of the five for the Ordinary Level.

Any meaningful study of the present Leaving Certificate syllabus necessitates a logical sequence. There is little point in studying topics such as Population (social geography) or Agriculture and Tourism (economic geography) without understanding the physical environments in which they operate. Hence the text begins with Physical Geography – a study of the physical forces which have shaped and continue to shape our physical surroundings. Once these forces are understood, variations in socio-economic activities, i.e. social and economic geography become easier to explain.

Chapter 22 of the text deals with Ordnance Survey maps and aerial photographs. In spite of the fact that OS maps form Question 1 on both the Higher Level and Ordinary Level papers, any fruitful study of map reading can only be undertaken when physical, social and economic geography have been studied. This becomes obvious when we realise that any OS map is essentially a detailed study of a small portion of the earth's crust. As such it has a physical dimension, i.e. relief and drainage. Superimposed on and interacting with this environment are people – Communication, Settlement and Land Use – hence the necessity for an understanding of social and economic geography. In Chapter 22, particular attention has been paid to answering techniques as existing textbooks fail to adequately address this question, given that it is a compulsory question in the Leaving Certificate examination.

Chapter 23 deals with political 'hotspots' – areas of conflict in the world. While this chapter does not pretend to be comprehensive, it is hoped that the brief notes might form the basis for group or individual project work.

Leaving Certificate Geography also contains a glossary of physical geographical terms and two world 'atlas' maps, one physical and one political.

Glossary of Physical Geographical Terms

Acid Rain: Weak acid that falls as rain caused by a mixture of oxides of sulphur (SO2) and nitrogen (NO) that are released into the atmosphere when coal, oil and natural gas are burned, and mixed with water vapour in the atmosphere. Acid rain, which adversely affects plants (especially trees) and marine life is a major form of environmental pollution.

Anticline: An upfold.

Aquaculture: The 'farming' of fish and shellfish for food products in rivers, lakes and coastal waters.

Aquifer: An underground reservoir of water held within a porous, water-bearing layer of rock with impervious rock below.

Atmosphere: A thin blanket of air that surrounds the earth, without which life as we know it would cease to exist.

Aureole: A zone or halo of contact metamorphism found in the rock surrounding an igneous intrusion.

Bajada (Bahada): An apron of sediment along a mountain front created by the coalescence of alluvial fans.

Barchan Dune: A solitary crescent-shaped sand dune with its tail pointing downwind.

Base Level: The level below which a stream cannot erode.

Basin: A circular downfolded structure.

Batholith: A large mass of igneous rock that formed when magma crystallised slowly at depth, and was subsequently exposed by erosion.

Baymouth Bar: A sandbar that completely crosses a bay, sealing it off from the main body of water.

Bedding Plane: A nearly flat surface separating two beds of sedimentary rock. Each bedding plane marks the end of one deposit and the beginning of another.

Belt of Soil Moisture: A zone in which water is held as a film on the surface of soil particles and may be used by plants or withdrawn by evaporation. The uppermost subdivision of the zone of aeration.

Block Mountain, also Horst: An elongated, uplifted block of crust bounded by faults.

Breccia: A sedimentary rock composed of angular fragments that were lithified.

Caldera: A large depression typically caused by collapse or ejection of the summit area of a volcano.

Capacity: The total amount of sediment a stream is able to transport.

Carbonation: Process of chemical weathering whereby carbon dioxide dissolved in rainwater forms a weak acid.

Cirque, also Cwm, Coom: An amphitheatre-shaped basin at the head of a glaciated valley produced by frost wedging and plucking.

Clint: Ridges in a limestone pavement.

Column: A feature found in caves that is formed when a stalactite and stalagmite join.

Competence: The ability of a river to carry its load.

Cone of Depression: A conical dip in the water table immediately surrounding a well.

Conservative Plate Boundaries: Where two plates move parallel to each other and the crust is not created or destroyed.

Contact Metamorphism: Changes in rock caused by the heat from a nearby magma body.

Continental Drift: Alfred Wegener's theory that all present continents once existed as a single supercontinent (Pangaea) that began to break into smaller continents which 'drifted' into their present positions.

Continental Glacier: A massive accumulation of ice that covers extensive land areas and whose flow is not usually controlled by the underlying topography.

Convergent Boundary: A boundary in which two plates move together, causing one of the slabs of lithosphere to be consumed into the mantle as it descends beneath an overriding plate.

Glossary of Physical Geographical Terms

Coom: See cirque.

Corrasion: The erosion of the banks and bed of a river by its load.

Crevasse: A deep crack in the brittle surface of a glacier.

Crust: Outermost, thinnest layer of the earth on which we live.

Deflation: The lifting and removal of loose material by wind.

Delta: An accumulation of sediment formed where a stream enters a lake or ocean.

Dendritic Pattern: A drainage system that resembles the branches of a tree.

Denudation: The combined effect of weathering and erosion.

Desert Pavement: A layer of coarse pebbles and gravel created when wind removed the finer material.

Destructive Plate Boundaries: Where two plates converge and the crust is destroyed as one plate is pulled under the other.

Dip: The angle at which a rock layer is inclined from the horizontal. The direction of dip is at a right angle to the strike.

Discharge: The quantity of water in a stream that passes a given point in a period of time.

Dissolved Load: That portion of a stream's load carried in solution.

Distributary: A section of a stream that leaves the main flow.

Diurnal Range: The range of temperature experienced daily.

Divergent Boundary: A boundary in which two plates move apart, resulting in upwelling of material from the mantle to create new sea floor.

Divide: An imaginary line that separates the drainage of two streams – often found along a ridge.

Dormant Volcano: Volcano which has erupted in the recent past but is not erupting at present.

Drainage Basin: The land area that contributes water to a stream.

Drift: The general term for any glacial deposit.

Drumlin: A streamlined asymmetrical hill composed of glacial till. The steep side of the hill faces the direction from which the ice advanced.

Dry Climate: A climate in which yearly precipitation is less than the potential loss of water by evaporation.

Dune: A hill or ridge of wind-deposited sand.

Dyke: Vertical walls of rock cutting across the bedding planes of the surrounding rock: formed when magma rises through vertical cracks or fissures.

Dynamic Metamorphism: The formation of metamorphic rock by pressure alone.

Earthflow: The downslope movement of water-saturated, clay-rich sediment.

Earthquake: Vibration of the earth produced by the rapid release of energy from within.

Emergent Coast: A coast where land formerly below sea level has been exposed either by crustal uplift or a drop in sea level or both.

End Moraine: A ridge of till marking a former position of the front of a glacier.

Entrenched Meander: A meander cut into bedrock when uplifting a meandering stream.

Epicentre: The location on the earth's surface that lies directly above the focus of an earthquake.

Erosion: The breakdown and transportation of material by a mobile agent, such as water, wind or ice.

Erratic: A large boulder deposited by a glacier.

Esker: Long ridge composed largely of sand and gravel deposited by a stream flowing in a tunnel beneath a glacier near its terminus.

Estuary: A funnel-shaped inlet of the sea that formed when a rise in sea level or subsidence of land caused the mouth of a river to be flooded.

Glossary of Physical Geographical Terms

Eustatic Movement: Refers to a rise or fall of sea level.

Eutrophication: Process in which lake waters become enriched with mineral nutrients. Much of this form of pollution occurs when chemical fertilisers, used in farming, find their way into water bodies. It causes an excessive growth of algae which reduces oxygen levels in water and this kills fish and plant life.

Exfoliation: Weathering of rock by repeated expansion and contraction of outer layers by the heat of the sun.

Extinct Volcano: Volcano which has not erupted in historic time.

Fault-Block Mountain: A mountain formed by the displacement of rock along a fault.

Fiord, also Fjord: A steep-sided inlet of the sea formed when a glacial trough was partially submerged.

Firn: Granular recrystallised snow, a transitional stage between snow and glacial ice.

Fissure Eruption: An eruption in which lava is extruded from narrow fractures or cracks in the crust.

Flood Plain: The flat, low-lying portion of a stream valley subject to periodic inundation

Focus (Earthquake): The zone within the earth where rock displacement produces an earthquake.

Foreset Bed: An inclined bed deposited along the front of a delta.

Fossil: The remains or traces of organisms preserved from the geologic past.

Frost Wedging, also Frost Blasting: The mechanical breakup of rock caused by the expansion of freezing water in cracks and crevices.

Fumarole: A vent in a volcanic area from which fumes or gases escape.

Geomorphology: The study of the topography or landscape of an area.

Geosyncline: A huge basin in which thousands of metres of sediment have accumulated.

Geyser: A fountain of hot water ejected periodically from the ground.

Graben: See rift valley.

Graded Stream: A stream that has the correct channel characteristics to maintain exactly the velocity required to transport the material supplied to it.

Gradient: The slope of a stream, generally measured in feet per mile.

Granite: Large-crystalled igneous rock formed by the slow cooling of magma.

Greenhouse Effect: Carbon dioxide and water vapour in a planet's atmosphere absorb and re-radiate infrared wavelengths - effectively trapping solar energy and raising the temperature.

Grikes: Furrows in a limestone pavement.

Ground Moraine: An undulating layer of till deposited as the ice front retreats.

Groundwater: Water in the zone of saturation.

Groyne: A short wall built at a right angle to the seashore to trap moving sand.

Hanging Valley: A tributary valley that enters a glacial trough at a considerable height above the floor of the trough.

Headward Erosion: The extension upslope of the head of a valley due to erosion.

Horn: A pyramid-like peak formed by glacial action in three or more cirques surrounding a mountain summit, e.g. the Matterhorn.

Horst: See block mountain

Hot Spot: Volcanic activity occurring in the middle of tectonic plates.

Hum: Residual mass of limestone in an old age limestone region.

Hydration: Process of chemical weathering whereby rocks absorb water and lose strength, hastening their breakdown.

Hydrocycle: The unending circulation of the earth's water supply. The cycle is powered by energy from the sun and is characterized by continuous exchanges of water among the oceans, the atmosphere, and the continents.

Glossary of Physical Geographical Terms

Hydrolysis: A chemical weathering process in which minerals are altered by chemically reacting with water and acids.

Hydrosphere: The body of water which rests on the surface of the earth and which covers 71 per cent of the total surface area.

Igneous Rock: A rock formed by the crystallisation of molten magma.

Infiltration Capacity: The maximum rate at which soil can absorb water.

Inner Core: The solid innermost layer of the earth, approximately 1216 kilometres (754 miles) in radius.

Inselberg: An isolated mountain remnant characteristic of the late stage of erosion in a mountainous arid region, e.g. Ayers Rock in Australia.

Intrusive Rock: Igneous rock that formed below the earth's surface.

Isostatic Movement: Rise or fall of land level.

Joint: A fracture in rock along which there has been no movement.

Kame: A steep-sided hill composed of sand and gravel originating when sediment collected in openings in stagnant glacial ice.

Karst: A limestone topography – as in the Burren.

Kettle-holes: Depressions created when blocks of ice become lodged in glacial deposits and subsequently melt.

Knickpoint: The point below which rejuvenation is effective.

Laccolith: A massive, igneous body formed when magma forces the overlying rock into a dome.

Landslide: Rapid downward movement of rock debris, often triggered by undercutting at the base of slope by rivers, sea or human activity.

Lateral Moraine: A ridge of till along the sides of an alpine glacier composed primarily of debris that fell to the glacier from the valley walls.

Lava Dome: A bulbous mass associated with an old-age volcano, produced when thick lava is slowly squeezed from the vent. Lava domes may act as plugs to deflect subsequent gaseous eruptions.

Lava: Magma that reaches the earth's surface.

Limestone: Sedimentary rock composed mainly of shells and skeletons of minute sea creatures. It is the most common rock in Ireland.

Lithosphere: The solid rock on which the oceans rest and on which we live.

Loess, also Limon: Deposits of windblown silt, generally buff colored, forming a highly fertile and productive soil, e.g. the Paris Basin.

Longitudinal Dunes, also Seif dunes: Long ridges of sand, oriented parallel to the prevailing wind.

Longhore Drift: The transport of sediment in a zigzag pattern along a beach caused by the uprush of water from obliquely breaking waves.

Magma: A body of molten rock found at depth, including any dissolved gases and crystals.

Magnitude (Earthquake): The total amount of energy released during an earthquake.

Mantle: The 2,885 kilometre (1,789-mile) thick layer of the earth located below the crust.

Marble: Metamorhic rock formed by regional metamorphism of limestone.

Mass Wasting: The downslope movement of rock, regolith, and soil under the direct influence of gravity.

Meander: A looplike bend in the middle or lower course of a stream.

Mechanical Weathering: The physical disintegration of rock, resulting in smaller fragments.

Medial Moraine: A ridge of till formed when lateral moraines from two coalescing alpine glaciers join.

Mercali Intensity Scale: A 12-point scale originally developed to evaluate earthquake intensity based upon the amount of damage to various types of structures.

Glossary of Physical Geographical Terms

Metamorphic Rock, also Thermal Metamorphism, Regional Metamorphism and Dynamic Metamorphism: Rock formed by the alteration of pre-existing rock, e.g. marble which is formed from limestone.

Mud Flow, also Earth Flow: Downslope movement of earth at speeds of up to 15 kilometres per hour.

Natural Levees: The elevated landforms composed of alluvium that parallel some streams and act to confine their waters, except during a flood.

North Atlantic Drift: A very broad, slow-moving current that moves from south-west to north-east across the North Atlantic. Since it originates in sub-tropical areas, this current washes the shores of Western Europe as a relatively warm ocean current. It therefore keeps the coastline ice-free in winter.

Outer Core: A layer beneath the mantle which is approximately 2,270 kilometres thick which has the properties of a liquid.

Outwash Plain: A relatively flat, gently sloping plain consisting of materials deposited by meltwater streams in front of the margin of an ice sheet, e.g the Curragh in Kildare.

Ox-bow Lake: A curved lake produced when a stream cuts off a meander.

Oxidation: A process whereby metals in rock, especially iron, mix with the air and weaken the structure of the rock.

Pangaea: The supercontinent which 200 million years ago began to break apart and form the present continents and landmasses.

Parasitic Cone: A volcanic cone which forms on the flank of a larger volcano.

Parent Material: The material upon which a soil develops.

Pater Noster Lakes: A chain of small lakes in a glacial trough that occupy basins created by glacial erosion.

Pediment: A sloping bedrock surface fringing a mountain base in an arid region – formed when erosion causes the mountain front to retreat.

Peneplanation: The process of making flat, the ultimate outcome of denudation is a flat, or almost flat landscape.

Perched Water Table: A localised zone of saturation above the main water table created by an impermeable layer (aquiclude).

Permafrost: Layer of permanently frozen subsoil. Usually found in the subarctic and arctic regions.

Permeable Rocks: Rocks which allow water to pass through.

Plate Tectonics: The theory which proposes that the earth's outer shell consists of individual plates which interact in various ways and thereby produce earthquakes, volcanoes, mountains, and the crust itself.

Plate: A rigid sections of the lithosphere that moves as a unit over the material of the asthenosphere.

Playa: The flat central area of an undrained desert basin.

Plucking (Quarrying): The process by which pieces of bedrock are lifted out of place by a glacier.

Poljes: Large depression, often up to several kilometres in diameter, found in limestone areas formed either from the joining together of uvulas or the downfaulting of limestone.

Porosity: The volume of open spaces in rock or soil.

Porous Rocks: Rocks which contain pores which allow water to flow through.

Pot Hole: A depression formed on the bed of a stream by the abrasive action of the water's sediment load.

Radial Drainage: A system of streams running in all directions away from a central elevated structure: such as a volcano.

Radio Carbon Dating: Method of dating material up to 70,000 years old by measuring the proportion of carbon-14 (which diminishes over time) to carbon-12 (which remains constant).

Rainshadow: A dry area on the leeward side of a mountain range.

Recessional Moraine: An end moraine formed as the ice front stagnated during glacial retreat.

Glossary of Physical Geographical Terms

Regional Metamorphism: Where the formation of metamorphic rock was caused by great pressure (and sometimes heat) over a great area, e.g. in areas surrounding plate boundaries.

Rejuvenation: A term applied to rivers that regain their erosive power through a change in base level.

Richter Scale: A scale of earthquake magnitude based on the motion of a seismograph.

Rift Valley, also Graben: A valley formed by the downward displacement of a fault-bounded block.

Roche Moutonnée: An asymmetrical knob of bedrock formed when glacial abrasion smoothes the gentle slope facing the advancing ice sheet and plucking steepens the opposite side as the ice overrides the rock.

Rock Flour: Ground-up rock produced by the grinding effect of a glacier.

Rock: A consolidated mixture of minerals.

Rockslide: The rapid slide of a mass of rock downslope along planes of weakness.

Rotational Slumping: Type of landslide where material not only slides downwards but moves rotationally at the same time.

Runoff: Water that flows over the land rather than infiltrating into the ground.

Salinity: The proportion of dissolved salts to pure water, usually expressed in parts per thousand.

Saltation: Transportation of sediment through a series of leaps or bounces.

Sand Spit: An elongated ridge of sand that projects from the land into the mouth of an adjacent bay.

Scree, see Talus.

Sea Arch: An arch formed by wave erosion when caves on opposite sides of a headland unite.

Sea Floor Spreading: The hypothesis first proposed in the 1960s by Harry Hess which suggested that new oceanic crust is produced at the crests of mid-ocean ridges.

Sedimentary Rock: Rock formed from the weathered products of pre-existing rocks that have been transported, deposited, and lithified.

Seismic Sea Wave: A rapid moving ocean wave generated by earthquake activity which is capable of inflicting heavy damage in coastal regions.

Seismograph: An instrument that records earthquake waves.

Seismology: The study of earthquakes and seismic waves.

Shield Volcano: A broad, gently sloping volcano built from fluid basaltic lavas.

Sill: Horizontal sheet of rock formed when magma forces its way into bedding planes of the surrounding rock.

Soil Creep: The slow downhill movement of soil and regolith, often less than one centimetre a year.

Soil Profile: A vertical section through a soil showing its succession of horizons.

Solifluction: Slow (5-20 centimetres per year), downslope flow of water-saturated materials common to permafrost areas.

Solution: Method of river transportation where solids are dissolved in water.

Spring: A flow of groundwater that emerges naturally at the ground surface.

Stalactite: Long spear-like structure that hangs from the ceiling of a cavern.

Stalagmite: Column-like form that grows upward from the floor of a cavern.

Strata: Parallel layers of sedimentary rock.

Stratified Drift: Sediments deposited by glacial meltwater.

Stream Piracy: The diversion of the drainage of one stream resulting from the headward erosion of another stream.

Stream: A general term to denote the flow of water within any natural channel. Thus, a small creek and a large river are both streams.

Striae (Glacial): Scratches or grooves in a bedrock surface caused by the scraping of rocks embedded in a glacier.

Leaving Certificate Geography

Glossary of Physical Geographical Terms

Submerged Coast: A coast whose form is largely the result of the partial drowning of a former land surface either due to a rise of sea level or subsidence of the crust: or both.

Subsoil: A term applied to the B horizon of a soil profile.

Surface Soil: The uppermost layer in a soil profile - the A horizon.

Suspended Load: The fine sediment carried within the body of flowing water or air.

Suspension: Transportation of material, on or near the top of river by hydraulic action or turbulence.

Swallow Hole, also Sink Hole, Slugga or Doline: A depression produced by hydraulic action or turbulence.

Syncline: A linear downfold in sedimentary strata, the opposite of anticline.

Talus, also Scree: An accumulation of rock debris at the base of a cliff.

Tarn: A small lake in a cirque.

Tectonics: The study of the large-scale processes that collectively deform the earth's crust.

Terminal Moraine: The end moraine marking the farthest advance of a glacier.

Traction: Transportation of material by rolling and dragging.

Transpiration: The release of water vapour to the atmosphere by plants.

Trellis Drainage: A system of streams in which nearly parallel tributaries occupy valleys cut in folded strata.

Truncated Spurs: Triangular-shaped cliffs produced when spurs of land that extend into a valley are removed by the great erosional force of an alpine glacier.

Tsunami: The Japanese word for a seismic sea wave.

Ultimate Base Level: Sea level - the lowest level to which stream erosion could lower the land.

Uvula: Large depression found in limestone regions formed when two or more swallow holes join together.

V-shaped Valley: Steep sided valley formed by the downcutting action of a river.

Ventifact: A cobble or pebble polished and shaped by the sandblasting effect of wind.

Vulcanicity: All processes by which solid, liquid or gaseous materials are forced into (intrusive) or on to (extrusive) the earth's crust.

Wadi: A desert stream course that is typically dry except for brief periods immediately following rainfall.

Water Table: The upper level of the saturated zone of groundwater.

Wave-Cut Cliff: A seaward-facing cliff along a steep shoreline formed by wave erosion at its base and mass wasting.

Wave-Cut Platform: A bench or shelf along a shore at sea level, cut by wave erosion.

Weathering: The disintegration and decomposition of rock in their original position.

Well: An opening bored into the zone of saturation.

Wind Gap: An abandoned water gap. These gorges typically result from stream piracy.

Xerophyte: A plant highly tolerant of drought.

Yazoo Tributary: A tributary that flows parallel to the main stream because a natural levee is present, named after the Yazoo river which flows parallel to the Mississippi.

Zone of Accumulation: The part of a glacier characterised by snow accumulation and ice formation. The outer limit of this zone is the snowline.

Zone of Saturation: Zone where all open spaces in sediment and rock are completely filled with water.

CHAPTER 1 — *The Planet*

Fig. 1.1 ›
Planet Earth from Space

This is the view of a planet which is vibrant, and dynamic in spite of its age – now estimated to be over 4,500 million years. At present it is home to over five billion human beings, a population which is expected to double within the next 50 years. While this population explosion is undoubtedly the most serious problem facing the human race and essentially belongs to the study of socio-economic geography, the problem is inevitably linked to the study of physical geography. The ability of any environment to support its population depends on a delicate balance between that population and the exploitation of the resources of that region. Over-exploitation of resources inevitably leads to a diminished quality of life and ultimately, in a global sense, to annihilation (total destruction). It is essential that we understand the source of those resources – our physical environment, i.e. physical geography.

Planet Earth is generally divided into three parts:

1. *The atmosphere* – a thin blanket of air which surrounds the earth and without which life as we know it, would cease to exist.

2. *The hydrosphere* – the body of water which rests on the surface and which covers approximately 71 per cent of the earth's total surface area.
3. *The lithosphere* – the solid rock on which the oceans rest and on which we live.

Changes in the atmosphere like wind and rain are everyday experiences and will be dealt with in a later chapter. Movements in the hydrosphere, such as waves, tides and currents will be examined in Chapter 6. Changes in the lithosphere are generally less obvious but can sometimes be more dramatic. While volcanoes and earthquakes may be the most dramatic and violent manifestations of the earth's ever-changing nature, the fact is that the landscape around us is constantly changing. No landform is exactly the same today as it was yesterday. Less than 200 million years ago, many of the great mountain ranges you see in your atlas did not exist. Even the continents occupied a different location on the surface of the earth.

Geomorphology is the study of the topography or landscape of an area. It should not be confused with geology which is the study of the origin, composition and nature of rocks. The processes or series of events which alter or change the earth's surface over time and which are responsible for our present landscape can be divided into two categories:

1. *Internal Forces*
   These forces operate inside or beneath the earth's crust and are responsible for earthquakes, volcanoes, folding and faulting.
2. *External Forces*
   These forces operate outside (i.e. on or above) the earth's crust. Their main effect is to wear away the landforms produced by internal forces. In so doing they often produce spectacular features such as cliffs and waterfalls. Their ultimate aim however is to reduce the landscape to sea level. This is why these forces are sometimes referred to as forces of *denudation* – as they are trying to 'denude' or make the landscape bare of all land forms. It can then be said that the object of denudation is peneplanation (almost flat).

> Denudation is the combined effect of weathering and erosion and will be studied in later chapters.

INTERNAL FORCES

As can be seen from the diagram, the earth is made up of three distinct, if uneven layers.

The Core

With a radius of 3,486 kilometres, this dense interior of the earth is larger than the planet Mars. The core is actually comprised of two parts; the inner which is thought to be solid due to intense pressure is surrounded by a more mobile liquid shell known as the outer core.

The Mantle

The mantle is a layer of red hot rock which separates the crust from the core. Its composition is extremely complex and seems to exhibit characteristics of

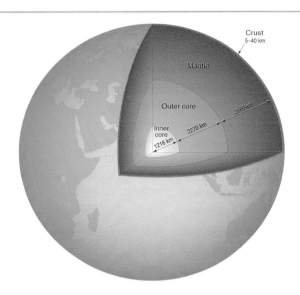

Fig. 1.2

The Structure of Planet Earth

both solid and liquid material. This is particularly true of the asthenosphere – the part of the mantle directly below the crust where the material is generally classified as 'plastic'.

The Crust – Lithosphere

The crust is the thinnest layer of the earth's structure and varies in thickness between 35 to 60 kilometres. The most common elements in the surface rocks are silicon (SI) and aluminium (AL) which explains the term SIAL which some geologists give to the continental part of the earth's surface. Below the SIAL is a layer of SIMA – where silicon (SI) and magnesium (MA) are the dominant elements.

Continental Drift – Plate Tectonics

Fig. 1.3

The notion that continents 'fit together' began the theory of plate tectonics

The idea that continents, particularly South America and Africa, fit together like a giant jigsaw puzzle originated with the improvements in world maps. It received little attention however until a German geophysicist Alfred Wegener published his work *The Origin of the Continents and Oceans* in 1912. In this work he argued that all the continents were once joined together to form a super-continent called *Pangaea* ('all land'). He then suggested that about 200 million years ago, this supercontinent began to break into smaller continents which then drifted into their present position. His argument was based on a number of observations including:

- Both South America and Africa 'fit together'.
- Similarities in the continents' ancient climates.
- Similarities in rock structure and fossils.

Wegener's inability to prove precisely how the continents actually moved meant his idea received a cool reception and were dismissed by others as utter 'damned rot'. It was not until the early 1960s when Harry Hess of Princeton University put forward his theory of sea-floor spreading that Wegener's ideas were seriously reviewed.

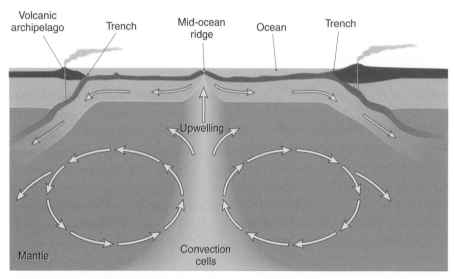

< Fig. 1.4
Sea-floor spreading

Sea-floor spreading involves the idea that the ocean-floors are being pulled apart along great cracks centred on mid-ocean ridges. Rising material from the mantle now fills the cracks forming a new crustal surface. The existing sea-floor is carried in a conveyor belt fashion away from the ridge crest, to be gradually consumed, as it descends into the mantle. Otherwise the earth would continue to expand which is not the case. In 1968, the concepts of continental drift and sea-floor spreading were united into a more comprehensive theory known as *plate tectonics*.

This theory states that the outer rigid shell of the lithosphere consists of several individual segments called plates. About 20 plates of various sizes have now been identified – the largest being the Pacific Plate. These plates are said to 'float' on the more 'plastic' asthenosphere underneath, and are driven by convectional currents. Though the movement of plates is very slow, about two centimetres per year in the Atlantic, three types of plate boundaries are thought to exist.

1. Constructive Plate Boundaries (Margins)

A constructive plate boundary is where two plates move away or diverge from one another and new crust is created. Initially the movement may cause huge rift valleys but molten rock or magma from the mantle rises to fill any gaps between the plates. The mid-Atlantic ridge was formed in this way, and more recently the island of Surtsey off the Icelandic coast in 1963.

2. Destructive Plate Boundaries (Margins)

When two plates converge or collide, they form destructive plate boundaries. As they compress, the heavier oceanic plate is pushed and pulled under the more stable continental plate. This now forms a subduction zone and is often associated with an ocean trench. As the crust descends the increase in pressure and friction helps to generate heat which converts the crust back into magma.

The resulting pressure often marks these destructive plate margins as areas of intense vulcanicity and may also trigger off major earthquakes. Where two continental plates collide the crust is forced up into fold mountains, e.g. Himalayas, Rockies and Andes.

3. Conservative Plate Boundaries

Conservative plate boundaries occur when two plates are forced to slide parallel to each other and land is neither created nor destroyed. Perhaps the best known example is the San Andreas Fault in California, scene of the recent earthquakes.

Plate Movements – Summary

Type of Plate Boundary	Description of Change	Examples
1. Constructive Margins	Two plates move away from each other forming mid-oceanic ridges and volcanoes.	Mid-Atlantic Ridge where the American Plate moves away from the Eurasian and African Plates.
2. Destructive Margins	Plates converge resulting in subduction producing deep sea trenches and volcanoes. Continental plates collide to form fold mountains.	Nazca Sinks under S. American plate. India collides with Eurasian forming Himalayas.

The advance in our knowledge since the 1950s of the structure of the earth has rightly been called a 'revolution'. Most *seismic* and *tectonic* activity is now known to be restricted to narrow zones already referred to as plate boundaries. The more important results of this activity are:

1. Earthquakes
2. Volcanoes
3. Folding and Faulting

Earthquakes

An earthquake is a vibration or tremor of the earth's crust. It is caused when a slow build-up of pressure within crustal rocks is suddenly released resulting in a jerking movement – in the same way as a ruler, slowly bent to breaking point suddenly snaps.

It is from the study of earthquakes (seismology) that much of our knowledge about the earth's interior has been deduced. The actual point where an earthquake begins is known as the *focus*. From the focus, seismic waves carry the shock or force of the quake to the earth's surface. The epicentre is the point directly above the focus which often receives the worst effects. Moving away from the epicentre, the intensity of an earthquake generally decreases. The force of an earthquake is recorded on a seismograph. Lines which join places receiving the same intensity of an earthquake are known as *isoseismal lines*.

MEASURING EARTHQUAKES

The intensity of an earthquake may be measured on the Richter Scale or Mercali Scale.

Richter Scale

This scale was designed by Richter, a German scientist and is a more scientific measurement of earthquake magnitude than the Mercali Scale. The magnitude is assigned a number. Most earthquakes range from force one to force eight. It is important to remember that each number on the scale signifies a force ten times stronger that the previous one, so that a force five earthquake is ten times more powerful than a force four earthquake.

The Mercali Scale

This allows people, without instruments, to judge the intensity of an earthquake by observation. Force one would go unnoticed while force 12 would be applied to total destruction.

EARTHQUAKE DESTRUCTION

Many factors determine the level of destruction caused by earthquakes, the most obvious being the magnitude of the earthquake, the proximity of the quake to a populated area and the degree of structural development of that region. During an earthquake, the region within 20 to 50 kilometres of the epicentre will experience roughly the same degree of ground shaking, but beyond this the vibrations decrease rapidly.

- Buildings, bridges and dams may collapse.
- Roads, railways, water, gas pipes and sewage pipes may crack or be displaced.
- Flooding, fires and disease (typhoid and cholera) may result.
- Landslides and avalanches may be triggered off.

- *Tsunamis* may result. These, sometimes inaccurately referred to as tidal waves, are caused by seismic sea waves. Travelling at speeds between 500 to 800 kilometres per hour they are known to reach heights of over 30 metres. Generally the first warning sign of an approaching tsunami, is a rapid withdrawal of water away form the shore. Between five to 30 minutes later this retreat of water is followed by a surge capable of extending hundreds of metres inland.

EARTHQUAKES: PREDICTION AND CONTROL

It is now known that major earthquake activity is largely confined to plate margins or plate boundaries, such as the Pacific Rim. Scientists are less able however to predict when an earthquake will strike. Some advance has been made with:

- Strainmeters – which measure the expansion of crusted rocks.
- Laser Beams – bounced off satellites to measure plate movements.
- Tiltmeters which reflect changes in the slope of the land.
- Radon gas is normally locked within rocks but is released during a build-up of stress.
- The strange behaviour of animals too such as dogs and snakes prior to the occurrence of earthquakes is being examined in countries such as China and the USA.

Volcanoes and Vulcanicity

Fig. 1.5

The term vulcanicity includes all processes by which solid, liquid or gaseous materials are forced 'into' (intrusive) or 'on to' (extrusive) the earth's crust. Volcanic activity is normally associated with plate margins. It is estimated that only about one per cent of the world's volcanic activity occurs in the middle of plates. Such areas are called *hot spots* and are believed to have caused island groups such as Hawaii and the Azores.

The Planet

EXTRUSIVE VULCANICITY – VOLCANOES

Depending on their level of activity, volcanoes are often said to be:

(i) Active – volcanoes which are erupting at present.

(ii) Dormant – volcanoes which have erupted in the past and are now said to be asleep (dormant).

(iii) Extinct – volcanoes which have not erupted in historic time and are thought to be 'dead'.

TYPES OF VOLCANOES: A CLASSIFICATION

Magma in the mantle is forced to the surface due to the build-up of pressure (at plate boundaries). Once magma is forced onto the surface it is known as lava (magma minus its gases).

Unfortunately there is no universally accepted method for the classification of volcanoes from a geographer's point of view. The most valuable method of classification is based on shape.

MAJOR VOLCANIC LANDFORMS

1. Fissure Eruption

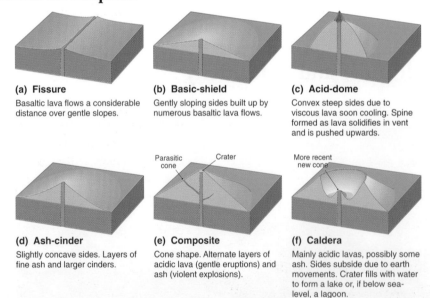

(a) **Fissure**
Basaltic lava flows a considerable distance over gentle slopes.

(b) **Basic-shield**
Gently sloping sides built up by numerous basaltic lava flows.

(c) **Acid-dome**
Convex steep sides due to viscous lava soon cooling. Spine formed as lava solidifies in vent and is pushed upwards.

(d) **Ash-cinder**
Slightly concave sides. Layers of fine ash and larger cinders.

(e) **Composite**
Cone shape. Alternate layers of acidic lava (gentle eruptions) and ash (violent explosions).

(f) **Caldera**
Mainly acidic lavas, possibly some ash. Sides subside due to earth movements. Crater fills with water to form a lake or, if below sea-level, a lagoon.

< Fig 1.6
Classification of Volcanoes

Fissure eruptions occur when lava wells up along a single fissure or series of parallel fissures. Normally the lava wells out quietly and being composed of basalt is more 'fluid' and so covers a large area. The plateau of Antrim in north-east Ireland was formed in this way and its columnar jointing produced by slow cooling of the lava provides a major tourist attraction known as the Giant's Causeway.

2. Basic or Shield Volcanoes

Here lava flows out through a central vent and, being fluid, spreads over a wide area before 'solidifying'. The result is a cone with a central vent and

Volcanoes and Vulcanicity

Fig 1.7

gently sloping sides. Mauna Loa on the island of Hawaii is one of the best examples.

3. Acid or Dome Volcanoes

Acid or dome volcanoes are composed of acid lava which, being more viscous (thick), solidifies more quickly. As most lava builds up near the vent, the result is a steep-sided convex cone, e.g. Mt Pelée in Martinique.

4. Composite Cones

Composite are probably the most common and typical of all volcanoes and include most of the highest volcanoes in the world. The main cone is built up of layers of ash and lava fed from the main pipe. Later explosions may blow off the top of the cone and so form a much larger crater within which a secondary cone may develop and parasite cones may grow on the flanks or sides. One of the best examples is Stromboli in the Lipari Islands off the 'toe' of Italy. The frequency of its eruptions (often at hourly intervals) has earned it the name 'the lighthouse of the Mediterranean'.

5. Ash and Cinder Cones

Ash and cinder cones form when fragments of solid material accumulate round a vent to form a cone whose shape is usually concave due to the spreading outwards of material at its base. One of the better known examples is Paricutin in Mexico which first erupted in 1943.

6. Calderas

Calderas occur when huge explosions remove the summit of the original cone. This often causes the sides of the crater to subside which widens the opening often to several kilometres.

This happened in the case of Krakatoa (1883) and Santorini. In both cases the enlarged crater was flooded by the sea resulting in lagoons in which later eruptions produced smaller cones.

The Planet

Volcanic activity has long ceased on the Irish landscape. The remnants of volcanic activity however can still be seen between Limerick and Tipperary where Derk Hill and Kilteely Hill are now little more than volcanic plugs.

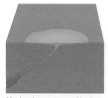

Mud volcano caused by hot water mixing with mud and surface deposits.

Solfatara is when gases, mainly sulphurous, escape onto the surface.

A **geyser** is when water is heated by rocks in the lower crust, turns to steam, pressure increases and the steam and water explode onto the surface.

Fumaroles result from superheated water turning to steam as pressure drops as it emerges from the ground.

< **Fig 1.8**
Minor Volcanic Landforms

INTRUSIVE FORMS OF VULCANICITY

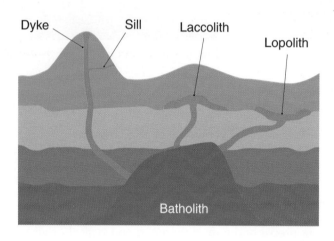

< **Fig 1.9**
Intrusive forms of Vulcanicity

7. Dykes

Dykes form when the magma rises through vertical cracks or fissures and solidifies to form 'vertical walls' of rock cutting across the bedding planes of the surrounding rock. They often occur in swarms as in the islands of Mull and Aran in Scotland.

8. Sills

Sills form when the magma forces its way into bedding planes to form horizontal sheets. They may vary in thickness from a few to a few hundred kilometres. Fair Head and Scrabo Hill near Newtownards are excellent example of sills in Ireland.

9. Laccoliths

The word laccolith comes from two Greek words meaning 'rock cistern'. They result when viscous magma forces the overlying strata (rock layers) into a dome.

10. Batholiths

Batholiths are formed by deep-seated movements on an enormous scale so that masses of magma cool slowly to form large-crystalled rock such as granite. These large masses of rock which often form the core of mountains which were exposed over time by prolonged denudation to form massive upland areas. The Wicklow mountains which form the Leinster batholith and the uplands of Brittany in France are some of the best known examples.

Effects of Volcanoes and Earthquakes – A Summary	
Benefits	**Hazards**
Lava and ash weather into fertile soils ideal for farming, e.g. the region surrounding Etna.	Earthquakes destroy buildings and result in loss of life.
Igneous rock contains minerals such as gold, copper, lead and silver.	Violent eruptions with blast waves and gas may destroy life and property, e.g. Mt Pelée, Mt St Helens.
Igneous rock is used for building purposes, e.g. Naples, Aberdeen	Mudflows may be caused by heavy rain and melting snow, e.g. Armero, Colombia.
Extinct volcanoes may provide sites for defensive settlement, e.g. Edinburgh.	Tidal waves/tsunamis, e.g. Krakatoa.
Geothermal power is being developed, e.g. Iceland, New Zealand.	Ejection of ash and lava ruin crops and kills animals.
Geysers and volcanoes are tourist attractions, e.g. Yellowstone National Park, generating revenue for local communities.	Interrupts communications.

Fig 1.10 >
Effects of Volcanoes and Earthquakes on Human Activity

FOLDING AND FAULTING

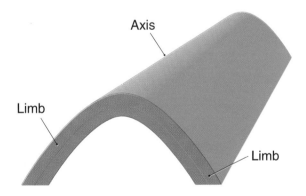

Fig 1.11 >
Simple Fold

Folding is associated with destructive plate margins where plates collide forcing the earth's crust to buckle or wrinkle, i.e. to fold. All the great mountain ranges of the world were formed in this way. Sediments were laid down on the ocean floor. When these plates were uplifted due to compression giant fold mountains, composed of sedimentary rock, were formed. This uplift was also aided by the upwelling of magma into the crust. The Andes which hug the Southern American coastline were formed by subduction while the Himalayas were formed by the collision of continental rocks.

Anticline	This is the upfold, the top of which is known as the crest.
Syncline	This is the downfold or valleys on either side of the anticline.
Limb	The sides of fold mountain are known as the limbs while the actual slope of the limb is known as the dip.

TYPES OF FOLD MOUNTAINS

Symmetrical

Fold mountains whose limbs have the same angle of dip are said to be symmetrical.

Asymmetrical

Variations in rock type and structure, together with variations in intensity of plate movement means that folding is more likely to be asymmetrical where the limbs are unequal.

Overfold and Recumbent

Where continued compression causes one limb of the anticline to be pushed over the other limb, overfolds and recumbent folds develop.

FOLDING AND THE IRISH LANDSCAPE

Caledonian Folding

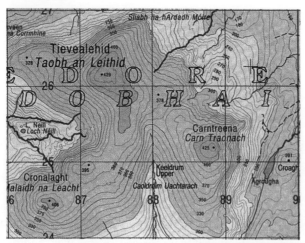

‹ Fig. 1.12

Caledonian fold mountains in Donegal

Caledonian folding produced the old fold mountains associated with the west and north-west of Ireland and include the Sperrin, Derryveagh, Bluestack, Maamturk, Mweelrea, Ox and Partry as well as the Twelve Pins. They run in a north-east to south-west direction.

The direction or trend of a mountain range is also known as the *strike*. The word Caledonian is derived from the Roman word for Scotland (Caledonia) where these mountains are also represented.

The distinct trend of the Scandinavian peninsula owes its origin to this mountain building movement.

Armorican Folding

Fig. 1.13 > Amorican fold mountains in Kerry

Less than 300 million years ago, the Armorican or Hercynian mountain building movement affected the south of Ireland and the mountains of Cork, Kerry and Waterford form the *'ridge and valley province'* – and have a distinct 'east to west' strike. This folding extended northwards to produce the Galtees, Silvermines and the Burren table-land.

Alpine Folding

This folding formed the young fold mountains in mid-tertiary times about 30 million years ago. While this period of mountain building did not affect Ireland, it was responsible for the great mountain ranges of the Alps, Himalayas, Rockies and Andes.

FAULTING

Like folding, faulting is also associated with plate boundaries where surface rocks are subjected to stresses, strains and compression.

Faulting produces two main features on the landscape:

(i) *Rift valleys* and (ii) *Block mountains*, which are also called *horsts*.

(i) Rift Valleys

Rift valleys or *graben* form when parallel faults develop as a result of tension (stretching) in the earth's surface and the land between the faults gradually slips down under the influence of gravity. In this way long and extensive valleys with steep sides known as fault scarps are formed. The Rift Valley of Scotland was formed in this way. In fact these fault lines entered into Northern Ireland running north and south of Lough Neagh until they eventually almost merge near Killary Harbour.

Other examples of rift valleys include; the Gulf of Aden and the Jordan.

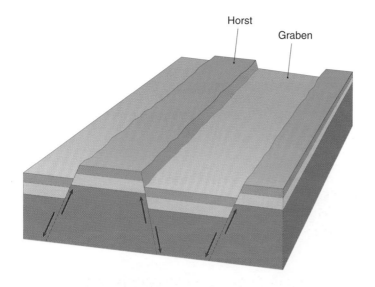

Fig. 1.14
Formation of Rift Valleys (graben) and Block Mountains (horst)

(ii) Block Mountains

Block mountains are the very opposite of fold mountains. Where parallel faults have developed the surface between the faults may be uplifted or alternatively the surface on the other sides of the faults may sink down. In either case a block mountain with steep fault scarps is the result. Single block mountains are known as horsts of which the Black Forest (Germany), Vosges (France) and Ox Mountain in Mayo provide excellent examples.

Types of Rocks – A Classification

Rocks are generally classified into three main groups depending on how they were formed:

1. Igneous Rocks 2. Sedimentary Rocks 3. Metamorphic Rocks

Igneous Rocks

These rocks result from the cooling of magma either inside – intrusive, or outside – extrusive, the earth's crust.

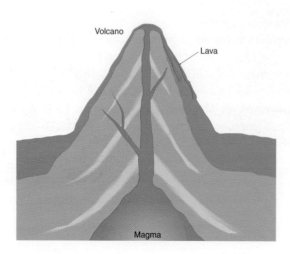

Fig 1.15
Formation of Igneous rocks

Intrusive Igneous Rocks

These rocks formed when magma cooled underground and are sometimes called plutonic rocks. Because they cooled slowly, large crystals were allowed to form, giving the rock a coarse texture.

Extrusive Igneous Rocks

When the magma escaped on to the earth's crust it cooled more rapidly resulting in small crystals. Extrusive rocks are also known as volcanic rocks.

Igneous rocks may also be classified according to the percentage of silica in the rock. Rocks with 65 per cent silica or over are termed *acidic*; rocks whose silica content is under 55 per cent are known as *basic* while rocks between 55 per cent and 65 per cent are called *intermediate*.

A Description of Some Common Igneous Rocks in Ireland

Granite

Granite is a coarse-grained, acidic (a high percentage of silica) plutonic rock composed mainly of quartz, felspar and mica. It is generally greyish in colour and is common in the Leinster chain, Donegal and the Mourne mountains.

Basalt

Basalt is the most common of the volcanic or extrusive rocks. Because it cooled rapidly it has small crystals. It is basic in composition, having a low silica content and has a blackish appearance. It is common in Antrim where it formed the famous Giants' Causeway.

Sedimentary Rock

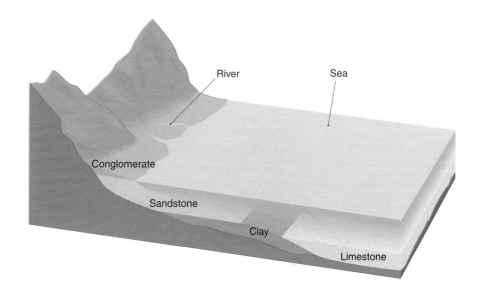

Fig 1.16 **Sedimentary rocks**

The Planet

Sedimentary rocks constitute nearly three quarters of the rock which appears on the earth's surface. They were formed from the destruction by weathering and erosion of igneous rocks and the accumulation of resulting debris in vast basins known as *geosynclines*. Later, other sedimentary rocks were formed from organic material such as decaying vegetation and animal organisms. As the material was deposited, usually in layers known as *strata*, the additional weight of the material on top compressed the lower layers into rock. In some cases this process was aided by the presence of a cementing agent such as calcium carbonate.

SEDIMENTARY ROCKS ON THE IRISH LANDSCAPE

Limestone

1. Limestone is laid down in **strata** (horizontal layers). The divisions between the layers are called **bedding planes**.

2. Limestone is **permeable**, which means that water can pass through the joints and bedding planes.

3. Vertical cracks or joints also occur in limestone.

4. Limestone is **easily weathered** (worn away). The rainwater which passes through it is a weak carbonic acid which dissolves the calcium carbonate that makes up the limestone.

5. Limestone may contain **fossils**. A fossil is the preserved remains of a plant or animal.

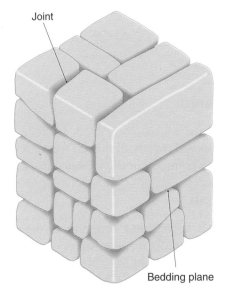

< Fig 1.17

The Structure of Limestone

Limestone is the most common rock in Ireland. Most limestones were formed from the shells and skeletons of minute marine animals which accumulated on the sea floor over millions of years.

Limestones are composed mainly of calcium carbonate but impurities in the sea-water during formation give them various colours, e.g. grey, blue-grey etc. Though limestones underlie most of the Central Plain of Ireland they only occasionally came to the surface on any large scale, e.g. the Burren, Co. Clare (see Chapter 3). Elsewhere they are covered by a blanket of peat bog and debris deposited by glaciers.

Coal

Coal was formed by the accumulation of decaying vegetation. During the carboniferous period, Ireland was experiencing a warm tropical climate, probably somewhat resembling that of the Amazon basin today.

As the tropical forests died and decayed they were compressed by layers of sand and mud which were deposited on them. Over time these deposits changed into peat and coal. While peat is widespread in Ireland today, due to post-glacial decay, coal deposits from the carboniferous period can be found at Arigna (Co. Leitrim), Castlecomer (Co. Kilkenny) and Coalisland (Co. Tyrone).

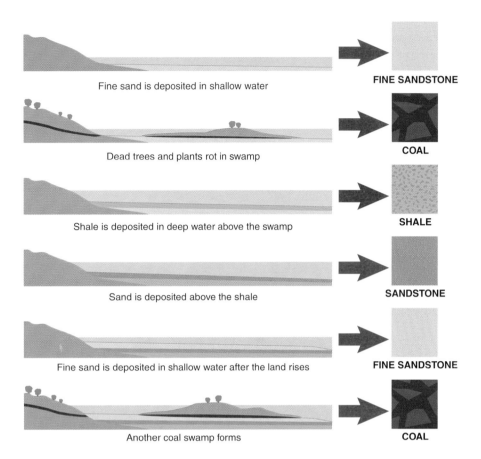

Fig 1.18
The formation of coal seams in sedimentary rock

METAMORPHIC ROCKS

The word metamorphic is derived from the Greek word meaning 'to change' and this name is given to both igneous or sedimentary rocks which have been changed by either heat, pressure or both. They are sometimes subdivided into three groups.

i) Thermal Metamorphism

Thermal metamorphism is where heat alone was responsible for the change. Here the intrusion of magma heated the local rock and it recrystallised. This is sometimes called *contact metamorphism*.

ii) Regional Metamorphism

This occurs over a wider area as a result of great pressure (and sometimes heat). It is generally associated with areas of plate boundaries where compression and subduction resulted in immense regional pressure.

iii) Dynamic Metamorphism

This occurs due to pressure alone and is generally associated with fault planes where rocks are crushed due to earth movements.

Marble

Marble is a coarse-grained rock formed by regional or contact metamorphism of limestone. In its pure state marble is white, being formed from pure limestone. Impurities however may result in marble having a black colour (due to carbon), a green colour or a reddish colour (iron oxide).

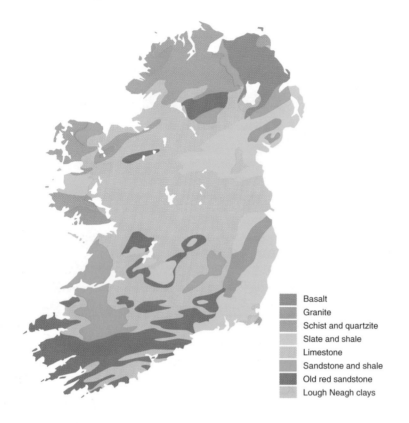

Fig 1.19
Main rock types in Ireland

ROCK DATING

Until recently, geologists could only tell the relative age of rocks but now thanks to modern technology the absolute age of rocks, i.e. their age in years can be determined.

1. Relative Age

The Law of Superposition

This states that in any rock strata the lowest layers are the oldest and the highest layers are the youngest. The exception to this is when earth movements may have overturned the rock strata.

2. Absolute Age

Radio Carbon Dating

This technique can now be used for dating material up to 70,000 years old. When a plant or animal dies the amount of carbon 14 and carbon 12 present is exactly the same. However while carbon 12 remains constant the amount of carbon 14 diminishes at a specific rate of half every 5,730 years. Thus by comparing the amount of carbon 14 with carbon 12, the age can be calculated.

Questions

LEAVING CERTIFICATE

Ordinary Level

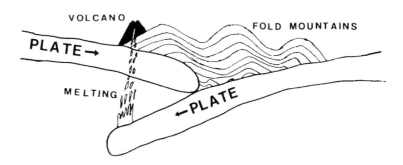

1. The above diagram shows a zone where plates meet on the Earth's crust.
   (i) Describe and explain the formation of **each** of the following, referring in your answer to examples you have studied:
   • Fold mountains
   • Volcanoes *(40 marks)*
   (ii) Earthquakes also occur in similar zones. Explain how earthquakes occur **and** describe their effects on a built-up area. *(40 marks)*
   80 marks

   Leaving Cert. 1996

2. Write a paragraph on – There is a wide variety of rock types in Ireland. *(20 marks)*

   Leaving Cert. 1996

3. Write a paragraph on – Major earthquakes are associated with certain regions of the world. *(40 marks)*

   Leaving Cert. 1995

4. (i) Fold Mountain, Rift Valley, Volcano:
   Select any **two** of the above landforms and for **each** one you select:
   • Name one example of the landform.
   • Describe and explain, with the aid of a diagram how it was formed. *(50 marks)*
   (ii) 'Earthquakes can have major effects on human society and on the physical landscape'.
   Explain this statement, with reference to examples you have studied. *(30 marks)*

   Leaving Cert. 1994

The Planet

5. (i) A study of patterns in the worldwide distribution of volcanoes and earthquake zones can help us to understand the causes of these events.
   Examine this statement, with reference to examples which you have studied. *(50 marks)*

   (ii) Describe some of the effects on human societies of the occurrence of a volcanic eruption or a major earthquake in populated areas.
   (30 marks)

   Leaving Cert. 1993

Higher Level

1. (i) Discuss how the Theory of Plate Tectonics helps us to explain the occurrence of volcanic activity and of earthquakes. *(60 marks)*

   (ii) Examine some of the immediate **and** long-term effects of **either** a volcanic eruption **or** an earthquake which you have studied.
   (40 marks)

   Leaving Cert. 1996.

2. (i) The processes of folding, faulting and volcanic action have produced distinctive landforms worldwide.
   With reference to **one** landform in **each** case, examine how these processes have helped to shape the surface of the earth.
   (60 marks)

   (ii) In the case of **one** of these landforms, examine **one** positive and **one** negative example of how human societies have interacted with it. *(40 marks)*

   Leaving Cert. 1995

3. (i) Earthquakes and volcanoes occur in quite predictable locations on the globe. Examine the theoretical basis for this statement.
   (60 marks)

   (ii) The frequency of occurrence of earthquakes and volcanic eruptions is much more difficult to predict than their location. Assess the accuracy of this assertion with reference to examples you have studied. *(40 marks)*

   Leaving Cert. 1993

4. 'The shudder which shook buildings in Dublin and most towns and villages of Leinster early yesterday morning marked the most severe earthquake registered in this country since scientific records began early in the present century. The earthquake – registering 5.5 on the Richter scale – demolished the myth that dangerous earthquakes cannot happen in this part of the world; in Britain, it has given rise to a new debate about the safety of nuclear power stations.'

   [Irish Times, 20 July, 1984]

Questions

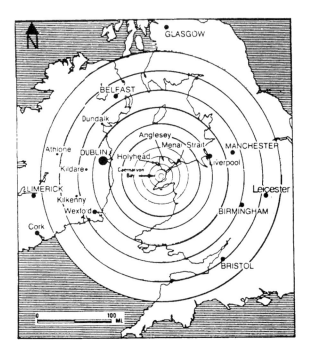

(i) In the passage above, the writer refers to the 'myth' that dangerous earthquakes cannot happen in Ireland or Britain. Examine the current theories about the causes of earthquakes and explain why this part of the world has always been regarded as being safe from such events.

(50 marks)

(ii) Explain some of the possible effects – including those referred to in the passage – of a severe earthquake happening in these islands and outline what steps governments could take in order to lessen the effects. *(50 marks)*

Leaving Cert. 1992

CHAPTER 2
Weathering and Mass Movement

< Fig 2.1
Avalanche

Rocks, and hence landforms, that form the landscape around us are constantly being attacked by various processes. The work of these processes which break up and remove the earth's land surface is known as denudation. *Denudation* is achieved in two ways – *weathering* and *erosion*.

> Weathering is the disintegration or breakdown of rocks 'in situ', i.e. in their original position.
>
> Erosion on the other hand is the breakdown of rock particles and their removal to a new location.

Weathering is the first stage of denudation and is one of the most fundamental of the geomorphical processes. It produces rock material for transport by mass movement and by such agents of erosion as running water, ice, wind and the sea.

While the degree of weathering depends upon the structure and mineral composition of the rocks it also reflects the local climate, vegetation and the length of time the weathering processes operate.

Types of Weathering

1. Mechanical	2. Chemical	3. Biological
i) Sun	i) Hydrolysis	i) Plants
ii) Frost	ii) Hydration	ii) Animals
	iii) Oxidation	iii) Humans
	iv) Carbonation	

1. MECHANICAL OR PHYSICAL WEATHERING

Mechanical or physical weathering is the disintegration of rocks into smaller particles without any change in the chemical composition of the rock. It is generally associated with areas which have little or no vegetation cover such as deserts and high mountainous regions.

i) Sun

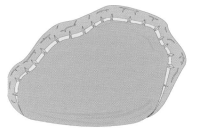
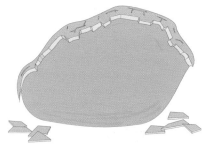

Fig. 2.2
Exfoliation (Onion Weathering)

Expansion and contraction produces cracks

Rock layers peel off

Like all solids, rocks expand when heated and contract when cooled. In hot deserts like the Sahara, the cloudless sky results in a high diurnal (daily) range of temperature i.e. very hot days but cold nights; the range can often exceed 40°C.

This daily rapid expansion and contraction sets up stresses and strains which lead to rock disintegration. Temperature changes like these essentially affect the outside of the rock leaving the interior relatively unaffected. As a result the surface peels off in layers. This 'onion weathering' is called *exfoliation*.

Laboratory experiments now prove that this heating and cooling alone do not produce exfoliation. Where water is present, exfoliation takes place rather rapidly. In hot deserts, the presence of atmospheric moisture, heavy dews and occasional downpours are believed to be sufficient to provide the required moisture.

ii) Frost

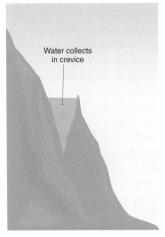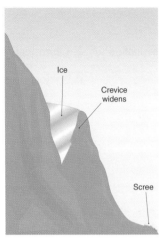

Fig. 2.3
Frost Wedging

In mid to high latitudes and in mountainous regions, freezing water is an effective agent of weathering. Water enters cracks and joints in rocks during the day. At night the water may freeze and its volume expands by nine per cent. This exerts pressure in the joints which wedges the rock apart. This freeze-thaw action is also known as *frost wedging* or frost blasting. When it occurs on steep slopes it can produce large quantities of loose angular rock fragments which accumulate to form *talus* or scree. Quite extensive talus heaps occur on Croagh Patrick (Mayo), Errigal Mountain (Donegal) and the Sugar Loaf (Wicklow).

Fig. 2.4
Mount Errigal

2. CHEMICAL WEATHERING

Chemical weathering is the decomposition of rock minerals and cannot take place without moisture being present in the atmosphere. Consequently, it is most active in humid regions.

i) Hydrolysis: This is the process by which felspar is broken down to a clay (e.g. kaolin clay from which chinaware is made). Hydrolysis is an important process in the weathering of all igneous rocks, especially granite.

ii) Hydration: Certain minerals (and so rocks) can absorb water as a result of which they expand and lose strength thus helping in the eventual destruction of rocks, e.g. sandstone.

iii) Oxidation: This is best seen in the case of rocks which contain iron. In humid conditions the oxygen in the atmosphere or in rainwater combines with the iron to form iron oxide or rust which crumbles easily causing the rock to disintegrate. The process is often visible, as the shell of the rock has a reddish-brown appearance. As a result the outside of the rock crumbles and the process commences all over again, somewhat similar to onion weathering.

iv) Carbonation: As rainwater passes through the atmosphere it dissolves small quantities of carbon dioxide to form a weak carbonic acid. This acid is particularly effective on rocks such as limestone and chalk as they contain large proportions of calcium carbonate.

3. BIOLOGICAL WEATHERING

The work of plants, animals and humans is known as biological weathering. Although vegetation cover actually protects rocks from atmospheric influences it also tends to assist weathering by mechanical and chemical means.

i) Plants: As plants grow, their roots may widen cracks already formed in rock and may eventually prise it apart. Water and air then enter and further destruction takes place. Rotting vegetation releases organic acids and carbon dioxide that aid the chemical disintegration of the rock.

ii) Animals: Various animals such as worms and rabbits can also contribute to the weathering process by boring through and loosening the surface thereby making material available to agents of erosion such as wind and running water.

iii) Humans: People can play a major role in the weathering process in particular regions. By overcropping, overgrazing and the removal of the protective covering provided by forests they have turned large areas of Africa into 'man-made deserts'. Also by quarrying, mining and building (roads and houses), underlying rock is broken down.

THE WEATHERING PROCESS IN DIFFERENT CLIMATIC ZONES

Fig. 2.5
Equatorial Rainforest

Mechanical, chemical and biological weathering do not act independently of each other and it is sometimes difficult to discern which is the dominant process in any given region.

Equatorial Regions

The high temperature and rainfall in areas such as the Amazon Basin would suggest that the chemical process dominates here. These climatic characteristics however also lead to rapid growth so that the contribution made by plants (and animals) also deserves careful consideration.

Hot Desert Regions

In regions such as the Sahara, the mechanical process of exfoliation caused by high diurnal ranges in temperature dominates. However the presence of water, if only in the atmosphere in the form of vapour is essential to explain the rate of disintegration. So even in hot deserts, chemical and mechanical processes combine in the destruction of rocks.

Fig. 2.6 Hot Desert

Temperate Regions

While chemical action is a dominant process in humid temperate regions such as Ireland, freeze-thaw action is responsible for producing large quantities of *scree* in the more mountainous areas of such regions. The water which enters the rock also has a chemical effect (before it freezes). Recent research has also tended to support the importance of chemical weathering in these areas as carbon dioxide, the principal agent of chemical weathering, is more soluble in water at low temperature than at high temperature.

Fig. 2.7 Irish Mountains

Mass Movement

When surface material such as soil, stones and rocks (regolith) moves downslope as a result of the pull of gravity, mass movement (mass wasting) is said to take place. Mass movement may be slow or rapid.

TYPE OF MASS MOVEMENTS

1. Slow	2. Fast
i) Soil Creep	i) Earth flows and Mud flows
ii) Solifluction	ii) Landslides
	iii) Rotational Slumping

Slow Movements

i) Soil Creep

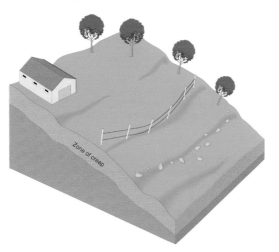

Fig. 2.8
Soil Creep

Soil creep is the slowest type of mass movement, so slow in fact that its movement may be almost imperceptible. Soil accumulation on the upslope side of fences, hedges and walls, however, provides easily observable evidence of soil creep. The downhill tilting of poles and fences are often the only tell-tale signs of such movement which is often less than one centimetre a year.

ii) Solifluction

This word, actually meaning soil flow, is a slightly faster movement averaging between five and 20 centimetres a year. When water is present there is a greater likelihood of movement. This is because water adds to the weight of the material, causes some of the clay minerals to swell and it lubricates the soil and lessens its resistance to gravity. It is most active under periglacial conditions (see Chapter 5) where surface meltwater cannot infiltrate downwards and temperatures are too low for much effective evaporation.

2. Fast Movements

i) Earth flows and Mud flows

Where slopes are steeper and more water is available the speed of movement will obviously increase. In humid regions then where the rock has been deeply weathered, masses of earth may move downslope at speeds up to 15 kilometres per hour. The movement of material may produce a concave scar at the origin of the slip and a convex bulge below. Earth flows may also be caused on river banks due to undercutting or by headward erosion. Mud flows occur when soils containing a high percentage of clay particles become saturated with water. they are often associated with periods of heavy rainfall and can reach speeds of over 80 kilometres per hour.

ii) Landslides

Landslides are rapid downslope movements of accumulated rock debris. They often occur on steep slopes such as glaciated mountain areas especially where the jointing or bedding planes are roughly parallel to the slope. The movement can be triggered off by the undercutting at the base of a slope by rivers, sea or human activity.

iii) Rotational Slumping

This is a type of landslide which involves not only the sliding downwards of material but its rotational movement at the same time. It is common on hillsides and coastal areas which are made of softer materials like clay or sandstone.

Weathering and Mass Movement

SOME FACTORS WHICH AFFECT MASS MOVEMENT

1. Relief

(i) Areas of bare rock are more susceptible to mechanical and chemical weathering which results in the formation of regolith.

(ii) Steep slopes are more likely to encourage mass movement than gentle slopes.

(iii) The presence of joints and bedding planes can encourage movement especially where they are parallel to the slope.

2. Climate

(i) Humid climates can encourage the weathering process leading to the formation of regolith.

(ii) Heavy rainfall increases the volume and lubricates the soils, which encourages mass movement.

(iii) Snow adds weight which can encourage rapid movement.

In early spring the downward rapid movements of large quantities of unstable snow in mountains like the Alps are known as avalanches.

3. Vegetation

The lack of vegetation means that there are fewer roots to bind the soil. It also encourages more surface run-off which also adds to surface instability.

4. Human Influence

Deforestation can dramatically encourage the rate of mass movement. Undermining slopes by quarrying or road building can have similar effects.

Questions

ORDINARY LEVEL

1. Write a paragraph on – The influence of gravity on mass movement has helped to shape and modify many land forms. *(40 marks)*

   Leaving Cert. 1993

HIGHER LEVEL

1. The processes of weathering, together with gravity, are important factors in shaping landscapes.
   (i) Explain this statement, referring to **three** weathering processes.
   (ii) Examine how human activities can accelerate or intensify any **one** of the weathering processes referred to above. *(25 marks)*

   Leaving Cert. 1995

CHAPTER 3
Underground Water and Limestone Regions

Fig. 3.1

In spite of the fact that as little as only half of one per cent of the world's water is found underground, the actual amount stored in the rocks and sediments beneath the earth's surface is vast. It has been estimated that the amount of water in the upper 800 metres of the continental crust is about 3,000 times greater than the volume of water in all rivers at any one time.

Source of Underground Water

THE WATER CYCLE – METEORIC WATER

When rain falls, some of the water evaporates, some runs off and the remainder soaks into the ground. The actual amount which percolates downwards depends on a number of factors such as:

- The intensity and frequency of the rain.
- The type and amount of vegetation.
- The slope of the ground – the greater the slope the greater the amount of run-off.
- The type of rock in the region – whether it is permeable or impermeable.

Impermeable rocks are rocks which do not allow water to pass through them. *Permeable* rocks are rocks which allow water to pass through either because they contain pores, i.e. porous rocks or have joints and bedding planes, i.e. *pervious* rocks.

THE WATER TABLE

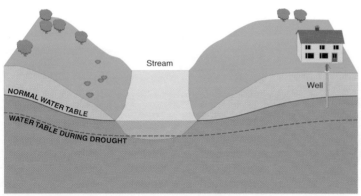

Fig. 3.2
The Water Table

The rain water which seeps into the surface rocks will eventually move downward until it reaches a layer of impermeable rock, unable to seep down any further it saturates the overlying layers filling all the pores and crevices (e.g. joints bedding planes etc.). This layer of rock which now holds water is called an *aquifer*.

> The water table is the line below which the rock is saturated.

In the wet season the water rises to form the temporary water table and in the dry season it falls to form the permanent water table. Contrary to what we might expect the water table is rarely level and generally follows the shape of the surface topography. A number of factors contribute to this irregular surface of the water table. Variations in rainfall and permeability from place to place leads to uneven infiltration and so to differences in the level of water table. More important, however, is the fact that ground water (underground water) moves very slowly. Because of this water tends to 'pile up' beneath high areas between stream valleys. Similarly where water has been extracted from the ground by the use of wells a dip occurs in the water table as a result of this drawdown effect. The result is a depression in the water table known as a *cone of depression*.

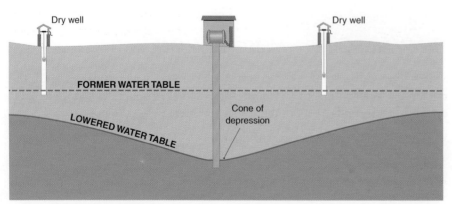

Fig. 3.3
Cone of Depression

A cone of depression in the water table often forms around a pumping well.
If heavy pumping lowers the water table, the shallow wells may be left dry.

In the wet season the rising water-table (temporary) may fill hollows in the surface to form lakes. In the dry season however the water table falls and the lakes may dry up. Disappearing lakes like these are known as turloughs and are common in the west of Ireland, e.g. Turlough Mór in Co. Galway.

SPRINGS

Where the water table intersects the earth's surface, a natural flow of ground water occurs which is known as a spring, so why do springs occur in one area and not another? The answer to this question may be found in the following diagram.

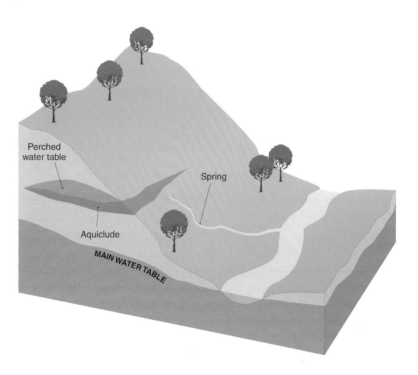

Fig. 3.4 › The Origin of a Spring

As water percolates downwards a portion of it is intercepted by the *aquiclude* thereby creating a localised zone of saturation called a perched water table.

> An aquiclude is an impermeable layer of material such as clay which may hinder or prevent water movement.

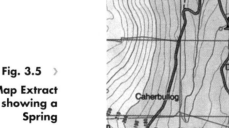

Fig. 3.5 › Ordnance Survey Map Extract of the Burren showing a Spring

The quality of spring water varies considerably with the depth at which it occurs and the rock type through which it passes. Water which passes through limestone is said to have a 'hard' quality. The best ground water comes from deep sandstone aquifers. Sandstone allows water to flow freely over long distances and this filtering effect removes all traces of organic matter.

Artesian Wells

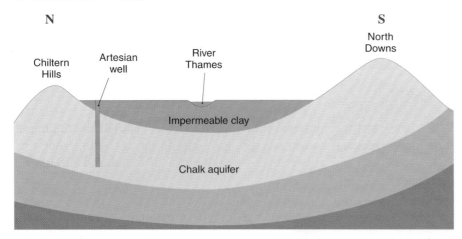

< Fig. 3.6
Artesian Well

For an artesian well to exist two conditions must be present:

(i) Water must be confined to an aquifer that is inclined so that one or both ends can receive water, and

(ii) Aquicludes (impermeable layers) both above and below the aquifer must be present to prevent the water from escaping.

As the water seeps into the *syncline* (downfold) through gravity it produces enough hydraulic pressure to flow up a well shaft. These wells receive their name from the Artois region in France where they were first used in the 12th century. The largest Artesian Basin in the world is the Great Basin of Eastern Australia, which extends over a staggering 1,500,000 square kilometres. In England the use of water in the London Basin has reduced the level of the water table by 17 metres. Over-exploitation of ground water where withdrawal exceeds recharge can result in major problems. In the USA, and California in particular, where ground water is literally being 'mined', subsidence and water shortages for agriculture and industry are the inevitable consequences, while closer to the coast, the lowering of the water-table allows supplies to be contaminated by salt water.

Limestone or Karst Regions

The most spectacular results of ground water are to be found in limestone or karst regions. The term karst is derived from a plateau region located along the north eastern shores of the Adriatic Sea in the border area of Italy and former Yugoslavia.

In Ireland this limestone topography is best developed in Northern Clare in a region known as the Burren, meaning 'Rocky Place'.

The reason why exposed limestone forms such a unique landscape is due both to its composition and structure. Limestone regions are subject to differential erosion which produces an extraordinary and often fascinating topography.

Limestone or Karst Regions

Fig. 3.7 >
The Burren, Co. Clare

COMPOSITION

When rainwater (H_2O) falls through the atmosphere it absorbs carbon dioxide (CO_2) to form a weak carbonic acid H_2CO_3. This acid changes the calcium carbonate ($CaCO_3$) in the limestone to a soluble bicarbonate, $Ca(HCO_3)_2$ which is easily removed in solution.

STRUCTURE

Limestone is a well jointed rock with many lines of weakness which allows water to pass freely through it.

SURFACE FEATURES

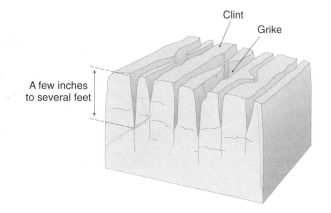

Fig. 3.8 >
Limestone Pavement

1. Limestone Pavement – Clints and Grikes

Because limestone is well jointed, rainwater tends to follow these lines of least resistance. As a result, chemical disintegration is more concentrated along the joints. Over time these joints are enlarged to form a pattern of deep narrow furrows *(grikes)* separated by ridges *(clints)*. Because the surface has a paved appearance it is sometimes referred to as a limestone pavement.

2. Swallow Holes

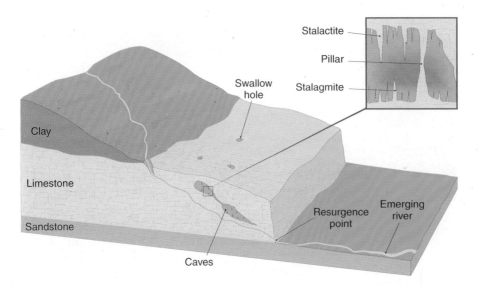

Fig. 3.9

Swallow holes are also known as *sink holes* or *slugga* and occasionally as *dolines*. They are conical shaped openings in the bed of rivers which flow over the limestone rock and may owe their origin to the chemical effect of rainwater or to the collapse of the roof of an underground cave. Their size can range up to several meters in diameter. Rivers and streams often disappear through swallow holes to follow underground passages only to re-appear again at a *resurgence point*. In the Burren, Poll na gColm and Poll an Phuca provide good examples of swallow holes, while the Aille River in Clare and the Gort in Galway flow underground for part of their course.

3. Uvulas

When two or more swallow holes coalesce (join together) a much larger surface depression is formed known as an *uvula*. They may also owe their origin to the caving-in of underground caves and can be up to 300 metres in diameter.

4. Poljes

Poljes are bigger depressions, again often up to several kilometres in diameter. They generally have steep sides with flat floors. They may originate either from the coalescence of uvulas or from down-faulting of the limestone. The Carran Depression in the Burren is a good example of a polje.

5. Dry Valleys

Dry valleys were formed by rivers that have since disappeared. They may owe their origin to glacial melt water when the ground underneath was frozen and so impermeable. However they may also have been formed when the climate was much wetter and the rainfall maintained a higher water table.

UNDERGROUND FEATURES IN LIMESTONE REGIONS

Fig. 3.10
Limestone Cave

1. Caves

As water passes through the limestone it seeps underground to form streams. Most caves or caverns are believed to be formed at or below the water table in the zone of saturation. Here the ground water follows lines of weakness/joints and bedding planes. Over time, the limestone is dissolved by the chemical process already discussed, slowly creating cavities and gradually enlarging them into caves. The Aillwee Caves in the Burren and the Mitchelstown caves in Tipperary are good examples of this feature.

2. Stalactites and Stalagmites

The continuous seepage of water through the ceiling of the cavern results in constant dripping and evaporation, where cracks occur. When evaporation takes place, tiny specks of calcite remain. Deposition occurs in a ring around the edge of the water drop. Over a considerable length of time a hollow limestone tube is created. Water now moves through the tube and remains suspended for a moment at the end of the tube contributing another tiny ring of calcite. Such features are aptly called soda straws.

While many of these break off, some eventually fill up to form stalactites.

Much of the seeping water falls to the floor of the cavern and evaporates there to form calcite. Over time this calcite grows upwards to form a stalagmite. Because the waters splash over the ground, stalagmites do not have a central tube and are usually larger in appearance than stalactites. When stalactites and stalagmites join together a column or pillar is formed. When water passes through narrow fissures in the ceiling and evaporates, the calcite takes on a 'curtain' appearance.

The Cycle of Erosion in Limestone Regions

Like all landscapes, karst regions go through their life cycle of youth, maturity and ultimately to old age when peneplanation is reached.

Youthful Stage

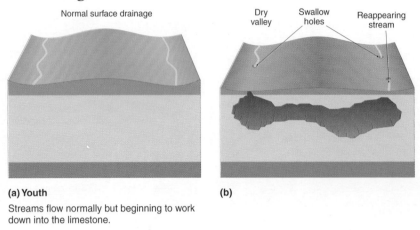

(a) Youth
Streams flow normally but beginning to work down into the limestone.

(b)

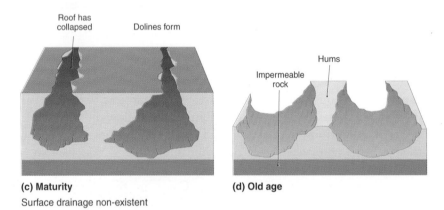

(c) Maturity
Surface drainage non-existent

(d) Old age

< Fig 3.11

In the youthful stage, the agents of denudation wear away the overlying bedrock (e.g. shale in the Burren) until the underlying limestone is exposed. Surface streams and rivers continue to erode into the limestone opening up the joints and bedding plains. As the rivers continue with the vertical erosion they eventually disappear underground through swallow holes.

Mature Stage

At this stage the underground water has eroded tunnels and caves. Over time the roofs of these caves collapse to form wider depressions known as dolines. The surface is gradually lowered in this way.

Old Age Stage

At this stage the original limestone surface has been lowered by the formation of poljes and uvulas. As a result all underground caverns disappear and the underground drainage now appears as surface drainage on the underlying impermeable rock which is now exposed.

Here and there, residual masses of limestone known as *hums* may rise above the general level of the landscape.

Fig. 3.12
Dolmen at Poulabroune, Co. Clare

The Burren was once described, by a soldier in Oliver Cromwell's army as a place where:

> *'There is not enough water to drown a man, not wood enough to hang him, nor earth enough to bury him.'*

The Burren is undoubtedly one of the most distinctive regions of Ireland. In many ways it resembles a desert. Surface water and vegetation are scarce and the density of population is very low. While the lack of surface water has been explained above, it is more difficult to explain the absence of soil cover. Many geologists believe that the region was stripped of its soil cover due to the scouring action of the ice sheets during the last Ice Age. Yet in spite of this, sufficient pockets of soil do exist in sheltered grikes where, aided by a mild maritime climate, Alpine and Mediterranean plants flourish.

Botanists from all over the world visit the region every year. The mild climate and the workable soil also helps to explain the early settlement of the region, which dates from pre-historic times. The unique flora has also attracted a unique fauna and certain species of butterflies found here are believed to be found nowhere else in the world. This region is a magnet for tourists while the maze of underground passages and caverns form an obvious attraction for speleologists.

Questions

Ordinary Level

1. Soluble Limestone or karst regions, such as the Burren, contain a great variety of landscape features, both over and under the surface.
   (i) Name any **three** of these features. *(15 marks)*
   (ii) For **each** feature named above, describe and explain, with the aid of a diagram, how it was formed. *(45 marks)*

(iii) The Burren is Ireland's best known soluble limestone region. Briefly explain its importance referring to some of the following:
- Heritage and Tourism
- Flora and Fauna.

(20 marks)

Leaving Cert. 1995

2. Examine the following statement in some detail.
   'Limestone regions have distinctive underground and surface features.'
   (40 marks)

   Leaving Cert. 1994

HIGHER LEVEL

1. (i) 'The karst topography of the Burren region exhibits many karstic landforms both above and below the surface'.
   Select **two** features above and **two** below the surface and explain with the aid of diagrams how each of these features was formed.
   (80 marks)

   (ii) Give your views on recent plans for an interpretative centre in the Burren region. *(20 marks)*

   Leaving Cert. 1993

CHAPTER 4 — Rivers and River Basins

Fig. 4.1

We depend upon rivers for transport, energy and irrigation while their fertile flood plains have nurtured human progress since the dawn of civilisation. They are unquestionably the most dominant agent in the whole process of denudation and of all the external agents of erosion have had the greatest impact on people.

N.B. River action is often referred to as *fluvial action* (Latin *fluvius*, a river).

The Origin of Rivers

When rain falls some of the water evaporates, some runs off and the remainder soaks into the ground. Rivers then get their water from run-off and in some cases from the water which soaks down and may later emerge as springs (see Chapter 3). As an agent of landscape alteration, rivers have three basic functions – to erode, transport and deposit.

Rivers and River Basins

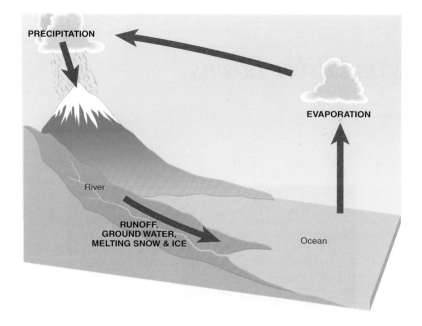

Fig. 4.2
The Hydrocycle

EROSION

The essential function of a river is to erode the land over which it flows. Erosion will continue until base level is achieved. This is a key concept in the study of stream activity.

> Base level is defined as the lowest elevation to which a stream can erode its channel.

Base level accounts for the fact that most streams have low gradients near their mouths (in the lower course) because the streams are approaching the elevation below which they cannot lower their beds.

The idea of base level was put forward in 1875 by an American geologist John Wesley Powell. He stated:

> *'We may consider the level of the sea to be a grand base level below which the dry land cannot be eroded; but we may also have, for local and temporary purposes, other base levels of erosion ....'*

RIVER VALLEY

Erosion is extremely active in the upper course. To erode, a river needs energy which it derives from its volume and velocity. It is important to note that the volume and velocity of any stream or river can change in a relatively short period of time. In the wet season and particularly in times of flooding, with increased volume and velocity, the energy of a river can dramatically increase giving it greater powers of erosion and transportation.

The Origin of Rivers

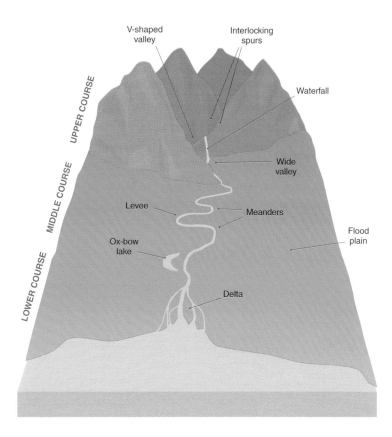

Fig. 4.3 > River Basin

THE PROCESSES OF RIVER EROSION

1. Hydraulic Action

Hydraulic action is the term used to describe the sheer force of running water which can be very effective near the banks on bends in a river. The process of *cavitation* is closely associated with hydraulic action. This is when bubbles of air collapse and the resultant vibration can loosen softer sands and clays at the sides of the channel.

2. Chemical Action – Solution

This occurs when the water in the river dissolves the rock both on the bed and sides. The chemical action involved will depend on the local rock type and so may be hydrolysis, hydration, oxidation or carbonation (see Chapter 2).

The combination of hydraulic and chemical action help to provide the river with a *load* (rock fragments etc.).

3. Corrasion

Corrasion is the wearing away of the banks and bed of a river by its *load*. The ability of a river to carry its load (which depends on the volume and velocity) is known as its *competence*. This is the most effective process of erosion especially in the wet season. This continual scraping, rubbing and bumping is also known as *abrasion*.

Leaving Certificate Geography

4. Attrition

As the river carries its load the particles are constantly colliding with each other and the bed of the river. As a result these particles become progressively smaller in size as they move downstream. This also explains why boulders and pebbles are rounded and smooth in appearance.

While vertical erosion is the most obvious form of erosion as the river erodes to base level it should be remembered that a river also erodes laterally, i.e. the sides of its banks, and headward as the river eats back through its source. This is also known as spring sapping.

These processes then explain how a river erodes to base level. The eroded material is now transported downstream as follows.

THE PROCESS OF TRANSPORTATION

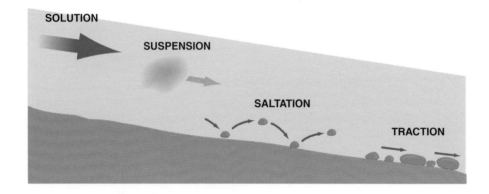

< Fig. 4.4
How a River Transports its Load

1. Suspension

Most of the fine particles of clay and silt are transported on or near the top of a river. Initially they may be lifted by hydraulic action and then kept there by the turbulence of the water.

2. Solution

Most rivers carry material in solution, though this method of transportation is not usually visible. Streams flow over rocks which contain soluble minerals and dissolve them. This is especially true in limestone and chalk areas.

3. Saltation

Saltation applies to the load where the particles are too heavy to be carried in suspension. They are lifted by hydraulic action but because of their weight they fall to the bed. The action is repeated so the load is said to bounce or jump (Latin, *saltus*, 'jump').

4. Traction

This applies to the load whose particles are too heavy to bounce. Instead they are rolled and dragged along the bed, especially in the wet season. The processes is sometimes referred to as '*bed load drag*'.

THE PROCESSES OF DEPOSITION

A river deposits its material due to a reduction in energy. This loss of energy can be caused by:

1. *A decrease in volume which may be due to:*
    (i) A period of drought or the ending of the wet season, e.g. summer in Ireland.
    (ii) A river flowing through a desert, e.g. Nile.
    (iii) A river flowing over a porous rock.

OR

2. *A decrease in velocity due to:*
    (i) A reduction in gradient.
    (ii) Entering a lake.
    (iii) Reduction in volume.

The Development of a River Valley

It is possible to compare the courses in the profile of a river with a river's stages of development:

Course	Stage
1. Upper	1. Young
2. Middle	2. Mature
3. Lower	3. Old

1. Upper Course

The river is said to be in its youthful stage in the upper course as it is primarily involved in vertical erosion.

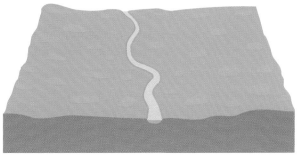

Fig. 4.5

YOUTHFUL STAGE

2. The Middle Course

The river is said to be in its mature stage as vertical erosion slows down, mainly due to a decrease in velocity.

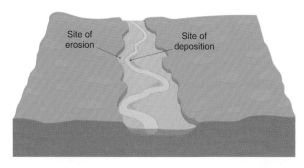

< Fig 4.6

MATURE STAGE

3. The Lower Course

The river is said to be in its old stage because vertical erosion has almost ceased as the river reaches base level. The main activities in the lower course are lateral erosion and deposition.

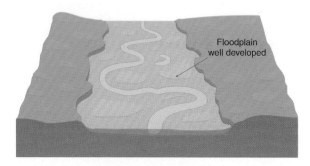

OLD STAGE

< Fig 4.7

FEATURES OF RIVER EROSION

The Upper Course - Young River Valley

1. V-Shaped Valley

Example: Upper Liffey and Upper Barrow.

In its upper course, a river concentrates on vertical erosion and cuts down into its bedrock. In the early stages it carries little material so that corrasion and attrition are less effective than hydraulic and chemical action. Over time the hydraulic and chemical action, together with the weathering process, provides the river with a load which enables it to erode by corrasion and attrition (see Fig 4.1).

Precisely how steep-sided and narrow the valley will be depends on a balance between the speed of vertical erosion by the river and the speed of weathering (e.g. frost action) and mass movement on the valley sides. In areas where the bedrock is more resistant to weathering (especially true of arid regions), the weathering process is slow and vertical erosion remains the dominant force so that *gorges* and *canyons* are common. There is a well developed gorge on the Munster Blackwater where it turns south at Cappoquinn.

2. Interlocking Spurs

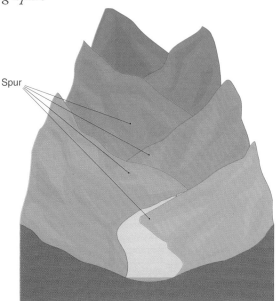

Fig. 4.8
Interlocking Spurs

Examples: Upper Liffey; Upper Barrow.

Rivers in their upper course rarely follow a straight course. The precise reason for this is not always clear. However in some areas variations in rock type either in the bed or the sides will encourage the river to follow a winding course.

Because the current tends to be strongest in the outside of a bend, these bends become more exaggerated. Over time sections of the valley sides project as spurs which are 'locked' into each other, hence interlocking spurs.

3. Waterfalls

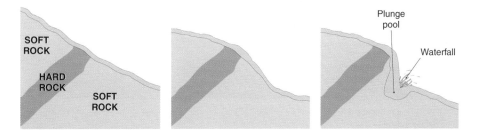

Fig. 4.9
Formation of a Waterfall

Examples: Torc Waterfall near Killarney, Co. Kerry; Powerscourt, Co. Wicklow.

Waterfalls are perhaps the most attractive and spectacular features produced by rivers. They owe their origin to the presence of a varied bedrock, i.e. where a band of resistant (hard) rock lies transversely across the valley of a river, so interrupting its progress towards a graded profile. The river erodes the softer rock downstream more quickly and so a waterfall forms over the resistant rock. The upstream less resistant rock is protected by the band of the resistant rock.

At the foot of the falls a plunge pool is formed, representing the increased hydraulic and corrasive power of the river at this point. Since all rivers will eventually produce graded profiles, waterfalls retreat upstream over time and eventually disappear. The Niagara Falls for example is retreating at the rate of one metre per year.

The formation of waterfalls is not exclusive to river erosion. They may also result from:

i) Glaciation	where hanging valleys produce a vertical fall of water (see Chapter 5).
ii) Plateaux	where a river plunges over the edge of a plateau e.g. Livingstone Falls on the Congo River (Africa).
iii) Faulting	the famous Victoria Falls on the Zambezi River owe their development in part to faulting.

4. Pot Holes

The formation of pot holes more than any other feature of erosion probably best illustrates the effectiveness of corrasion. Using its load as a 'grinding tool' the pebbles in the river are whirled round by eddies in hollows in the bed to scour and excavate circular depressions known as pot holes.

The Middle Course – The Mature River Valley

As a river enters its middle course (mature stage) lateral erosion becomes more effective than vertical erosion and gradually the valley develops a broad flat-bottomed appearance (in cross profile).

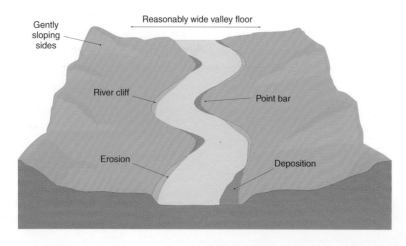

‹ Fig. 4.10
Mature River Valley

1. Meanders

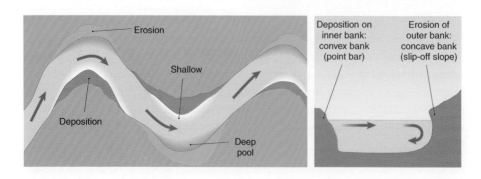

‹ Fig. 4.11
Formation of a Meander

The Development of a River Valley

Fig. 4.12 Meanders

Example: River Shannon: near Bunratty; River Lee near County Hall (Cork).

Meanders are pronounced curves or bends in the middle course of a river. As the river swings from side to side the current, being stronger on the outside, erodes the bank by lateral erosion to form a river cliff. Undercutting now produces a concave slope.

On the opposite bank however, the current flows more slowly and the material which is deposited forms a point bar, where the slope of the bank is more convex in shape. While this helps to explain the development of meanders it does not account for their initiation which is still not fully understood.

As the meanders migrate downstream they create an increasingly wide valley. Over time, each spur is completely removed having a cusp or bluff overlooking the valley floor.

2. Flood Plains – Levees

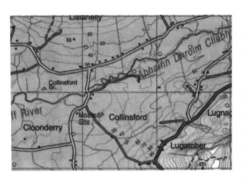

Fig 4.13

Examples: Lower Course Shannon near Limerick City; 'Lee Fields' near Cork City.

As the name implies, a flood plain is that part of a river valley that is covered by water during a flood. As rivers enter their middle-lower course a reduction in velocity reduces the river's ability to transport its load. As a result the river deposits gravel on its bed and so becomes shallower. These gravel deposits may split the river into numerous smaller channels and the river is now said to be braided. In times of heavy rainfall then, the river often bursts its banks and floods the surrounding land which is known as a flood plain. As the river overflows its banks onto the flood plain the water flows as a broad sheet. Since such a flow pattern significantly reduces the water's velocity the

Leaving Certificate Geography

Rivers and River Basins

coarser portion of the suspended load is deposited in strips bordering the channel to form natural *levees*.

As the water spreads out over the flood plain a finer sediment or alluvium is laid down over the valley floor when the water stagnates. The natural levees in the lower Mississippi rise six metres above the lower portions of the valley floor. The presence of levees often restricts the movement of water back into the river after a period of flooding. As a result these areas are often poorly drained and are known as *back swamps*. Levees can also hinder the entry of a tributary into the main channel. As a result, the tributary may be forced to flow through the back swamp zone parallel to the main river for many kilometres before it eventually joins the main river. Such streams are called *yazoo* tributaries after the Yazoo river which parallels the lower Mississippi. They are sometimes referred to as *deferred juncture streams*.

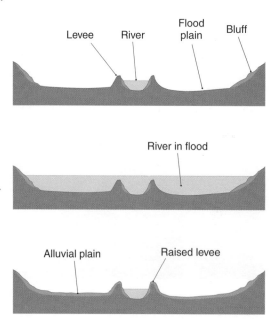

< **Fig. 4.14**

THE LOWER COURSE: OLD AGE

1. Ox-Bow Lakes

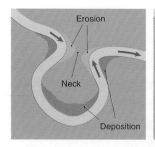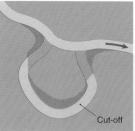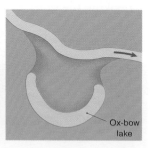

< **Fig. 4.15**

Formation of an Ox-Bow lake

Ox-bow (horse-shoe) lakes are normally associated with the lower course of old river valleys. As the meanders move downstream they may become more pronounced or exaggerated. In times of flooding the velocity of the river increases and the current cuts through the neck by lateral erosion. The meander is now isolated from the river. Over time, deposition at the sides of the channel finally results in the meander being cut off. Generally, ox-bow lakes have a short life span. They lose their water through evaporation and are often covered over by reeds and rushes.

These *mort lakes* or *meander scars* are easily recognisable in aerial photographs by the colour of the vegetation which contrasts sharply with that of the surrounding area.

2. Deltas

When a river enters a lake or the sea its rate of flow is checked and deposition takes place. Gradually a fan-shaped mass accumulates at the point of entry. Such deposits of sediments are known as deltas since they resemble the shape of the fourth letter of the Greek alphabet. Deltas may form in either lakes *(lacustrine deltas)* or the sea *(marine deltas)*. When a river enters a lake or sea it drops the coarse material first, i.e. the bed load. Further out, it drops its suspended load. The fine material settles as horizontal layers lying roughly parallel to the sea or lake floor. These deposits are called the *bottom set beds*. The coarser material deposited on top of the bottom set beds gradually build to form *foreset beds*. Over time the delta gradually builds seawards. The horizontal deposits resting on top of the foreset beds are known as *top set beds*.

Not all rivers form deltas. In many areas currents and tides are sufficiently strong to remove all the material brought down by the river. Deltas occur where the rate of deposition exceeds removal. A river tends to carry most of its load at the centre since velocity is highest there. Consequently when a river enters the sea, it builds a bank at the centre of the estuary.

The mouth of the river is now divided into two channels and the process is repeated. In this way the river keeps dividing up into smaller channels called *distributaries*.

Types of Deltas

In spite of their name not all deltas have the idealised shape. Differences in the nature of shoreline processes and in the shape of the coastline can result in deltas having very different shapes.

i) Arcuate Deltas

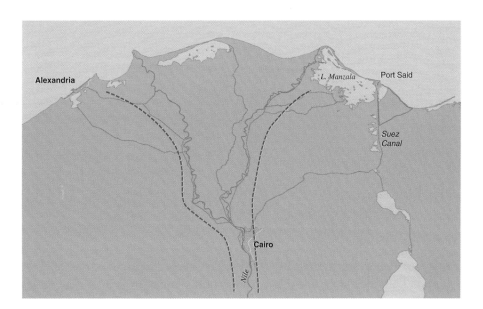

Fig. 4.16
The Nile Delta (Arcuate)

Arcuate deltas form the classic triangle shape and can be found at the mouths of the Nile and Po rivers. They are generally composed of coarse sands and gravels where sea currents are relatively strong.

ii) Estuarine Deltas

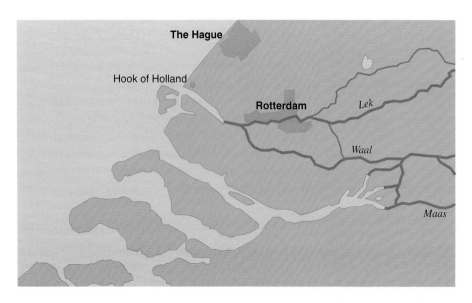

Fig. 4.17
The Rhine Delta (Estuarine)

Estuarine deltas form at the mouth of submerged rivers. The deposits form narrow estuarine fillings down both sides of the estuary. It is thought that this type eventually develops into an arcuate delta. The Rhine delta is a good example of an estuarine type.

iii) Bird's Foot Deltas

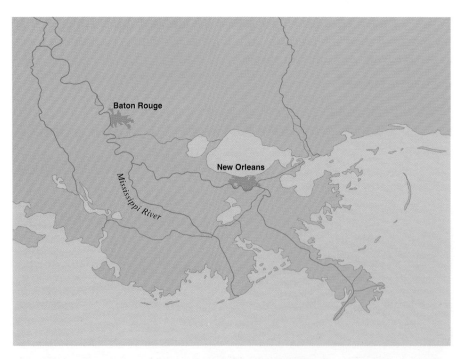

Fig. 4.18
Mississippi Delta (Bird's Foot)

Bird's foot deltas are formed by rivers which carry large quantities of fine material. The river deposits this material and divides in only a few large distributaries. It extends into the sea like toes – to form a bird's foot shape. The Mississippi is a good example.

The Development of a River Valley

RIVER REJUVENATION

> Rejuvenation means 'to be made young again' and is applied to a river when it regains its erosive power.

Causes of Rejuvenation

The most common cause of rejuvenation is a change in base level. This may be due either to a fall in sea-level or a rise in land level. In either case there is a change in the relative levels of land and sea and vertical erosion is renewed.

The Effects of Rejuvenation

(i) Knickpoints; (ii) Terraces; (iii) Incised meanders.

i) Knickpoint

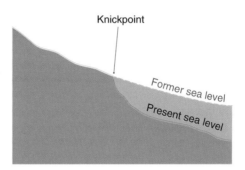

Fig 4.19

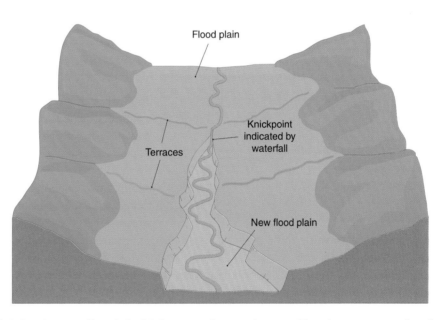

Fig 4.20
Rejuvenated River Valley

Knickpoints are 'breaks' which occur along a river profile when a new cycle of erosion is initiated. A number of knickpoints may occur on a river, an indication of the number of times the river was rejuvenated.

An example of a knickpoint can be found on the River Barrow at Muine Bheag.

ii) River Terraces

As a result of rejuvenation, vertical erosion is renewed and the river cuts into the alluvium it has already deposited on its flood plain. Over time the river

Rivers and River Basins

develops a new, if narrower, flood plain at a lower level within the older flood plain. The old flood plain now remains as terraces on either side of the river and are referred to as paired terraces. A number of terraces may occur on either side of the river indicating the number of times rejuvenation has occurred.

Examples of terraces can be found at Inniscarra on the Lee and in the Dodder (a tributary of the Liffey) at Terenure.

iii) Incised Meanders

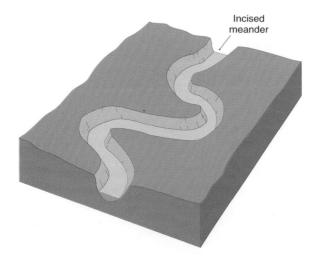

The word 'incised' means 'cut into' and refers to a meanders where the river has cut deep into its bed while maintaining its winding course. Incised meanders occur in the River Nore south of Thomastown and in the Barrow south of Graiguenamanagh.

‹ Fig. 4.21
Incised Meanders

Patterns of Drainage

Dendritic

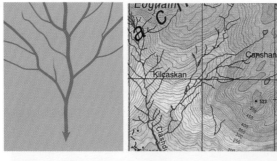

The development of a drainage basin begins with a number of main streams flowing directly down a slope to the sea. These are the result of, or consequent upon, that slope and are therefore called consequent streams. As they develop, tributaries flow towards these main valleys joining the parent river obliquely and in turn these are joined by minor tributaries. If the bedrock is homogenous (all the same), a dendritic (tree-like) pattern will develop, so called because the pattern resembles the branches of a tree; (*dendron* is the Greek word for a tree). The river Shannon provides a good example of this pattern.

‹ Fig 4.22
Dendritic Drainage Pattern

Trellised

When tributaries flow into the main river at right angles a trellised pattern is formed. This often develops in a varied bedrock.

Streams are joined where bands of resistant and less resistant rocks lie at right angles to the consequent streams.

The Development of a River Valley

Fig. 4.23
Trellised Drainage Pattern

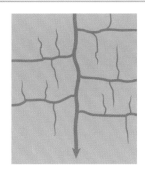 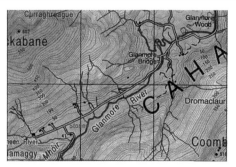

Radial

Where the upland topography is circular or oval shaped rivers and streams radiate from the watershed in different directions to form a radial pattern.

Fig. 4.24
Radial Drainage Pattern

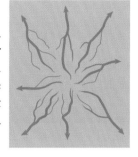 

Deranged

As the name suggests, rivers and streams have a confused or deranged appearance with no obvious direction of flow.

It often develops on deposits of glacial material as in the Newport-Westport lowlands.

RIVER CAPTURE

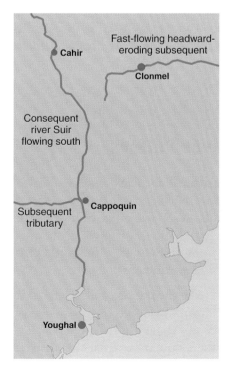 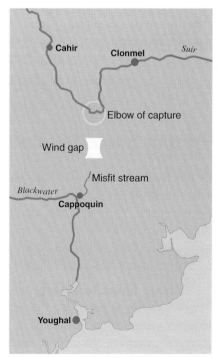

Fig. 4.25
River Capture

Leaving Certificate Geography

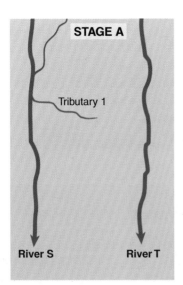
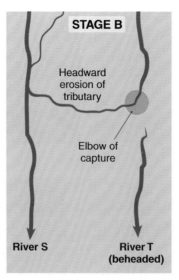

Fig. 4.26
River Capture

When two river systems (river basins) exist side by side the high ground which separates them is known as a *watershed*. By the process of *headward* erosion *(spring sapping)* rivers constantly eat back through their source. Because the tributaries of the more powerful river will be at a lower level (due to its increased vertical erosion) it *captures* the flow of the neighbouring stream. River capture or *stream piracy* is now said to have occurred. The place where capture occurred is usually marked by a bend in the river known as the *elbow of capture*. The captured stream is now beheaded and its reduced volume means that it is disproportionate to the size of the valley which it occupies. In consequence it is now referred to as a *misfit*. The dry valley which now exists between the elbow of capture and the source of the misfit stream is known as a *wind gap* or *dry gap*.

As a result of capture the volume of the subsequent stream (which captured its neighbour) can be dramatically increased giving the river increased erosional power. For the captured stream the geomorphic processes are proportionately decreased. River capture can be seen as a struggle for existence between rivers, a type of 'survival of the fittest' as streams battle to become the 'master-stream' in that area.

Human Societies and River Valleys

Since earliest times, people have been attracted to rivers and some of the world's earliest civilisations have developed along river valleys. This is understandable when we realise that apart from supplying two of the essential resources needed for human survival, i.e. water and soil (alluvium) they also provide food (fish) and provided a ready form of transport – the world's first highways – over which people first travelled long distances.

The essential difference however between early civilisations and people today is that societies adapted to the river's environment in the past whereas modern society sets out to change it. While this change has allowed people to successfully manage river environments, the long term repercussions are less clear and have proved catastrophic in some areas.

DAM BUILDING

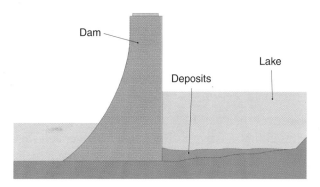

Fig 4.27

The physical effects of dam building in a river are quite obvious. Before dam building, vertical erosion was the main function of the river. As a result of the dam a new base level develops and deposition takes place behind the dam. This now interrupts the progress of the river towards a graded profile. The presence of the dam also deprives the river downstream of its load and in consequence the rate of vertical erosion is reduced.

While the geomorphological effects of dam building may be obvious the socio-economic effects of this type of *river management* are less clear as dam building can have major long-term effects for the river's environment.

H.E.P.

One of the more positive effects of dam-building is the production of hydro-electric power. This 'atmospheric friendly' fuel can be used for domestic and industrial purposes and often reduces the need to import fossil fuels. The inevitable consequence of dam building however is the creation of an artificial lake. Apart from their obvious recreational value such lakes can also be used for fish farming. Dam building allows for more effective control of downstream flooding too with obvious benefits for agriculture, etc.

Dam building however also has negative socio-economic effects. The resultant artificial lake means that from an environmental viewpoint a 'river ecology' has been replaced by a 'lake ecology' with obvious effects for the local flora and fauna. The ecology behind the dam is now submerged and consequently destroyed, often involving the loss of valuable agricultural land. In warm countries, the water stored behind the dam is subjected to evaporation and seepage. The stagnant waters also provide ideal breeding grounds for diseases like bilharzia. Downstream from the dam often fares no better. The absence of flooding deprives the flood plain of fertile alluvium, a lifeline to survival for millions in developing countries. The reduction in the discharge also increases the salinity of coastal waters which has all but wiped out the fishing industry along the Mediterranean coast in countries such as Egypt.

Artificial Levees

Sometimes artificial levees are built along rivers as means of flood control. They are generally easy to distinguish from natural levees as their slopes are much steeper. When a river is confined by levees during periods of high water it deposits material in its channel as the discharge diminishes. This is sediment that otherwise would have been dropped on the flood plain. So apart from depriving the flood plain of this 'natural fertiliser' the bottom of the channel builds up as deposits are left on the river bed each time there is a high flow. With the build-up of the bed, less water is required to overflow the levee.

As a result, the height of the levee must be raised to protect the flood plain. Artificial levees are not, therefore, a permanent solution to the problem of flooding. If protection is to be maintained the structure must be heightened periodically, a process that cannot go on indefinitely.

Short-term objectives then can lead to more serious, longer-term problems. It is now recognised that successful river basin management must involve the whole river – from source to mouth – a difficult task when major rivers often cross several countries whose peoples may have different priorities for the use of that basin and different levels of wealth to develop and conserve it.

Rivers then, apart from their geomorphological contribution, have been the lifeline of human society since the dawn of history which makes it all the more difficult to understand why modern society treats them like sewers – convenient dumping grounds for domestic, agricultural and industrial waste.

Questions

ORDINARY LEVEL

1. V-shaped valley, Ox-Bow Lakes, Delta, Interlocking spurs, Waterfall.
   (i) In the case of **each** of the above features, found along the course of a river, state whether it is formed by erosion or deposition.

   (18 marks)

   (ii) Select any **three** of these features and, with the aid of a diagram, describe how each was formed. *(48 marks)*

   (iii) 'Large-scale flooding by rivers has caused enormous problems for local communities.' Discuss the statement. *(14 marks)*

   Leaving Cert. 1996

2. Examine the following statement in some detail. Erosion by rivers has resulted in the formation of distinctive landscape features. *(40 marks)*

   Leaving Cert. 1995

3. (i) Erosion and deposition by rivers have helped shape the landscape. Name **three** landforms which result from river action.

   (15 marks)

   (ii) For **each** landform selected, describe and explain with the aid of a diagram how it was formed. *(45 marks)*

   (iii) 'Rivers are of great use to people and yet are often abused by people.'
   Explain this statement. *(20 marks)*

   Leaving Cert. 1994

4. (i) Explain, with reference to **three** typical landforms, how rivers help to shape the landscape. *(60 marks)*

   (ii) Examine briefly, using examples which you have studied, how human activities change the operation of natural processes in river valleys. *(20 marks)*

   Leaving Cert. 1991

HIGHER LEVEL

1. (i) Discuss with reference to **three** characteristic landforms, the natural processes at work in river valleys. *(75 marks)*

   (ii) Flooding in river valleys can be worsened by human activity. Examine this statement. *(25 marks)*

   Leaving Cert. 1997

2. (i) With reference to processes of erosion and processes of deposition, examine **three** ways in which rivers shape the Irish landscape. *(75 marks)*

   (ii) Examine briefly **two** examples of human management of rivers. *(25 marks)*

   Leaving Cert. 1996

3. (i) With reference to erosional processes **and** to depositional processes, examine some of the ways in which rivers shape the landscape. *(75 marks)*

   (ii) Human societies have always sought to control or manage river processes. Discuss this statement briefly, using **one** example which you have studied. *(25 marks)*

   Leaving Cert. 1994

4. Processes of erosion, transportation and deposition produce characteristic landforms. Select three landforms, one produced by glaciation, one by river action, and one by coastal processes. For **each** landform selected, explain with the aid of diagrams how it was formed. In your explanation include Irish examples of each landform. *(100 marks)*

   Leaving Cert. 1993

5. (i) Explain how the natural physical processes which are active in river valleys help to shape the landscape. *(75 marks)*

   (ii) Examine the effect which a drop in base level can have on a mature river system. *(25 marks)*

   Leaving Cert. 1992

CHAPTER 5
Glaciation and Landforms

< **Fig. 5.1**

In Chapter 3 we learned that the earth's water is in constant motion. However when precipitation falls in high elevations or high latitudes the water may freeze and become part of a large mass of ice known as a glacier. This is what happened about one million years ago during the Pleistocene period when global temperature fell and much of the earth's precipitation remained in a solid state to form glaciers which covered about 30 per cent of the earth's surface, compared to the present 10 per cent.

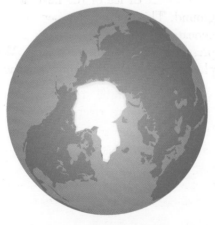

Europe during the Ice Age   **Europe today**

< **Fig. 5.2**

Leaving Certificate Geography

Types of Glaciers

CONTINENTAL GLACIERS

These are sometimes referred to as *ice sheets* and are massive accumulations of ice that cover large portions of a land mass such as Greenland and Antarctica.

ALPINE GLACIERS

These are smaller glaciers and are generally confined to mountain valleys. They are sometimes referred to as *valley glaciers*.

FORMATION AND MOVEMENT OF A GLACIER

Snow builds up when the amount that falls in winter *(accumulation)* exceeds that which melts in summer *(ablation)*. As the snow thickens, the lower layers are compressed into ice. This granular ice is known as firn or neve. As more snow collects, compression increases until all the air has been squeezed out and blue ice is formed. Once a glacier has formed it may advance, retreat or remain stationary. The actual movement depends upon the balance between accumulation and ablation. If accumulation exceeds ablation the glacier will *advance* which usually happens in winter and vice versa in summer.

Whatever their movement, glaciers are major geomorphic agents of landscape change and profoundly change the topography by erosion (as they advance) and deposition (as they retreat).

PROCESSES OF GLACIAL EROSION

Glaciers erode the land in two ways.

Plucking

Melt water is normally present at the sides and base of a glacier due to either the pressure of ice or the heat caused by friction as the ice passes over the ground. This melt water penetrates the cracks and joints of the bedrock beneath the glacier and freezes. As the water expands it exerts tremendous pressure that pries the rock loose. When the glacier moves on (advances), it plucks these rock fragments from the sides and floor of the valley.

Abrasion

Because of plucking, rock fragments become embedded into the sides and base of the glacier. The glacier now uses these to *scour*, rasp or *abrade* the rock surface over which it passes. This action often leaves long scratches or grooves in the bedrock known as glacial striae. While all glaciers erode by plucking and abrasion the extent of erosion will vary depending on:

- The rate of glacial movement.
- The thickness of the ice.
- The nature of the bedrock, i.e. whether it is resistant (hard) or less resistant (soft).

Features of Glacial Erosion

The effects of glacial erosion are best seen in upland areas which form collecting grounds for ice and from which the ice usually moves fairly rapidly downhill.

Cirque

Examples:

Coomshingaun in the Comeragh Mountains; Devils Punch Bowl, Mangerton Mountain, Kerry.

Cirques are large amphitheatre-like or circular hollows often arranged around mountain peaks. These great hollows are called cirques (French), cwms (Welsh), corries (Scottish), cooms (Irish).

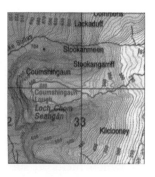

< **Fig. 5.3**
OS Map showing Coomshingaun

Cirques are where glaciers were first formed during the ice age. In the beginning snow accumulated in sheltered hollows usually on north and north-east facing slopes. The hollow was gradually enlarged by the process of nivation, i.e. where material was weathered by freeze-thaw action and removed by meltwater. As the snow accumulated in this hollow year after year, it was gradually converted into ice. As the ice was drawn downslope by gravity it deepened the hollow by plucking and abrasion. The ice pulled away from the backwall of the cirque and a

< **Fig. 5.4**
The Matterhorn

gap or crevasse called the *bergschund* appeared. While the freeze-thaw action continued on the backwall, rock fragments fell down into the base of the glacier which were then used to abrade the cirque floor. As the ice moved, it pivoted around a point scouring and deepening the floor – a process known as *rotational slip*. Today many of these hollows are occupied by lakes known as *tarns*. When two cirques cut back to back or side by side a process known as headward recession, a sharp ridge forms between them known as an *arête*. Three or more cirques cut back to back to form a peak known as a *pyramidal peak* or horn.

Features of Glacial Erosion

U-shaped Valley

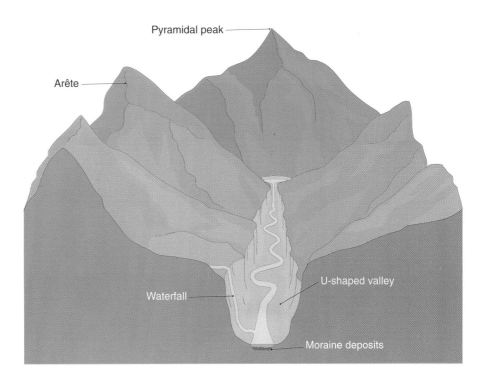

Fig 5.5

Fig 5.6
OS Map showing the Gap of Dunloe

Examples: The Gap of Dunloe (Kerry) and Glendalough (Wicklow) where the original lake has been divided into two by deposits carried down by rivers, are examples of U-shaped valleys.

Rather than creating their own valleys, glaciers take the path of least resistance by following the courses of pre-existing valley streams. Before glaciation the valleys were characteristically narrow and V-shaped because the streams were well above base level and so were involved in vertical erosion. During glaciation however the valleys are widened and deepened by the glacier creating a U-shaped valley or *glacial trough*. Because the glacier is less fluid than the former river it tends to remove the (interlocking) spurs of land that extend into the valley. The results of this activity are triangular shaped cliffs called *truncated spurs*. Apart then from straightening and widening the original river valley, glaciers often leave depressions in the valley floor, which may be due to the scouring of less resistant rock. When these hollows fill with water they form *rock basin lakes* or ribbon lakes. When they occur at intervals they are called *pater noster* (Our Father) lakes as they resemble a string of rosary beads. They are destined to disappear as a result of infilling by alluvium which is deposited by streams.

Leaving Certificate Geography

Glaciation and Landforms

OTHER FEATURES DUE TO GLACIAL EROSION

Hanging Valleys

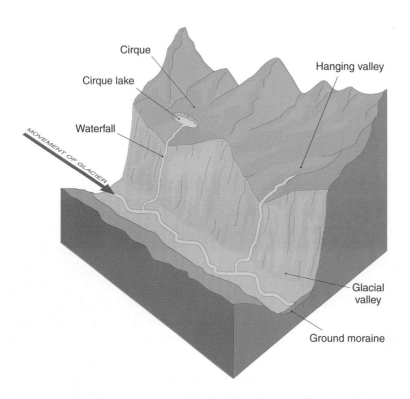

Fig. 5.7
Features caused by Glacial Erosion

Hanging valleys occurred when tributary valleys were not eroded vertically as fast as the main valley. This may be due to the fact that the tributary valley was occupied by a smaller glacier and so hadn't the same power of vertical erosion as the glacier in the main valley. As a result the tributary valley was left 'hanging' at the side of the main valley. Their presence is often marked by waterfalls as in Glencar, Co. Sligo.

Fig. 5.8
A Waterfall cascades from a Hanging Valley

Leaving Certificate Geography

Roches Moutonnées

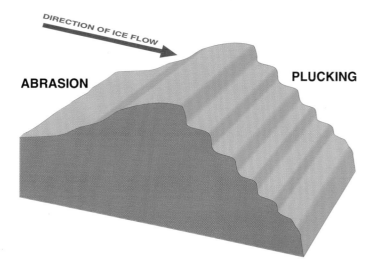

Fig 5.9 > Formation of Roches Moutonnées

When glaciers were unable to remove large rock outcrops in valley floors they simply moved over them. As the glacier approached the object, pressure resulted in melting which allowed the glacier to slide over the upstream side. However, once the pressure was released on the downstream side plucking took place resulting in a rough jagged appearance. They are so called because they resemble a type of wig which was fashionable in nineteenth-century France.

The feature known as 'crag and tail' is closely related to roches moutonnées. The crag is the plucked leeward side while the opposite side, the tail forms a gentle slope usually composed of glacial drift. One of the finest examples occurs in Edinburgh, Scotland where Edinburgh Castle is situated on the crag and the street, 'Royal Mile' situated on the tail. Carrigogunnel in Co. Limerick also forms a spectacular example.

Glacial Deposition

Unlike rivers which are capable of only carrying relatively fine debris, moving glaciers are capable of carrying all types of material varying from the finest 'rock flour' to boulders weighing several tonnes. Glaciers, for example, transported Galway granite south as far as Co. Cork. Rocks that are 'foreign' to the surrounding bedrock found in an area are termed *erratics*. All the material deposited by glaciers is known as *drift*.

When valley glaciers reach the zone of wastage, i.e. the point of melting they drop their load. This type of drift, deposited by the actual glacier is 'unsorted' and is called *boulder-clay* or *glacial till*.

As the glaciers melt, vast quantities of *fluvio-glacial* meltwater transport finer sediments. Unlike glacial till, this material is *sorted*.

It is possible then to distinguish two sets of landforms related to glacial deposition.

1. Landforms resulting from deposition by the glacier.
2. Landforms resulting from fluvio-glacial activity.

Landforms Resulting from Glacial Deposition

Moraines

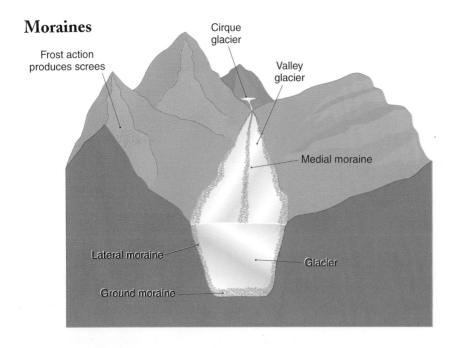

> Fig. 5.10
> **The Formation of Moraines**

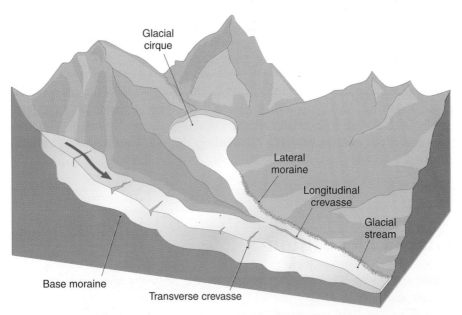

> Fig. 5.11
> **Crevasses and Moraines**

Moraines

Alpine glaciers (valley glaciers) produce two types of moraines that occur exclusively in mountain valleys – *lateral moraines* and *medial moraines*.

Lateral Moraines

As a glacier moves through a valley the ice erodes the sides of the valley. In addition large quantities of debris broken off by freeze-thaw action are added to the glacier's surface and collect on the edges of the moving ice. When the ice eventually melts this accumulation of debris is dropped next to the valley sides. The ridges of till running parallel to the valley sides are called lateral moraines.

Medial Moraines

Medial moraines are formed when two glaciers coalesce. The lateral moraines of the two glaciers join together to form a medial moraine. The till that was once carried along the edges of each glacier now forms a single dark strip of debris within the newly enlarged glacier. Medial moraines are one obvious proof that glacial ice moves because they could not form if the ice did not flow down the valley.

Terminal and Recessional Moraines

A terminal moraine is the outermost end moraine which marks the limit of the glacial advance. At the snout (front) all the material carried by the glacier like a giant conveyor belt is dropped to form an unstratified crescent shaped ridge of material known as a terminal moraine. Further up the valley 'temporary' terminal or *recessional* moraines mark the temporary halts in the final melting of the glacier. They may be intensive enough to block valleys and thus dam the waters of a river to form a moraine-dammed lake. The lakes in Glendalough, the Lower Lake in Killarney and many of the Alpine lakes such as Lake Como and Maggiore were formed in this way.

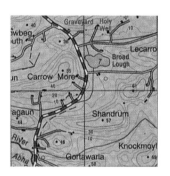

Fig. 5.12
A Road winds its way through Drumlin Country

Drumlins

Examples:

Leitrim – Monaghan (the county of little hills) Drumlin Belt; Clew Bay region, Co. Mayo (drowned drumlins).

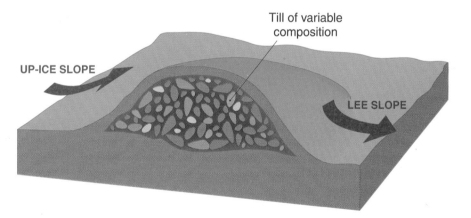

Fig. 5.13

Drumlins are oval shaped or egg-shaped elongated hills which can range in height from 15 to 50 metres and may be up to one kilometre long. The steep side of the hill faces the direction from which the ice advanced while the gentler, longer slope points in the direction the ice moved. Drumlins are found in swarms or clusters known as Drumlin fields and form a 'basket of eggs topography'. Drumlins are formed from boulder clay and their long axis lies in the direction of ice movement. While their origin is not yet fully understood, they are more commonly associated with continental rather than valley glaciation. They were probably formed when the ice was still advancing slowly but had begun to lose its power of transportation – possibly as a result of melting. The ice then deposited the boulder clay where friction between the clay and the underlying floor was greater than that between the clay and overlying ice. Its actual shape was then streamlined by the ice movement. In some cases the boulder clay was deposited on obstacles such as boulders to form false drumlins or rock drumlins.

< Fig. 5.14
Drumlins in Co. Louth

FLUVIO-GLACIAL LANDFORMS

Towards the end of the Ice Age, temperatures rose and vast amounts of meltwater flowed from the melting ice. These meltwater streams transported and sorted large quantities of stratified drift.

Outwash Plain

Examples: Curragh (Kildare); Wexford Lowlands

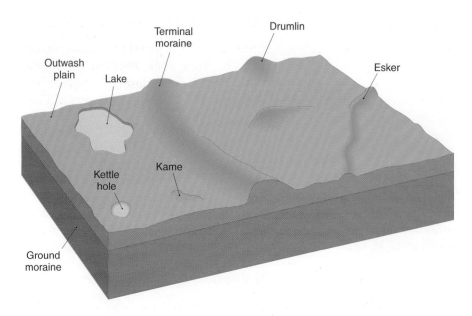

< Fig. 5.15
Fluvio-glacial Landforms

At the same time as the terminal moraine was forming, water from the melting glacier cascaded over and through the till transporting large quantities of material. As it spread over the relatively flat surface beyond the terminal moraine it rapidly lost velocity depositing the coarser material first and the finer drift further away to form an outwash plain, as in the Curragh in Kildare where fluvio-glacial deposits are up to 60 metres thick. Frequently blocks of stranded ice were buried under the outwashed deposits. When these blocks finally melted the overlying material slumped to form hollows known as kettle-holes, called after Kettle range in the USA where they are very common. They are common too in the Wexford lowlands (Curracloe) where they frequently form lakes.

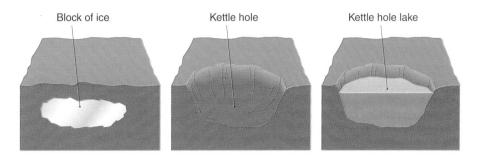

Fig. 5.16 › Formation of Kettle-Holes

Kames and Eskers

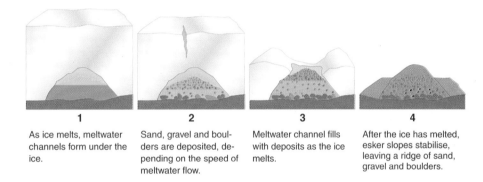

Fig. 5.17 › Formation of Kames And Eskers

1. As ice melts, meltwater channels form under the ice.
2. Sand, gravel and boulders are deposited, depending on the speed of meltwater flow.
3. Meltwater channel fills with deposits as the ice melts.
4. After the ice has melted, esker slopes stabilise, leaving a ridge of sand, gravel and boulders.

Examples: Clonmacnoise, Co. Westmeath; Athlone to Athenry (Galway).

Eskers are long winding ridges of stratified sand and gravel which run parallel to the direction of ice movement. In form they resemble a railway embankment which winds its way over a glaciated lowland without any relationship to the geology of the area. Their precise origin is uncertain. They may represent the beds of former streams which flowed in or under the ice. When the ice finally disappeared these beds of subglacial streams remained as eskers. The more probably explanation however is that they represent a continuously receding 'delta' formed at the edge of an ice-sheet or glacier.

Intermittent streams of melt-water emerging from the ice front formed isolated mounds of sand and gravel. These delta-like mounds are known as kames and are sometimes sufficiently large to allow development as sand and gravel workings.

Glacial Spillways

Examples of spillways on the Irish landscape include the Scalp and the Glen of the Downs in Wicklow and the Pass of Keimaneigh in Co. Cork.

As the glaciers melted (ablation) great quantities of melt water were released. When this melt water was trapped between the ice front and valley moraine (or a ridge of high ground) it formed a lake known as a *pro-glacial lake* (or

Glaciation and Landforms

inter-glacial lake). Over time the lake filled up and the water eventually escaped at the level of the lowest *col* (gap) in the ridge. The stream formed from this overflow now cut a deep valley across the obstructing high ground known as a glacial overflow channel or spillway.

Many of these are now dry while others are occupied by misfits.

OTHER CONSEQUENCES OF MELTING ICE

Apart from the landforms already discussed large quantities of drift of varying thickness covered the solid rock often forming ponds and small lakes. In some areas this uneven distribution of boulder clay gave rise to a confused drainage pattern, generally referred to as deranged drainage. Elsewhere, terminal moraines blocked the pre-glacial rivers forcing them to change direction as in the case of the river Bann. The great quantity of melt water too resulted in a rise in sea level. As a result, the sea invaded the lower courses of river valleys or glaciated valleys to form *rias* and *fiords* (see Chapter 6).

⟨ **Fig 5.18**
Geirangerfjord, Norway

Periglaciation

The term periglaciation strictly speaking means 'near to or at the fringe of an ice-sheet'. These areas have a climate which is not cold enough to cause glacier formation but do have long, cold winters just like the Arctic or Tundra regions of Northern Canada, Alaska and Northern Russia.

Permafrost

Permafrost is a layer of permanently frozen subsoil and weathered rock which can vary in depth but is thought to be up to 300 metres in Siberia and Alaska and at present covers about 25 per cent of the earth's surface. In these areas rocks had their joints and bedding plains sealed by ice and were thus rendered impermeable. Run-off meltwater then, unable to seep down, formed river valleys. When temperatures rose and the ice in the joints melted, the surface water disappeared. It was in these conditions that many of the dry valleys associated with chalk and limestone regions (Burren) are thought to have formed.

Active Layer

In areas where summer temperatures rise above freezing point the surface layer thaws to form an active layer. This active layer is often saturated because meltwater cannot infiltrate downwards through the impermeable permafrost.

Cryoturbation and Patterned Ground

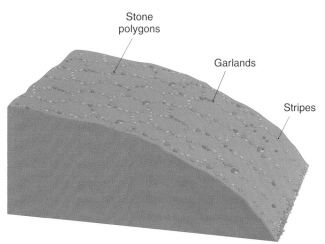

Fig. 5.19 > **Patterned Ground**

As temperatures fluctuate above and below freezing (0°C) the water in the active layer continually freezes and expands. This frost heaving or *cryoturbation* pushes stones sidewards and upwards to form a collection of stones. The larger stones with their extra weight move outwards to form, on almost flat areas, stone circles or stone polygons. Where this process occurs on slopes with a gradient in excess of 6° the stones will slowly move downhill under gravity to form elongated stone stripes or garlands.

This explains why a plot which was left stoneless in Autumn has become stone-covered by the spring following a cold winter.

Pingos

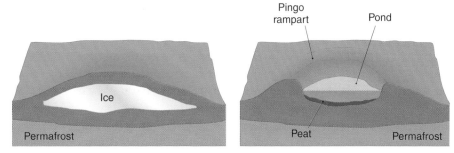

Fig. 5.20 > **Formation of a Pingo**

The land surface in many periglaciated regions is often irregular and hummocky. One of the most striking mounds or hillocks is a pingo which can have a diameter of up to 500 metres and may rise to 50 metres in height.

Pingos have a core of ice overlain by layers of surface debris. How the ice becomes trapped is not fully clear. One theory is that ground water is trapped between advancing underlying permafrost and a newly frozen surface. This water eventually freezes, expands and forces its way upwards in the same way as frozen milk lifts the cap off a bottle. When the ice eventually melts the mound collapses to form a circular depression often containing a pond. Ancient pingos occur west of Wexford town at Camaross. The centre of these mounds too have collapsed leaving a shape resembling a rath. Relative to Canadian pingos they are very small having a diameter of about one metre. Pingos are also known as *hydrolaccoliths*.

Because the permafrost layer prevents the infiltration of surface water the active layer often becomes saturated and very mobile where slopes exist. This 'soil flow' or solifluction has already been referred to in Chapter 3. Freeze-thaw action too is much more active than chemical weathering. On relatively flat upland surfaces, extensive coverings of large angular stones formed 'in situ' by frost action are known as blockfields or *felsenmeer* (literally a sea of stones). When meltwater removes this material the process is known as nivation.

Because high atmospheric pressure existed over the ice-sheets, strong winds blew outwards from these areas and carried with them fine materials which were deposited beyond the ice-sheet. This fine material with a high potential fertility is known as *loess* or limon (French).

DATING ICE AGE DEPOSITS

Radio Carbon Dating

Carbon 14 is formed in the upper atmosphere and reaches the ground as carbon dioxide in rainwater and so is present in living plants and animals. While the organism is alive the decaying Carbon 14 (radio carbon) is continually replaced. As a result, the ratio of Carbon 14 to Carbon 12 remains constant. However when the plant or animal dies the amount of Carbon 14 gradually decreases as it decays to Nitrogen 14. The rate of this decay is half every 5,730 years. Therefore, by comparing the amount of Carbon 14 to Carbon 12 it is possible to ascertain the approximate number of years it is dead. Because the remains of living organisms are found in glacial deposits then it is possible to estimate their actual time of deposition. New sophisticated analytical techniques has increased the usefulness of this 'clock' and it can now be used to date events as far back as 75,000 years.

The Counting of Varves

During the Pleistocene period, glacial lakes received rivers of meltwater in summer. The material carried by these rivers consisted mainly of sand, gravel and clay. Because the flow of melt-water was greatest in summer the river carried and deposited coarse material into the lake. As the flow of water was reduced in autumn to spring (as temperatures fell) finer silt was transported and deposited on top of the coarse material. This double layer of coarse and fine material is known as a *varve* and represents the total deposition for one year. Varve counting has revealed that the decay of the last European ice sheets commenced about 25,000 years ago and the ice-sheets finally disappeared from Ireland about 10,000 years ago.

Causes of Glaciation

Fig. 5.21

Although the theory of glaciation is now 150 years old no complete agreement exists as to the causes of this event.

THE THEORY OF CONTINENTAL DRIFT

(See Chapter 1)

Glacial features in present day Africa, Australia and India suggest that these areas experienced an Ice Age about 250 million years ago. It is now believed that during the geological past, continental drift accounted for many dramatic climatic changes as land masses moved to different latitudinal positions. While this may explain past ice ages however, the theory of plate tectonics is extremely gradual and cannot be used to explain the recent Pleistocene glaciation.

Milankovitch – The Astronomical Theory

Milankovitch, a Yugoslavian mathematician suggested a theory based on the belief that the climatic oscillations that characterised the Pleistocene period may be linked to variations in the earth's orbit. He deduced that ice ages resulted from changes in the amount of solar energy reaching the earth.

Obliquity

We know that the earth's tilt helps to explain the causes of seasons – summer and winter. However, every 41,000 years this tilt varies a few extra degrees.

Orbit

Every 100,000 years the earth's orbit stretches, and planet earth moves a little further from the sun.

Precession

The earth makes one complete wobble every 20,000 years.

Milankovitch calculated that the combined cycles of tilt, wobble and orbital stretch could change the distribution of incoming solar energy to cause cycles of ice ages.

There is now strong evidence to support this theory. An examination of sea-floor deposits such as fossil shells reveals changes in ocean temperature which match the ice-cycles predicted by Milankovitch.

GLACIATION AND HUMAN ACTIVITY

Agriculture

Glacial till usually improves the fertility of the soil. The boulder-clay deposits on undulating topography have made the Golden Vale in Munster a highly productive agricultural area.

In contrast, outwashed sands and gravels have limited agricultural value as in the case of the Curragh in Kildare while in areas like the Burren, glaciers scoured and removed much soil cover.

Communication

Glaciated valleys are generally straighter with flat floors in comparison to river valleys. This facilitates the construction of communication links such as roads and railways.

Building Materials

Apart from being used as natural routeways in ancient times, eskers today provide sand and gravel for the construction industry.

Energy Supplies

Waterfalls which flow from hanging valleys may be harnessed to provide hydro-electric power, while various glacial lakes provide natural reservoirs which help to maintain water supplies throughout the year.

Tourism

The rugged scenery associated with the west of Ireland in general and Donegal and Kerry in particular owes much of its beauty to the erosive power of ice. The Lakes of Killarney are rock-basin lakes though Lough Leane is ponded back by a moraine. Glacial erosion too has enhanced the beauty of Glendalough and the Wicklow region and the resulting tourist spin-off helps to off-set some of these regions' agricultural deficiencies.

Questions

LEAVING CERTIFICATE ORDINARY LEVEL

1. (i) Name any **three** features of glacial erosion. Describe and explain with the aid of diagrams how **each** feature was formed.
   (60 marks)

   (ii) 'Glaciated landscapes have been both an advantage and disadvantage to people'. Briefly discuss this statement. *(20 marks)*

   Leaving Cert. 1995

2. Examine the following statement in some detail. Glacial erosion creates many distinctive landforms. *(40 marks)*

   Leaving Cert. 1994

3. Examine the following statement in detail.
   Glacial erosion in highland areas has produced some of our most spectacular landscapes. *(40 marks)*

   Leaving Cert. 1993

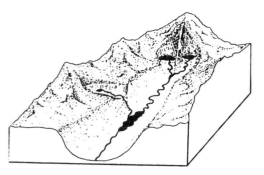

4. This diagram shows an area which experienced glaciation in the recent geological past.
   (i) With reference to landforms which are evident in the diagram, describe and explain how glaciation shaped the area. *(60 marks)*

   (ii) Examine briefly the human economic potential of landscapes which have been glaciated. *(20 marks)*

   Leaving Cert. 1992

LEAVING CERTIFICATE HIGHER LEVEL

1. (i) Discuss **three** ways in which glaciation has shaped the Irish landscape. *(75 marks)*

   (ii) Examine briefly how materials deposited by ice can be dated. *(25 marks)*

   Leaving Cert. 1997

2. (i) Identify *three* landforms produced by glaciation. In the case of *each* landform, describe and explain how it was formed. Use diagrams to support your answer. *(75 marks)*

Glaciation and Landforms

    (ii) Examine briefly how the results of glaciation of the landscape have been used to economic advantage by human societies.

(25 marks)

Leaving Cert. 1994

3. Processes of erosion, transportation and deposition produce characteristic landforms.

   Select **three** landforms, one produced by glaciation, one by river action, and one by coastal processes. For **each** landform selected, explain with the aid of diagrams how it was formed. In your explanation include Irish examples of each landform. *(100 marks)*

Leaving Cert. 1993

4. (i) Explain what basic conditions of climate and of topography are necessary in order to bring about a period of widespread glaciation of a landscape. *(30 marks)*

   (ii) Examine how moving ice shapes the landscape. Refer in your answer to:
   - Landforms formed beneath the ice and
   - Landforms formed at or beyond the ice front. *(70 marks)*

Leaving Cert. 1992

Leaving Certificate Geography

CHAPTER 6

Coastal Landforms – Marine Erosion and Deposition

Fig 6.1

In spite of the fact that the sea occupies over 71 per cent of our planet, its effects as an agent of landscape change are essentially confined to the coast.

Coast: The zone of contact between land and sea. The coast should not be confused with the shore or beach.

Shore: The land between the lowest tides and the highest point reached by storm waves.

Beach: This consists of an accumulation of sand, shingle and stones on the shore.

Most of the energy that shapes and modifies shorelines comes from wind-generated waves. Where the land meets the sea, waves that may have travelled unimpeded for thousands of kilometres suddenly meet a barrier that will not allow them to advance any further. As one writer put it:

'... the shore is the location where a practically irresistible force confronts an almost immovable object. The conflict that results is never ending and sometimes dramatic.'

The factors which influence the nature of the coastline are as follows:

MARINE	TECTONICS	CLIMATE
Waves, Tides, Currents	Movements of the Crust	Winds: Waves Currents
GLACIATION	**HUMAN**	**GEOLOGY**
Changes in sea level	Defence, Pollution Recreation	Rock Type and Structures

Leaving Certificate Geography

75

1. The Work of Water

The work of water is the work done by waves, tides and currents which act as agents of erosion, transportation and deposition.

2. The Nature of the Coastline

The nature of coastline refers to the type of rock – resistant or non-resistant – homogenous (all the same type) or varied; and whether the area is steep or low-lying.

3. Isostatic or Eustatic Movements

Isostatic and eustatic movement refers to changes in the relative levels of land and sea. Isostatic movement refers to movements in land levels. Eustatic movement refers to movements in sea levels.

4. Human Interference

Human interference might involve the construction of coastal defences against erosion such as groynes and rock armoury or the building of piers.

While the movement of waves, tides and currents all contribute to the actual shaping of coastlines, waves are the main agent of coastal destruction from an erosional viewpoint. Tides, which decide the actual position where waves break only play a minor role in the process of coastal disintegration.

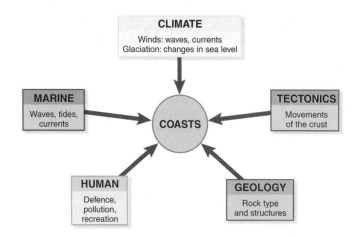

< Fig 6.2
Influences on the Coast

Coastal Erosion

How Waves Erode – The Processes of Erosion

1. Hydraulic Action

Hydraulic action is the shattering impact of the waves as they crash against the coast. It is particularly effective in stormy weather and it is estimated that Atlantic waves exert a force of nearly three tonnes per square foot on cliff faces in the west of Ireland.

2. Compressed Air

Fig. 6.3

Air is trapped in cracks, joints and bedding planes by the incoming wave (swash). This forces the air to compress which exerts enormous pressure on the rocks. When the wave retracts (backwash) the air expands. This sudden expansion too has an explosive effect which shatters rocks.

3. Corrasion

Corrasion is the term used to describe the pounding of the coast by the load in the waves (sand, pebbles, boulders). It is particularly effective during storm time and at high tide level when the breaking waves reach high up the cliff face. Corrasion is also known as *abrasion*.

4. Attrition

The rock fragments which are pounded against the coast and against each other are themselves broken down into smaller particles such as sand and shingle by a process known as attrition. Attrition is important because breaking the load down into smaller particles facilitates the process of corrasion by the swash. It also enables the material to be carried away by tides and currents.

5. Chemical

Sea water contains dissolved chemicals which have a solvent effect on rocks such as limestone and chalk.

Coastlines in Retreat – Features of Coastal Erosion

Cliff – Wave-Cut Platform

Example:

Cliffs of Moher, Co. Clare; Killiney, Co. Dublin; Ballybunion, Co. Kerry.

Coastal Landforms – Marine Erosion and Deposition

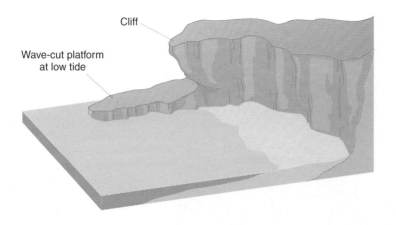

< Fig 6.4
Cliff and Wave-cut Platform

Cliffs and wave-cut platforms are common erosional features where the coast is exposed to destructive wave action. A notch is eroded near high-tide level by the processes of hydraulic, chemical, corrasion and attrition. Over time, the notch is enlarged until the overhanging rock collapses and a steep rock face or cliff is formed. As the notch continues to be cut, the cliff retreats inland and a platform cut by the waves, i.e. a wave-cut platform, is formed. The wave-cut platform slopes gently seawards. As it grows wider the depth of water decreases and there is a corresponding decrease in the erosive power of the sea. Eventually marine erosion will cease and the cliff retreats in response to other processes of sub-aerial denudation, e.g. freeze-thaw action. The actual profile of any cliff then (whether it is sloping or vertical) will depend on a balance between sub-aerial processes tending to reduce angle slope, sea erosion which tends to steepen the cliff face and the actual type and structure of the rock.

Caves – Arches – Stacks

Examples: Ballybunion, Co. Kerry; Kilkee, Co. Clare.

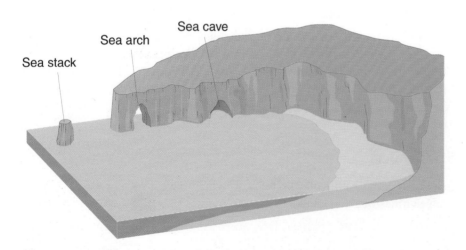

< Fig 6.5

Caves tend to form where the local rock has lines of weakness due to the presence of joints, bedding planes and faults. While all the processes are active in cave formation, compressed air is particularly effective. Lines of weakness are opened up and enlarged to form cavities. These, in turn, are enlarged to form caves. Sometimes air compression inside the cave may result in the formation of a chimney-like tunnel in the roof of the cave known as a *blow-hole*. In other areas, the roof of a cave may collapse leaving a long narrow sea inlet known as a *geo*. When sea caves form on the opposite side of headlands they may eventually meet to form a *sea-arch*, as can be found at Portsalon in Donegal and the Old Head of Kinsale in Cork.

Fig 6.6 > **Sea Arch**

Eventually the arch will collapse and the isolated part of the headland is known as a *sea-stack*. Eroded below sea-level it is known as a *sea-stump*.

Bays and Headlands

Examples: Wicklow Coast

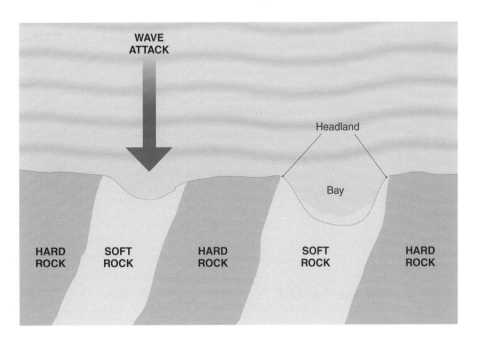

Fig 6.7 > **Formation of Bays and Headlands**

Bays and headlands occur where the coast is formed of a varied rock, i.e. bands of resistant and less resistant rock.

Because the less resistant rock is eroded more quickly (differential erosion) a bay is formed. The resistant rock then remains as headlands. Bays are funnel-shaped due to the reduction in wave energy. This reduction is best illustrated by the presence of a beach at the head of the bay known as a *bay head beach*.

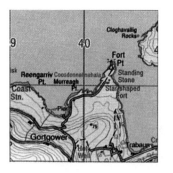

< Fig 6.8
OS Map showing Bay and Headland

MARINE DEPOSITION

Waves

The material eroded from the coast and brought in by rivers is transported in the sea by waves, tides and currents. This material is moved up the shore by the *swash* (on-shore wave) and seawards by the *backwash*. If the swash is more powerful than the backwash, material will accumulate on the shore so the waves are said to be *constructive*. If the movement of material down the shore by the backwash is greater, the waves are said to be *destructive*.

Longshore Drift

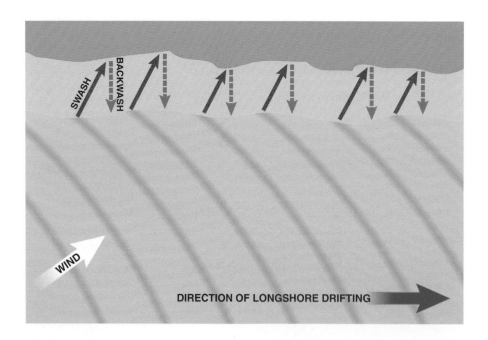

< Fig 6.9
Longshore Drift

When waves (driven by wind) approach the shore at an angle the material is moved obliquely up the beach by the swash. In contrast the backwash always returns at right angles so that the material is moved along in a zig-zag pattern. In this way the material drifts along the shore, i.e. *longshore drift*.

As a result, sand may accumulate in one part of the beach while other areas are denuded of sand. To prevent such uneven distribution of sand, groynes are often constructed to help to check such movement.

Features of Marine Deposition

Beach

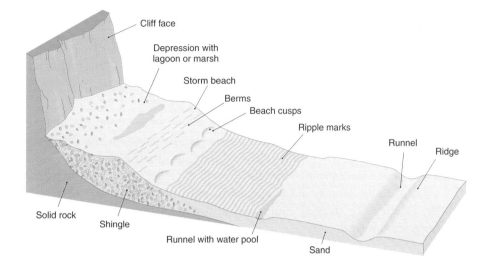

Fig 6.10 > Profile of a Beach

Fig 6.11 > OS Map showing a Beach

A beach is formed by the accumulation of material in response to constructive wave action between high and low tide. Sometimes storm waves may transport material above the normal level of high tide to form a *storm beach*. The beach may be non-existent on upland coasts when erosion is dominant, while a low sheltered bay between headlands usually has a crescent-shaped beach known as a bay-head or pocket beach.

Generally, the typical beach has a gentle concave profile and can be divided into two parts: (i) the upper beach or backshore, and (ii) the lower beach or foreshore area.

Upper Beach – This is composed of stones and other coarse material. The inland limit of the beach is in a form of a ridge composed of coarse material. This storm ridge or *berm* represents the material pushed up the beach by constructive waves at the highest tides.

Lower Beach – This is composed of smaller sand particles and shell fragments. A common feature of this part of the beach are ridges and runnels. The ridges are the result of constructive wave action running parallel to the coastline and are separated from each other by *runnels* – depressions which often contain water.

While most beaches conform to this general description, it must be remembered that a beach profile is constantly changing to adapt to varying conditions of tide, wave frequency, wave height, etc.

Sand Dunes

Example: Strandhill and Rosses Point, Co. Sligo; Portmarnock, Co. Dublin; Brittas Bay, Co. Wicklow

Sand dunes form on all coastal areas where wide expanses of sandy beaches are exposed at low tide and dried by the wind and sun.

On-shore winds then transport the sand off the beach by *saltation* and *surface creep* to form small mounds which are quickly colonised by vegetation such as marram grass. The root system stabilises the sand mounds while the grass itself traps more sand. Over time individual mounds grow larger to form sand dunes and sand hills. In some areas the growth of dunes is so rapid that they may overrun fertile land as has happened in parts of Wexford and Donegal. Human interference such as camp fires and pathways can also encourage dune migration inland.

Coniferous trees are often planted to stabilise the dunes as in Strandhill in Sligo and Curracloe in Wexford.

Sand Spit

Examples: Clew Bay (Co. Mayo); Rosslare (Co. Wexford); Dingle Bay (Co. Kerry).

Sand spits develop in areas where the coastline undergoes a sharp change of direction as in bays, inlets and estuaries. The headlands check the progress of longshore drift down the coast in the same way as groynes. As beach deposits are carried seawards they gradually build up to form a projecting ridge of material called a spit which is attached to the mainland at one end. The seaward end of a sand spit is often curved or hooked as a result of wave refracturing which causes the waves to swing round the end of the spit. Where extensive spits develop, sand dunes often develop on their landward side.

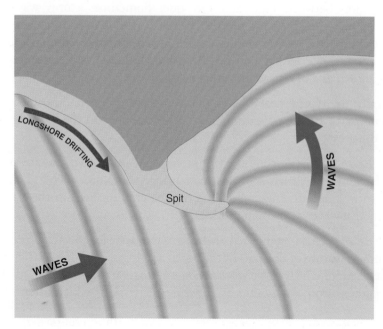

Fig 6.12
Sand Spit

Fig 6.13
Inch Strand and Sand Spit

Sand Bars

Sand bars may be divided into two types:

(1) Off-shore Bars; (ii) Baymouth Bars.

(i) Off-shore Bars

These form well away from, and roughly parallel to the shoreline. Their origin is disputed but they generally develop on very gentle sloping coastlines where there has been considerable off-shore deposition in the past. They may be due to wave action where the material which drifted along the shore is deposited just inside the line where the incoming waves first break (hence the term 'break-point bar') or they may form from material 'combed' directly down the beach by the backwash of the waves. The end result is the formation of a bar, behind which marshes, mudflats and lagoons can accumulate.

(ii) Baymouth Bars

As their name suggests, these are ridges of sand and shingle which lie across the mouth of a bay or estuary. They can originate as off-shore bars and migrate shorewards. In time they may cut the bay off from the sea to form a lagoon as in Tacumshin Lake (Wexford). Mud is deposited in the enclosed area (by the inflowing river). As the accumulation of mud continues, the lagoon becomes shallower. Over time the mud flats build up to form salt marshes as at Lady's Island Lake (Wexford) and north of Dublin Bay.

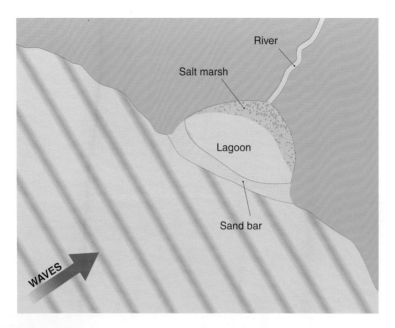

< Fig 6.14

Tombolo

Where a bar or a spit joins an island to the mainland it is known as a tombolo (an Italian word meaning 'connecting bar'). The Sutton tombolo (Dublin) is an excellent example. Howth Head was once an island. A submarine bar developed which joined the island to the mainland. Isostatic uplift after the Pleistocene was responsible for raising the tombolo to its present level. Castlegregory in Co. Kerry is another good example.

Classification of Coastlines

The presence of drowned drumlins in Clew Bay (submergence) and raised wave-cut platforms in Antrim (emergence) suggest that the relative levels of land and sea change over time.

1. Eustatic Movement

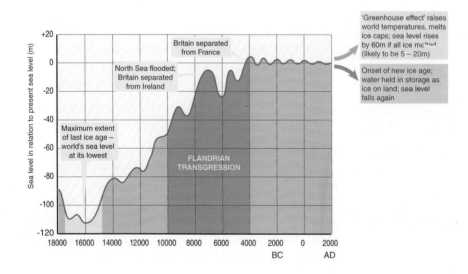

< Fig 6.15

Sea levels 18,000 BC – 2000 AD (projected)

The world's sea level was at its lowest 18,000 years ago when the ice was at its maximum. As temperatures began to rise and ice caps melted there was a universal rise in sea-level, i.e. eustatic movement. As the sea level rose, many coastal areas were submerged – hence drowned drumlins.

2. Isostatic Movement

In Chapter 1, we noted that the crustal 'plates' float on the more plastic asthenosphere underneath. As the ice sheets grew (during the Ice Age) the land underneath was pressed down, i.e. 'isostatic depression'. The subsequent melting of these ice sheets at the end of the Pleistocene meant that the land, with less weight began to uplift (emerge). Today, measurements suggest that parts of north-west Scotland are still rising by four millimetres a year and areas of the Gulf of Bothnia (Scandinavia) by 20 millimetres a year.

TYPES OF COASTLINES

It is convenient to classify coasts based on:

1. The nature of the movement, i.e. submergence or emergence.
2. The nature of the former coast, i.e. upland or lowland.

As a result four types of coastlines can be identified.

1. *Submerged Coasts:* (a) Upland (b) Lowland
2. *Emerged Coasts:* (a) Upland (b) Lowland

Features of a Submerged Upland Coast

i) Rias

ii) Fiords

iii) Dalmatian Coasts

Ria

Examples: Bantry Bay, Co. Cork; Dingle Bay, Co. Kerry.

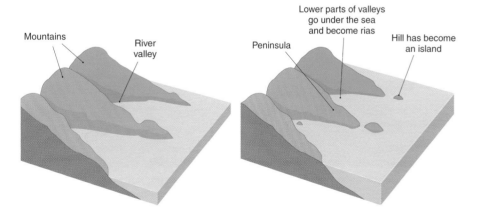

Fig 6.16
Formation of Rias; the Drowning of River Valleys

When ridges and valleys meet the coast at right angles, the coastline is said to be discordant.

Rivers occupied these valleys forming river valleys (see Chapter 4). At the end of the ice age the sea level rose and invaded the river valley to form a ria. A ria is a submerged river valley in a discordant coastline caused by a eustatic movement at the end of the Pleistocene Ice Age.

It is a funnel-shaped opening which increases in width and depth moving seawards.

Fiord

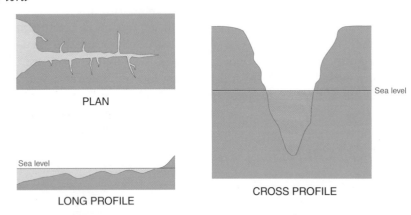

Fig 6.17
Profile of a Fiord

Example: Killary Harbour, Co. Mayo.

Like rias, fiords are also associated with discordant coastlines. While there is much controversy regarding their formation they generally possess many of the features of glaciated valleys which were submerged. They are U-shaped in cross-section and possess truncated spurs and over-deepened rock basins. Unlike rias their width is quite uniform and they possess a shallow seaward entrance known as a threshold. The origin of this 'lip' is debatable and while some represent terminal moraines, they most probably result from a reduction in a glacial erosion as melting of the glacier took place with rising temperatures near the sea.

Fiords then are submerged glaciated valleys in discordant coastlines caused by a eustatic movement at the end of the Pleistocene Ice Age.

Dalmatian Coasts

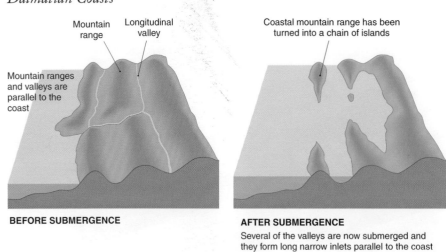

Fig 6.18
Formation of Dalmation Coasts

Examples: Cork Harbour; Pacific Coast.

Unlike rias and fiords, Dalmatian coasts are associated with concordant coasts. As the sea-level rises it invades the ridge which lies closest to the coast and the sea floods the valley to form a new coastline. The original coastal range is now represented as a series of off-shore islands which run parallel to the new coast.

Submerged Lowland Coasts

Example: Clew Bay, Co. Mayo.

Features: Drowned drumlins; Submerged forests

Emerged Coastlines

Coastlines of emergence are less common than coastlines of submergence because the melting of the Pleistocene ice-sheets was completed only relatively recently. It is only in areas then where the rate of isostatic response was greater than the post-glacial rise in sea-level that land which was formerly submerged was raised to form dry ground.

Features associated with emergence include raised beaches, raised sea cliffs, caves and sea-stacks.

COASTLINES AND HUMAN ACTIVITIES – INTERFERENCE AND MANAGEMENT

Coastal erosion is presently threatening over 1,500 kilometres of the Irish coastline. Destructive wave action can attack cliffs causing them to retreat inland resulting in the loss of farm land and residences. Coastal erosion forced the Dublin-Bray railway, for example to be moved inland at Killiney.

Longshore drift too can remove beach material and deposit it elsewhere often resulting in the silting of ports as in Dublin and Wexford.

Groynes

Example: Youghal, Co. Cork; Rosslare Strand, Co. Wexford.

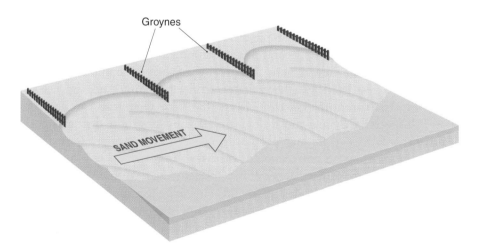

Fig 6.19 › Groynes

Groynes are built to trap sand that is moving parallel to the beach as a result of longshore drift. They consist of wooden fences which are built at right angles to the coastline. Their basic effect is to maintain or widen the beach which dissipates wave energy and so reduces the effect of coastal erosion further inland. However this action can have repercussions elsewhere, because places further along the coast will lose their supply of beach material and become more susceptible to erosion.

Sand Removal

Beaches dissipate (lessen) wave energy. When sand and shingle are removed from the beach, for agricultural use or for the construction industry, less energy is absorbed by the beach thus permitting larger and steeper waves to attack and erode cliffs and dunes. In 1887 over half a million tonnes of shingle were removed from a beach in Devon (England) to help in the construction of the naval dockyard at Plymouth. As a result the level of the beach fell by four metres. Erosion of the cliffs behind the beach was accelerated – six metres were lost between 1907 and 1957 so that today the local village of Haltsands is almost completely abandoned and left in ruins.

In Ireland the removal of beach material is now illegal and since 1990 the Department of the Marine has assumed responsibility for coastal protection.

Camping near beaches often results in the destruction of the local vegetation by camp fires and trampling.

Sea Walls and Breakwaters

The construction of sea walls with hydrodynamic curves dissipates wave energy and so protects the coast. Breakwater, built either at right angles to the coast or parallel to it are not aimed at halting waves but reducing their energy. *Gabaons* (rock armoury) have similar effects. In order to restrict silting in Dublin Harbour by south-flowing longshore drift, the Bull Wall, almost three kilometres long was constructed in 1819. This prevented sediments from entering and accumulating in Dublin Harbour. The building of sea walls and other defences however is expensive.

A cheaper and more effective solution is to add more sand, shingle and boulders in order to widen the beach and so dissipate wave energy.

Coastal Management

The development of fish farming or aquaculture along Irish coasts such as Cork and Mayo provides an example of successful coastal management. Here fish are reared in a natural, if controlled environment and provide a valuable source of income for these regions. The rearing of such large numbers of fish however in a restricted area necessitates artificial feeding and the use of chemicals. Apart from the negative effects that these chemicals can have on other fish species, they may move along the coast by longshore drift and pose health hazards to bathers and people involved in other water sports. The decrease in sea-trout off one section of the Mayo Coast is a case in point. In the late 1980s up to 12,000 catches of this species were recorded each year. By 1992 the catch had fallen to less than 500. This fall-off ruined the livelihood of one local hotelier who specialised in offering angling holiday breaks to German tourists. Many still attribute the decline in the fish catch to the growth of fish farming in the region, which can increase the incidence of fish lice.

Industries too often locate on estuaries not only for trading or extraction of water for cooling and washing but to use the sea as a dumping ground for their waste products.

Questions

ORDINARY LEVEL

1. Coastal landscapes are shaped by erosion and deposition.
   (i) Select any **three** landforms that result from action by the sea and, in the case of each that you have selected:
   - Describe and explain, with the aid of a diagram, how it was formed.
   - Name a specific location where the feature may be found.

   (60 marks)

   (ii) Describe any **two** ways by which human activities attempt to protect coastlines from erosion. *(20 marks)*

   Leaving Cert. 1996

2. Sea-cliff, Sea-arch, Beach, Lagoon, Blowhole, Sand spit.
   (i) In the case of **each** of the above coastal features, state whether it is the result of erosion or deposition. *(18 marks)*
   (ii) Select any **three** of the features listed above and in the case of **each**
   - Name a specific location where the feature may be found.
   - Describe and explain, with the aid of a diagram how it was formed.

   (48 marks)

   (iii) In recent years, coastal erosion has caused enormous damage to coastal areas. Describe **two** methods used to limit the damage.

   (14 marks)

   Leaving Cert. 1995

3. Examine the following statement in detail:
   Marine erosion has created spectacular coastal landscapes. *(40 marks)*

   Leaving Cert. 1994

4. The diagram shows an area where marine erosion has been active.

   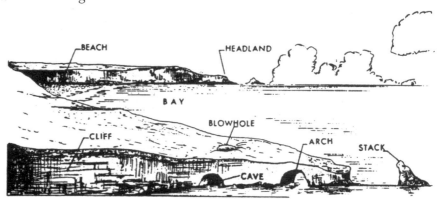

   (i) With reference to any **three** major landforms which are evident in the diagram, describe and explain how marine erosion has shaped them. *(60 marks)*

(ii) Materials produced by erosion at one part of a coastline are transported to other parts and deposited. Human action is sometimes taken in order to interfere with these natural processes. Explain why. *(20 marks)*

Leaving Cert. 1993

Higher Level

1. (i) Examine the processes which influence the formation of any **three** landforms found along the Irish coast. *(75 marks)*
   (ii) 'Coastlines are subjected to much human use and abuse.' Discuss this statement with reference to Ireland. *(25 marks)*

   Leaving Cert. 1996

2. The Irish landscape is famous for its scenic landforms.
   (i) Identify **three** such landforms – **one** formed by river action, **one** formed by marine action and **one** formed by glaciation.
   For each landform identified, explain how it was formed, referring to specific locations in Ireland. *(75 marks)*
   (ii) Examine **one** example of the way in which human recreational use of the landscape may cause environmental damage. *(25 marks)*

   Leaving Cert. 1995

3. Processes of erosion, transportation and deposition produce characteristic landforms. Select three landforms, one produced by glaciation, one by river action, and one by coastal processes.
   For **each** landform selected, explain with the aid of diagrams how it was formed. In your explanation include Irish examples of each landform. *(100 marks)*

   Leaving Cert. 1993

CHAPTER 7 — *Wind and Desert Landforms*

Fig 7.1

The above photograph may be the mental image which most people have of hot deserts, but it does little to convey the real desert landscape. While large areas of sand dunes do exist, only 10 per cent of the Sahara, the world's largest desert is actually covered in sand. A more accurate picture of hot deserts is one associated with the traditional 'Western' movies set in North America.

The traditional definition of a desert is an area receiving less than 250 millimetres of rain per year. Modern attempts to define deserts are more scientific and are specifically linked to the water balance. This approach is based on the relationship between input of moisture (precipitation) and output (evaporation).

N.B.   P = Precipitation   E = Evaporation

Fig 7.2
Scale of Precipitation and Evaporation

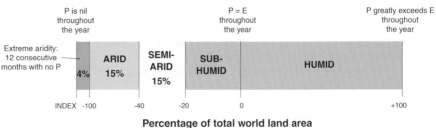

Deserts, therefore are areas where evaporation is always greater than precipitation.

Desert Landscapes

TYPES OF DESERT LANDSCAPES

Desert landscapes may differ due to variations in rock type and structure, in climate and time.

Generally three types of desert landscapes can be identified.

1. Erg

Erg are the large areas of sand and sand dunes and cover about 12 per cent of the world's deserts. The Sahara and the desert of north-west Australia provide examples of the type.

2. Reg

Reg are stony deserts where the surface is littered with boulders and pebbles. Parts of Algeria and Egypt provide good examples.

3. Hamada

Hamada is a desert with a bare rock surface where sand has been removed by wind. The northern Sahara in Libya provides a good example.

Wind is generally associated with hot deserts as an agent of landscape change because these regions provide the optimum conditions in which it can operate.

As a geomorphic agent, wind is most effective in hot deserts because these regions provide:

(i) Vast open spaces which allow wind to increase its speed.
(ii) Loose or unconsolidated material which the wind can then pick up and use to sandblast desert surfaces.

It must also be remembered that the effects of other processes, e.g. running water and ice are at a minimum.

The process of mechanical weathering in deserts, i.e. exfoliation or onion peeling has been dealt with in Chapter 3.

This *insolation weathering* or *thermal expansion* and *contraction* is particularly effective in boulders which possess joints and bedding planes and is termed 'block disintegration'. The term granular disintegration is used to describe rocks which break down into individual grains. As was pointed out in Chapter 3, the presence of water (moisture) is essential to explain the relatively rapid rate of disintegration.

THE PROCESS OF AEOLIAN (WIND) EROSION

1. Deflation

Deflation is the removal of fine material by wind. Although the effects of deflation are sometimes difficult to note, because the entire surface is being lowered at the same time, they can be significant. One of the most devastating effects of deflation was seen in the Dust Bowl (in the Western USA) in the 1930s, where vast areas of land were lowered by as much as one metre in only a few years.

2. Abrasion

Abrasion is the *sand blasting* effect which the material being transported by the wind has on the surrounding topography. Because the wind cannot lift particles very high, the action is confined to within one metre of the ground. Its general effect is to polish and smoothen rock surfaces in some areas while shaping others into distinctive landforms.

3. Attrition

Attrition involves the breaking down of particles which are being transported by the wind by constant friction with one another and with the ground.

HOW WIND TRANSPORTS MATERIAL

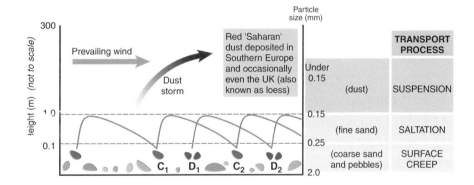

Fig 7.3 Methods of Wind Transportation

The ability of the wind to transport material is related to several factors including the velocity of the wind, the size of the particles and the nature of the surface over which it passes.

1. Suspension

Suspension is where the wind carries fine material with a diameter of less than 0.15 millimetres. The material can be raised to considerable heights and carried over vast distances. In the 1980s, red dust from the Sahara was deposited as 'red rain' over parts of Ireland.

2. Saltation

Saltation occurs where coarse material is bounced along in a series of hopping movements.

3. Surface Creep

Surface creep is when larger particles, too heavy to be lifted, move along the desert surface. It can be compared to traction in river transportation.

LANDFORMS PRODUCED BY AEOLIAN EROSION

Deflation Hollows

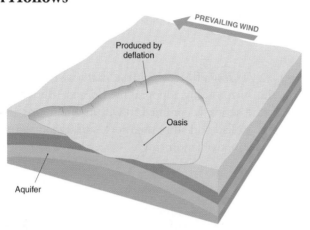

< Fig 7.4
Deflation Hollows

Example: The Qattara Depression in Egypt

The most notable results of deflation are shallow depressions which are sometimes called 'blow outs'. Once a small 'blow out' has been formed it is gradually enlarged and deepened by the eddying action of the wind. The factor which controls the depth of the depression is the water-table which acts as the local base level. Once the water-table has been reached a swamp or oasis is formed. While many deflation hollows are no bigger than small dimples, the Qattara depression in Egypt measures 320 by 160 kilometres and is 130 metres below sea level.

Elsewhere deflation is responsible for desert pavements where deflation removes the fine material until eventually only a continuous cover of coarse material remains.

Rock Pedestal

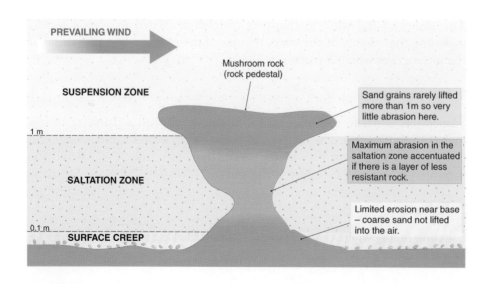

< Fig 7.5
Formation of a Rock Pedestal

Rock pedestals are some of the more spectacular structures produced by wind abrasion and are sometimes known as mushroom rocks or gour (plural, gara) in the Sahara. Because maximum abrasion occurs in the saltation zone (less friction) the base of the rock is undercut to produce a mushroom effect. The process is accentuated where there are layers of less resistant rock.

Chemical weathering due to the presence of water at the base is also believed to play a part in their formation.

Yardangs and Zeugens

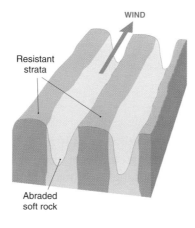

Fig 7.6 Formation of a Yardang

Examples: Atacama Desert in South America

When rocks of varying resistance occur in bands roughly parallel to the direction of the prevailing wind, the less resistant rock is eroded more quickly by the process of abrasion. The result is a 'ridge and furrow' topography where the ridges, the yardangs can stand up to six metres above the furrows.

Where a hard horizontal layer lies above a soft layer and weathering along the joints breaks the hard surface, wind abrasion will carry on until separate tabular masses known as zeugens are left standing upon softer rock.

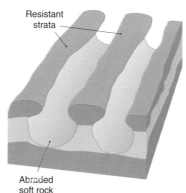

Fig 7.7 Formation of a Zeugen

Wind abrasion too is responsible for ventifacts (Latin: *ventus* wind, and *facere* to make). These are stones which are too heavy to be carried away by the wind and so are polished on several sides as a result of seasonal changes in wind direction. Ventifacts known as *Dreikanter* (German: three sided) are the most common.

While the processes of deflation and abrasion create actual landforms, attrition reduces the wind-blown material to a fine powder known as *'millet seed'* said to be the 'end product' of desert erosion.

LANDFORMS PRODUCED BY AEOLIAN DEPOSITION

As with running water, wind deposits its load when wind velocity falls and so wind energy diminishes. Thus sand begins to accumulate wherever an obstruction across the path of the wind slows down its movement. The most common form of deposition is in the form of a mound or ridge called a dune.

When the wind meets an object such as rock or clump of vegetation (or even a dead animal), the movement of the wind is impeded and the particles moving in saltation and surface creep, pile up behind the obstacle. As the accumulation of sand continues, the mound grows into a dune.

Barchan Dunes

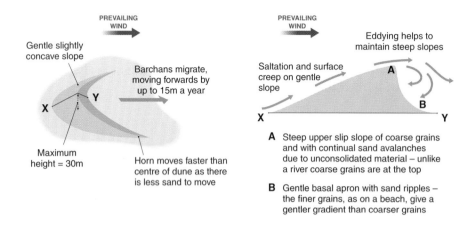

Fig 7.8
Formation of Barchan Dunes

Example: Sahara Desert

Barchan dunes are sand dunes shaped like crescents with their 'tails' pointing down wind. They generally form behind an object (rocks, vegetation, etc.) where the supply of sand is limited. As the sand piles up it forms a gentle concave slope on the windward side. As the dune grows sand moves over the crest (top) and slips down the leeward side to form a steep slope. Because the movement of wind is faster at the sides 'horns' or 'tails' develop. Barchans normally occur in groups or swarms and can reach a height of up to 30 metres. Occasionally barchans form without the presence of any object. Their origin then is thought to be due to variations in wind velocity and direction. Like other dunes, barchans are known to migrate at speeds of up to 15 metres a year.

Longitudinal Dunes

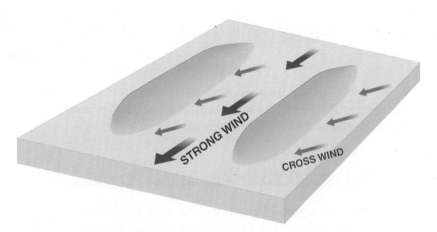

Fig 7.9
Origins of Longitudinal dunes

Example: South of the Qattara Depression in the Sahara.

Longitudinal dunes are elongated straight dunes which lie parallel to the direction of the prevailing wind. They may extend for over 100 kilometres and reach a height of 200 metres. While the exact origin of these *seif* dunes, as they are also known, is not fully understood, they seem to form where the prevailing winds cut troughs in the desert floor and pile the sand particles in symmetrical ridges on either side. A second theory suggests that they develop from barchans.

Cross winds may reorientate the horns (tails) so that they become elongated forming a ridge which lies at right angles to the wind and gradually individual barchans join together to form a series of parallel dunes.

THE EFFECTS OF WATER IN DESERTS

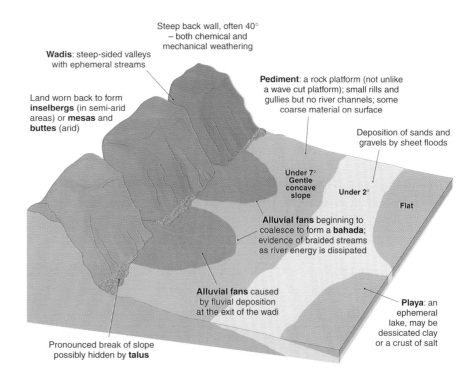

Fig 7.10 > Water Features in Deserts

Wadis

While rain seldom falls in hot deserts, ephemeral streams do flow after rainstorms. Although short-lived, these streams have high levels of discharge. This is due to the fact that the torrential nature of the rainfall exceeds the infiltration capacity of the ground and there is almost a complete lack of vegetation. As a result, these streams are effective agents of vertical erosion and the resultant gullies may over time develop into deep steep-sided ravines known as wadis or *arroyos*. The size and number of wadis suggests that they may have been formed in a past pluvial (wet) period. Today, they are dry channels which can become extremely dangerous after rainstorms.

Bajadas

At the mouth of a wadi, the river spreads out and its energy is dissipated. As a result it deposits the load in a cone-shaped alluvial fan. Where several wadis occur close to each other their alluvial fans may merge to form a continuous deposit of sand and gravel known as a *bajada* (bahada).

Bolsons

Most desert rivers drain into inland basins known as bolsons. They are often surrounded by mountains from which they receive their streams. Bolsons are occupied by temporary salt lakes during rain storms. These salt lakes are known as *playas* or *salinas*.

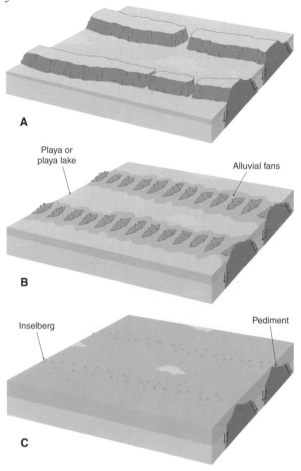

< Fig 7.11
Minor Desert Landforms

Occasionally isolated flat topped remnants of former highlands rise from the desert surface. They are known as *mesas* (Spanish, table) while smaller versions are known as *buttes*. Isolated hills of resistant rock on the other hand such as the famous Ayers Rock in Australia are known as *inselbergs* (German, island mountain).

At the foot of the highland region lies the *Piedmont* – a gently concave sloping area often covered by a thin veneer of debris.

< Fig 7.12
Ayers Rock

Wind and the Irish landscape

As already stated, to be an effective agent of erosion, transportation and deposition, wind needs:

i) Vast open expanses which allows the wind to build up speed

ii) Loose or unconsolidated material.

The presence of mountains, vegetation cover and human structures essentially rules out the occurrence of both of these requirements in Ireland. Even ploughed fields in spring are relatively safe from wind deflation due to their moisture content and texture.

The environment which best mirrors the topographical and climatic conditions of hot deserts are found on the beach. Here, exposed to the open sea, on-shore winds can built up impressive speeds. On dry days at low tide, the loose sand is easily transported inland by saltation and surface creep and may develop into sand dunes.

While sand dunes are the direct result of aeolian action, the wind also acts as an indirect agent of landscape change. Waves and currents, responsible for coastal features of erosion and deposition owe their origin to the direction and speed of the local winds (see Chapter 6).

Desertification

Fig 7.13

Desertification is the term used to refer to the spread of deserts. It is a process causing particular concern in areas like the African Sahel and the surrounding countries. While the frequency of drought in these areas has increased, it is also a fact that the situation has been aggravated by human activity. Timber for fuel and more land for farming as the population increases has led to deforestation. In the 1960s over 20 per cent of Ethiopia was covered in forests. Today that forest cover has shrunk to less than five per cent. The problem is further aggravated by overgrazing which bares the soil to the effects of wind and water. Overcropping in other areas quickly exhausts the soil which then is unable to support any protective covering. This leads to a reduction in evapotranspiration from plants which lowers the amount of atmospheric water and so reduces the chance of rain.

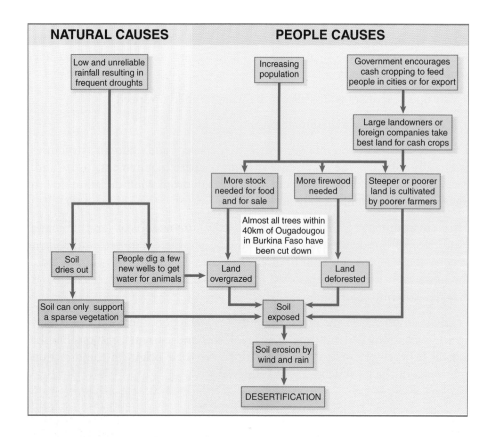

◁ Fig 7.14
The Causes of Desertification

Questions

ORDINARY LEVEL

1. Examine the following statement in some detail. The Sahara Desert is expanding southwards at over 50 kilometres each year.

   (40 marks)

   Leaving Cert. 1995

2. Examine the following statement in detail. By variations in the processes of transportation and deposition, wind produces several different types of sand dunes. *(40 marks)*

   Leaving Cert. 1993

HIGHER LEVEL

1. (i) Describe how wind shapes and modifies landscape features in hot desert regions. *(40 marks)*

   (ii) 'Desertification is a major problem of our time.' With reference to African deserts explain the main causes and consequences of desertification. *(60 marks)*

   Leaving Cert. 1993

CHAPTER 8

Weather and Synoptic Charts

Fig 8.1

The term *weather* is used to describe the conditions of the atmosphere at a certain place at a *specific time*, i.e. day to day or even hour to hour. The atmosphere is composed of large bodies of air known as *air masses*.

> An air mass is a large body of air which moves over the earth's surface. Atmospheric conditions such as temperature, pressure, humidity, etc., are generally uniform and as a result air masses are often said to be homogenous.

Air masses originate in what are termed *source regions* from where they get their characteristics. This is because when air remains stationary in an area for several days it tends to assume the temperature and humidity properties of that region. So while air masses may change somewhat as they move, they generally bring the atmospheric conditions of their source regions with them.

SOURCE REGIONS

The major areas of stationary air are the areas of high pressure, i.e.
- Polar Areas (P)
- Tropical Areas (T)

Weather and Synoptic Charts

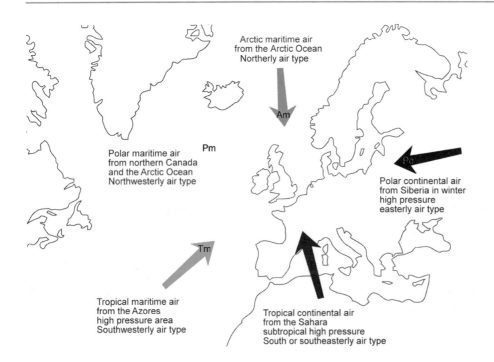

Fig. 8.2
The Air Masses that affect Ireland

The surface over which these air masses form is also important as it will determine their moisture content, i.e. Maritime (M); Continental (C).

Types of Air Masses

Polar Maritime (PM)

Polar maritime air masses originate over the North Atlantic and are very common over Ireland and Britain. They generally bring cool conditions and heavy showers especially in upland areas.

Polar Continental (PC)

Polar continental air masses are formed over the Arctic regions of Northern Eurasia. As a result these air masses are very cold and dry. In winter they usually bring clear weather and sometimes heavy snowfalls.

Tropical Maritime (TM)

Tropical maritime air masses are formed over the oceans in the tropical regions and as a result are warm and moist. As these air masses move northwards they are cooled resulting in a lot of rain. In winter they bring mild wet weather with thick cloud cover while in summer they are usually associated with thundery showers.

Tropical Continental (TC)

Tropical continental air masses form over hot-desert regions such as the Sahara. The lack of moisture means that the air is fairly stable. The resulting weather in winter is often dry and warm while heat wave conditions can be experienced in summer.

If you examine Figure 8.2 carefully you will see that Ireland lies in an area where air masses meet. This has a major influence on our weather, and because different air masses have different temperatures etc., they do not mix easily. The line which separates two air masses is called a *front*. The most important front is the polar front where the warm air from the tropics meets the cold air from the poles.

Warm Front

A warm front is where warm air is advancing and being forced to override cold air. A warm front is represented by a line of semi-circles.

Cold Front

A cold front occurs when advancing cold air undercuts a body of warm air.

In both cases the rising air cools resulting in the formation of clouds and later precipitation. It is at polar fronts that depressions (low pressure systems) form. Cold fronts are represented by a line of triangles.

HOW DEPRESSIONS ARE FORMED

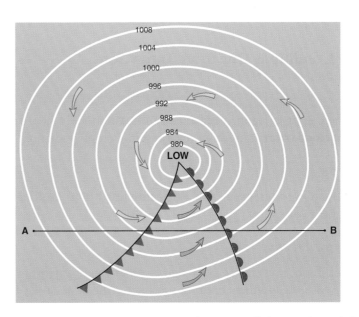

Fig 8.3
A Depression

The early stages of a depression begins when a small dent or 'wave' develops as the warm air mass bulges into the cold mass.

As the wave or bulge increases in size it forms a well-defined 'tongue' known as the 'warm sector'. As the cold air from the north-west pushes against the warm sector it forms a cold front.

In front of the warm sector, the warm air pushes against the cold air to form a warm front. Because the warm air in the warm sector is forced to rise, pressure falls and the inward blowing winds increase in strength. The rising warm air now cools to form clouds and eventually precipitation.

The depression eventually begins to die out when the cold front catches up with the warm front to form an occluded front. An occluded front is represented by a line of alternate semi-circles and triangles.

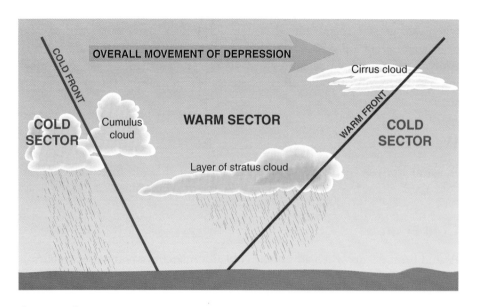

Fig 8.4
Cross-section of a Depression

Anticyclones

Anticyclones are the opposite to depressions. While depressions are centres of low pressure around which winds blow anticlockwise, anticyclones are centres of high pressure around which winds blow clockwise (in the northern hemisphere). They also tend to cover larger areas and tend to remain stationary for days.

Weather Conditions Due to an Anticyclone

Summer

Because there is little cloud cover, intense insolation gives hot, sunny days. The lack of cloud cover at night however results in rapid radiation.

Winter

Descending air currents mean that skies are clear often resulting in severe frost at night.

Ridge of High Pressure

A ridge of high pressure is a wedge or elongated extension from an anticyclone with the weather conditions characteristic of an anticyclone.

Trough

A trough is a wedge or elongated extension of a depression with similar weather characteristics.

Hurricanes

Intense low pressure systems with a diameter measuring hundreds of kilometres often form in tropical latitudes between 6° and 20° north and south of the Equator. They are known as hurricanes in the West Indies, typhoons in the China Seas, cyclones in the Indian Ocean and willy willies off the Australian coast. They have a very steep barometric gradient. The centre of the hurricane, (the eye) is an area of calm or light winds, around which wind whirls at up to 200 kilometres per hour.

They are usually accompanied by torrential rain and do enormous damage.

Synoptic Charts

Fig 8.5
Hurricane Damage, Darwin, Australia.

Synoptic Charts

In order to produce a synoptic or weather chart the relevant information must first be collected. This is done by a number of weather stations throughout Ireland which constantly supply information about the atmospheric conditions in their area to the Meteorological Office in Dublin. This information is then plotted in the form of a weather chart which has to be constantly updated. The plotting is based on a *station model*.

THE STATION MODEL

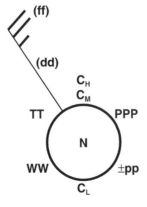

The Station Model
(simplified version)

dd = wind direction (from which wind is blowing) in degrees from true North in a clockwise direction. (090°=E, 180°=S, 270°=W, 360°=N).

ff = wind speed in knots (each full feather represents 10 knots; the half-feather 5 knots). Feathers appear on right hand of arrow shaft, looking into wind.

N = total cloud amount.

C_L = type of low cloud.

C_M = type of medium cloud.

C_H = type of high cloud.

PPP = barometric pressure in millibars (tens, units and tenths figures i.e. omitting the initial 9 or 10 and also decimal point).

pp = barometric tendency i.e. change in barometric pressure in past 3 hours (in tenths of millibar with decimal point omitted). Use plus (+) if current pressure value is above that of 3 hours ago, minus (−) if below. For no change use +00.

TT = air temperature to nearest whole degree Celsius. Preceded by minus (−) if below freezing level.

WW = present weather.

Fig 8.6

Wind - See dd and ff

Wind speed is recorded in *knots*. The actual speed is shown by feathers on the shaft. Each full feather represents 10 knots while a half feather represents five knots.

The shaft points in the direction from which the wind is blowing.

Clouds

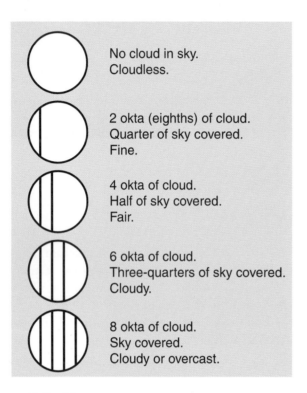

< Fig 8.7

Selected Symbols for Total Cloud Amount (N). This is plotted inside the Station Circle

Cloud Amount – See N.

The amount of cloud cover is indicated by the number of lines drawn in the centre circle of the station model. The sky is divided into eight parts of *oktas*. Total cloud cover is said to be eight oktas.

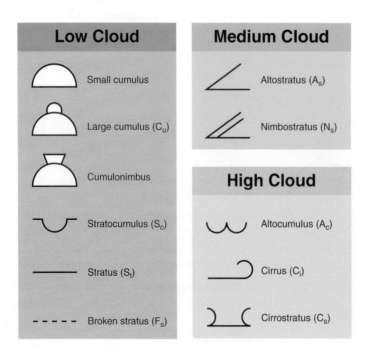

< Fig 8.8

Selected Symbols for Cloud Type

Cloud Types – See CL, CM, CH.

The actual type of cloud cover present at a weather station is indicated by its appropriate symbol.

Barometric Pressure – See ppp

Barometric pressure is measured in millibars (mb). Generally pressure varies between 950 mb. and 1,050 mb. In a station model the first numbers are dropped and the decimal point omitted so that a pressure of 985.6 mb. would be shown as 856 and a pressure of 1025.6 mb would be shown as 256.

Isobars are lines on a map which join places of equal barometric pressure.

Barometric Tendency – See PP

Barometric tendency shows the change that has occurred in the pressure over the previous three hours.

Temperature – See TT

Temperature is measured by a thermometer in a Stevenson's Screen.

Present Weather

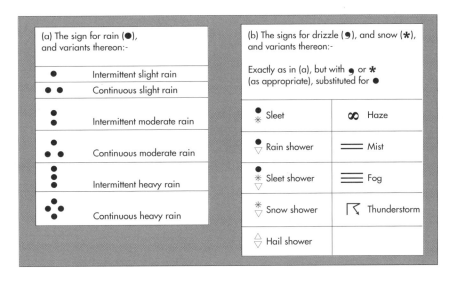

Fig 8.9
Selected Present Weather Symbols

Present weather refers to the type of precipitation (if any) which the weather station is experiencing and is recorded using the appropriate symbol.

THE IMPORTANCE OF WEATHER FORECASTING FOR CERTAIN OCCUPATIONS

Agriculture

As an economic activity, farming is particularly sensitive to variations in weather conditions.

Arable Farming

< Fig 8.10

The importance of the weather to arable farming, i.e. sowing, spraying and harvesting is outlined in the following charts.

\multicolumn{5}{c}{Conditions for Sowing and Harvesting}				
Operation	Element	Cereals	Sugar Beet	Potatoes
Sowing/ Planting	Temperature (air/soil)	>4.5°C	>7°C	>7°C
		(sustained for a period of a number of days)		
	Weather	2-3 dry days	Dry, especially on day	Dry on day, not too wet before
	Soil moisture	Moist to dry but not too dry	Moist (not wet) at seed depth	Moist to dry but not too dry
Harvesting	Temperature	—	>0°C in soil	>8°C (avoids bruising)
	Weather	Bright and dry not humid; delay ops. after dew/rain	Preferably dry; some rain acceptable	Dry on day not too wet previously
	Soil Moisture	Dry soil; at least below field capacity	Moist to moderately dry soil	Moist to dry but not too dry*
\multicolumn{5}{l}{* Heel end rot or serious tuber discolouration may occur if haulm desiccants are applied when crop is drought stressed (i.e. SMD> 60-80mm). Mechanical damage (bruising) can also occur in drought conditions.}				

< Fig 8.11
Weather Conditions for Sowing and Harvesting

Synoptic Charts

Suitable Conditions for Spray Application			
	Before	**At time of**	**After**
Weather	Dry on day (min. 1-3 hr)	Dry (or a light dry mist)	Dry for 1-6 hr (depends on agrochemical) Heavy rain causes crop damage
Temperature	Sufficient for growth	Sufficient for growth: In high temperature and intense sunlight plants are susceptible to leaf scorch	Susceptible to frost damage after some herbicides
Wind	–	Light, 2-3 m/s; strong wind causes spray drift and uneven distribution	Light/moderate (calm wind slows drying or increases risk of frost

Fig 8.12 › Weather Conditions for spraying

Pastoral Farming

Pastoral farming too is no less sensitive. The weather can have direct or indirect effects on animal health. Stress caused by heat and cold may reduce the animals' resistance to infection.

Fig 8.13 ›

Disease	Animals Involved	Weather Relationship
Foot & Mouth	Cattle, pigs & sheep	High humidity, rain and light winds
Liver Fluke	Cattle, sheep	Plenty moisture and temperature > 10°C.

Fig 8.14 › Weather and Animal Disease

The Meteorological Service co-operates with the veterinary authorities in this country in the preparation of special forecasts relating to the spread of a number of diseases. These forecasts predict the timing and intensity of the disease so that farmers may take preventative action with specific treatments.

Transport

Aircraft and shipping are particularly sensitive to visibility, i.e. the occurrence and intensity of fog and the direction and strength of winds.

Leaving Certificate Geography

Weather and Synoptic Charts

Questions

ORDINARY LEVEL

1. Examine the following statement in some detail.
   In Ireland, weather forecasts are important but not always reliable.
   (40 marks)

   Leaving Cert. 1995

2. Examine the following statement in some detail.
   Accurate weather forecasts are important for certain occupations.
   (20 marks)

   Leaving Cert. 1994

3.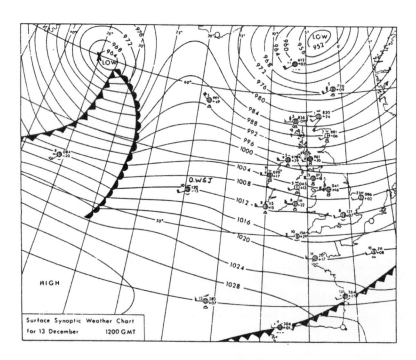

   (i) Give an explanation for the cloud type and cover recorded on this chart at station OWSJ. *(15 marks)*

   (ii) Describe in detail and give an explanation for the weather report recorded on the chart in the warm sector of depression 'X'.
   (20 marks)

   (iii) Explain in some detail the occurrence of any **three** of the following:
   (a) Thunderstorms in periods of hot summer weather. *(15 marks)*
   (b) Very cold sunny and dry periods of weather over Ireland in Winter. *(15 marks)*
   (c) Land and sea breezes. *(15 marks)*
   (d) Fog in river valleys in the early morning. *(15 marks)*

   Leaving Cert. 1992

HIGHER LEVEL

1. The changeable nature of Irish weather is the product of a variety of air masses which affect us.

    (i) Give a concise definition of the term 'air mass'. *(10 marks)*

    (ii) Identify **two** different types of air mass which affect Ireland. Describe and explain the major weather characteristics of **each** type. *(60 marks)*

    (iii) Examine briefly the impact of advancing technology on weather forecasting. *(30 marks)*

    Leaving Cert. 1994

2.

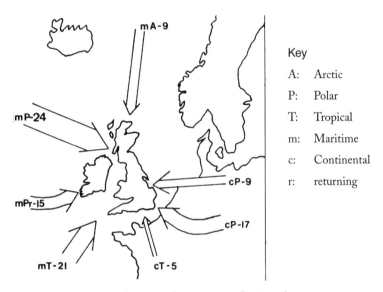

An approximate picture of the percentage frequency of air masses over Britian and Ireland in January. 'mPr-15' stands for 'polar maritime returning 15%'.

Examine the map above and answer the following:

    (i) Explain what you understand by the term 'air mass'. *(10 marks)*

    (ii) Select **three** of the air mass sources shown on the map, and explain the major characteristics of the weather patterns associated with **each**. *(90 marks)*

    Leaving Cert. 1988

CHAPTER 9 — Climate and Natural Vegetation

< Fig 9.1

Climate can be defined as the average weather condition experienced in an area over a considerable period of time, i.e. 25 to 30 years.

By studying weather records over long periods it is possible to identify certain unique characteristics that an area possesses in relation to temperature, pressure, precipitation and winds. In some parts of the world the 'climate' is harder to describe than others and the changeable nature of the atmospheric conditions in Ireland and the British Isles might well suggest that we experience 'weather' rather than 'climate'. A general description of the equatorial climate on the other hand will apply with a high degree of accuracy to almost any day of the year.

While the average of all atmospheric conditions is important in identifying a climatic type, temperature, and to a lesser extent precipitation, is widely accepted as the main reason for the different climatic regions.

Climatic Divisions Based on Temperature

Because latitude is the most important influence on temperature, it is possible to divide the world into basic climatic divisions, i.e. (i) Tropical (hot); (ii) Temperate (Moderate); and (iii) Cold climates.

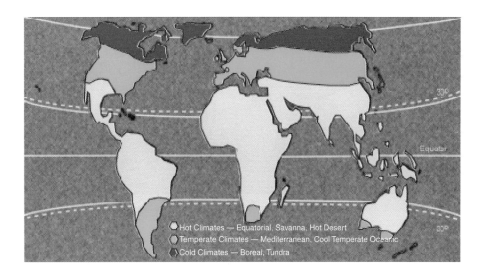

Fig 9.2 >
Main Climatic Areas

This basic division can be modified when we take 'effective' precipitation, i.e. total precipitation less evaporation into consideration.

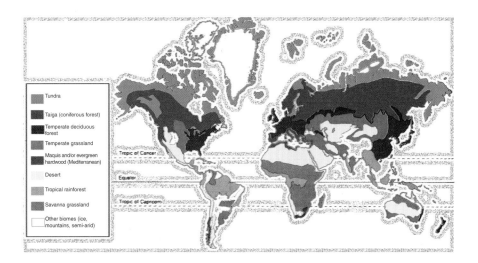

Fig 9.3 >
World Biomes. A Biome is the Largest Ecosystem Unit based on Dominant Vegetation, Soil Type and Fauna

Apart from the actual description of a climatic type, geographers also attempt to explain the climatic conditions and their effect on the fauna (animal) and flora (plant) of the region.

Climate and Natural Vegetation

Hot Climates

1. Equatorial

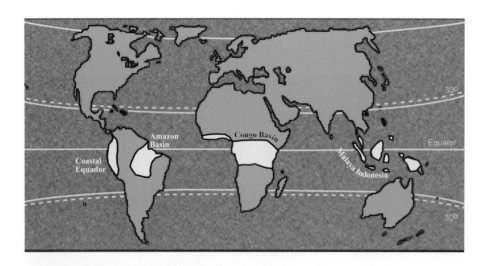

< Fig 9.4
Areas of Equatorial Climate

Climate

The equatorial climate is located within 5° on either side of the equator.

Because the sun is high in the sky, throughout the year, temperatures are constantly high with an annual range of only 2°C to 3°C.

The mean monthly temperature also ranges from 26°C to 28°C. Diurnal (daily) ranges are also small with evening temperature rarely falling below 22°C due to the presence of cloud cover.

If you examine Figure 9.6 carefully you will notice that the region is dominated by low pressure. This results from the high temperatures which lead to rising air currents. Rising air currents in turn result in cloud formation and precipitation with heavy convectional rainfall being experienced each afternoon, often accompanied by thunder and lightning. Total amounts can be up to 2,000 millimetres. The constant high humidity produces a sticky heat which is extremely oppressive for Europeans.

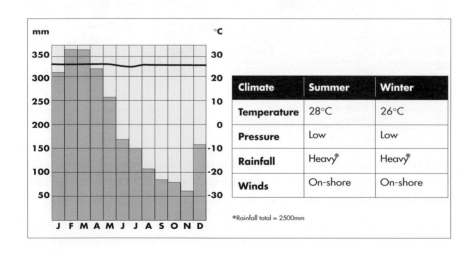

< Fig 9.5
Temperature and Rainfall Chart for Equatorial Climate

Climate	Summer	Winter
Temperature	28°C	26°C
Pressure	Low	Low
Rainfall	Heavy*	Heavy*
Winds	On-shore	On-shore

*Rainfall total = 2500mm

Climatic Divisions Based on Temperature

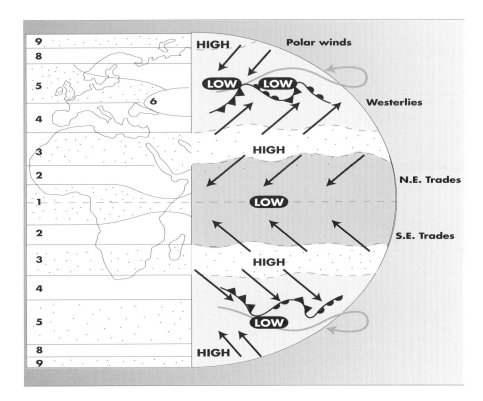

Fig 9.6 > Band 1 shows the Low Pressure which, together with the Equatorial Location produce the Equatorial Climate

Vegetation

Fig 9.7 > Equatorial Forests

The high temperature and humidity produce a dense covering of forest vegetation generally referred to as tropical rain forests or *selvas*. The variety of species is bewildering and it is estimated that one square kilometre in Amazonia can have up to 300 species, including rosewood, mahogany, ebony and rubber. Although the trees are mainly deciduous, they are evergreen as there is a continuous growing season. Generally three tiers can be distinguished. In the tallest layer, emergents can reach up to 50 metres. Below these is the canopy which can absorb up to 70 per cent of the light and forms the habitat for most of the birds, animals and insects of the rainforest. Below the canopy is the under-canopy which consists of trees which can reach up to 25 metres. Below this is a shrub layer which consists of small trees which have adapted to the shade.

The absence of light often means that the actual forest floor is almost clear of vegetation. Elsewhere on low-lying muddy coasts where the water is calm, mangrove forests occur.

2. Tropical Climate – Savanna Type

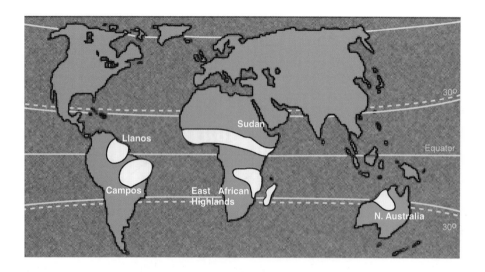

Fig 9.8
Areas of Savanna Climate

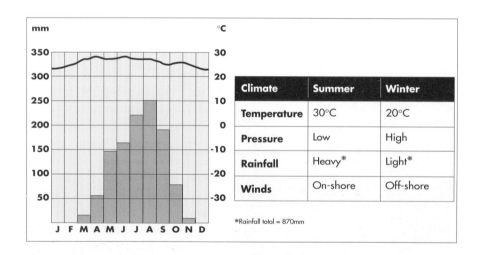

Fig 9.9
Temperature and Rainfall Chart for Savanna-type Climate

Climate

The tropical, savanna-type climate is generally found between 5° to 15° North and South of the equator. This region is essentially a transitional zone between the equatorial low pressure which dominates in summer and the sub-tropical high pressure which dominates in winter. The main characteristic of this climate then is the alternate wet and dry season. The wet season occurs in summer when the sun moves overhead and with it the equatorial low pressure belt. The result is heavy convectional rainfall which can last for up to five months and can account for 80 per cent of the total of 870 mm. In contrast the winters are dry as the low pressure moves away and the area comes under the influence of the trade winds. Distance from the equator is best illustrated by the increase in the seasonal range of temperature, as differences between summer (30°C) and winter (20°C) insolation become more pronounced with increasing latitude.

Climatic Divisions Based on Temperature

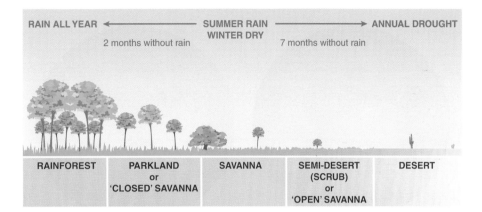

Fig 9.10 >
Transition from Rainforest to Desert

Vegetation

The transitional nature of this region is probably best reflected in the vegetation as it changes from rainforest to desert. There is still sufficient rainfall nearer the equator to support forest cover though it is less dense and more obviously deciduous as it loses its leaves in winter. Trees tend to be *xerophytic* or drought resistant in response to the dry season (winter). Further away (from the equator) the trees are replaced by grasslands, the Savannas or tropical grasslands where elephant grass can reach for a height of up to five metres. This is the so called 'big game' country, inhabited by over 50 different species of herbivores and carnivores such as the wildebeest, zebra and lions.

Fig 9.11 >

3. Monsoon Climate

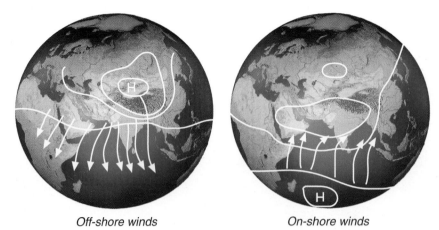

Off-shore winds   On-shore winds

Fig. 9.12
High Pressure results in Off-shore Winds bringing the Wet Season across the Indian Ocean

Location: South-east Asia especially India and Northern Australia.

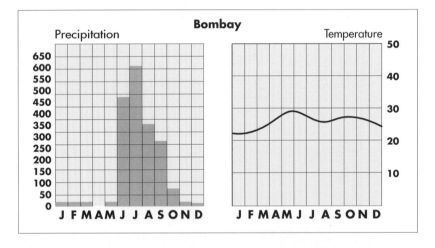

Fig. 9.13
Temperature and Rainfall Charts of Monsoon Climate

Climate

The word monsoon is an Arabic word meaning 'season' and refers to the change in wind direction brought about by the heating and cooling of great land masses. Summer in the northern hemisphere results in low pressure over Asia (as the land heats up) but high pressure over Northern Australia (winter).

From May to October the prevailing wind is from Australia approaching India from the south-west. As they cross the Indian Ocean, the winds pick up huge quantities of moisture which they drop over South-East Asia as they approach the mountains. The reverse happens in winter as the sun moves south over Australia resulting in high pressure over Asia. Now the winds are off-shore giving South-East Asia its dry season and Northern Australia its wet season (see Figure 9.12). While rainfall is essentially seasonal, actual amounts vary with relief. Bombay on the west coast of India can receive up to 2,000 millimetres in the four months from June to September. Because India is protected from the cold northerly winds by the Himalayas, average winter temperatures are about 10°C while summer temperature can average 30°C.

Vegetation

Because the monsoon lands of South-East Asia are home to almost half the world's population, most of the natural vegetation has been cleared for farming and human habitation. Where the natural vegetation does exist it reflects variations in rainfall and is often deciduous shedding its leaves in the dry season. The most frequently occurring trees are teak, bamboo and acacia. Teak is by far the most valuable and large stands are to be found in Burma, which supplies the world with over half its requirements, and in Thailand. This durable hardwood is now carefully conserved and enjoys extensive replanting.

4. Hot Deserts

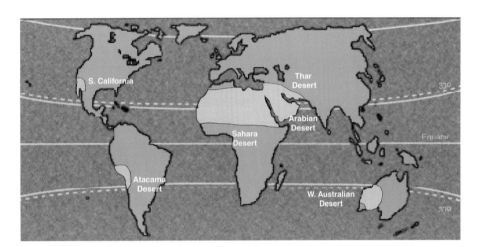

Fig 9.14
Areas of Hot Desert Climate

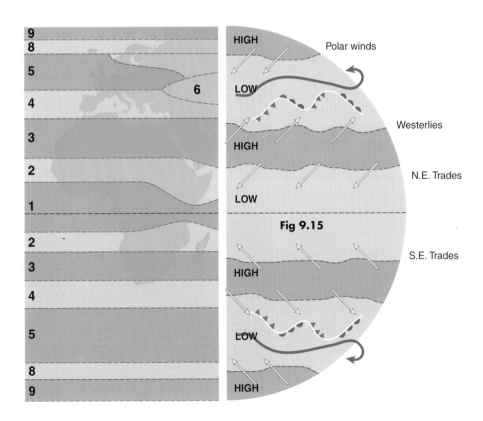

Fig 9.15
Band 3 shows Hot Desert Regions

Climate

Hot desert climates are usually found on the west coasts of continents between 15° and 30°C north and south of the Equator. Temperatures are characterised by their extremes. The annual range can be over 20°C while the diurnal (daily) range can exceed 50°C, hence the saying 'night is the winter of the desert'. Daytime temperature are intense in summer with the overhead sun provide maximum insolation because of the lack of cloud cover. It is the same lack of cloud cover however which explains the often subzero night-time temperatures as a result of maximum radiation. The low rainfall is the result of high pressure with descending air currents. The air heats up as it descends thereby increasing its capacity to hold moisture. The winds also travel over long stretches of land and generally move from cool to warmer regions.

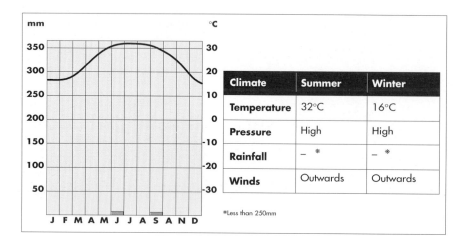

< Fig 9.16

Temperature and Rainfall Chart for Hot Desert Climate

Vegetation

While all deserts suffer an acute shortage of water none are truly arid (dry). When rain does fall it is unreliable and often heavy.

The result is a rapid surface run-off with low infiltration and high evaporation rates. Vegetation, where it does exist, has to be *xerophytic* (drought resistant). Many plants like the cactus are *succulents*, i.e. they can store water in their tissue.

They also have thick barks and a waxy coating to limit transpiration (loss of water). Other plants like the acacia survive courtesy of long tap roots – up to 15 metres – while others spread over the surface to take maximum advantage of any rain or dew. Vegetation in hot desert regions tends to be widely spaced to avoid competition for water. Fauna are also restricted by the high temperature and lack of water. Many animals are nocturnal, burying themselves in the daytime from the oppressive heat.

< Fig 9.17

Cacti in the Arizona Desert

It is the fringes of these deserts with their delicately balanced ecosystem that are threatened by desertification as in the African Sahel (see Chapter 7).

Temperate Climates

1. Mediterranean Climate

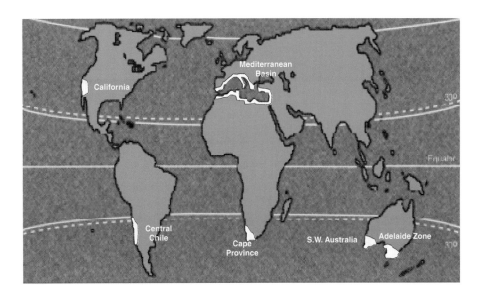

Fig 9.18
Areas of Mediterranean Climate

Climate

The Mediterranean climate is located on the west coat of continents between 30° and 40° north and south of the equator, i.e. Mediterranean Europe, California, Central Chile, Cape Province (South Africa) and Southern Australia.

> The Mediterranean climate is noted for its hot, dry summers and mild, wet winters.

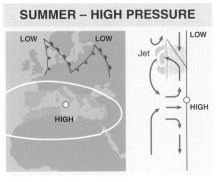 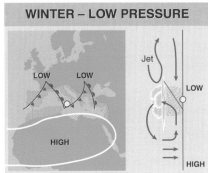

Fig 9.19
Pressure Belts over the Mediterranean

In summer high pressure dominates and with the sun shining directly on the nearby tropic, cloudless skies result in hot days with temperature well into the high 30s. Descending air currents and trade winds bring arid conditions with the length of the dry season increasing towards the desert margins. In winter the low pressure returns allowing the westerlies to dominate. While these ensure a mild winter (10°), they are also responsible for rainfall which varies in amount depending on the local relief. Rome, on the western windward side of the Apennines can receive over 700 millimetres, almost twice that of

Climate and Natural Vegetation

the leeward (rainshadow) east coast. The Mediterranean region is also noted for its local winds. The hot dry winds of the *Sirocco* can raise temperatures to over 40°C, while the cold *Mistral* from the Alps can wreak havoc on the vine crop in the Rhone Basin (France).

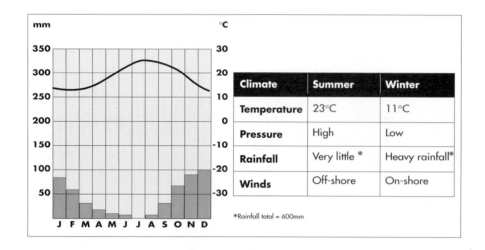

Fig 9.20
Temperature and Rainfall Chart for Mediterranean Climate

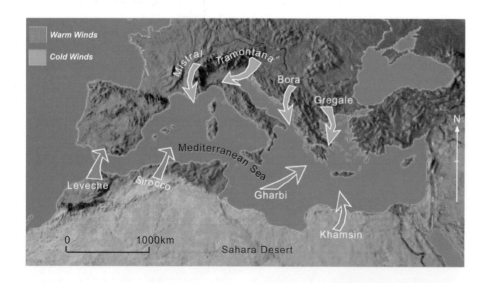

Fig 9.21
Mediterranean Winds

Vegetation

Mediterranean regions were originally covered with open woodlands most of which consisted of evergreen oak and cork oak. Most of this natural vegetation has been removed by human activities and widespread fires during the summer drought. The natural vegetation then that does exist is xerophytic or drought resistant and is referred to as *maquis* or *garrigue* in Europe and *chaparral* in California. The cork oak, for instance, then has a thick bark to reduce transpiration while the olive and eucalyptus have long tap roots to reach underground water supplies. Elsewhere forests of conifers such as pines, firs, cypresses and cedars occur.

2. Maritime: Cool Temperate Western Margin

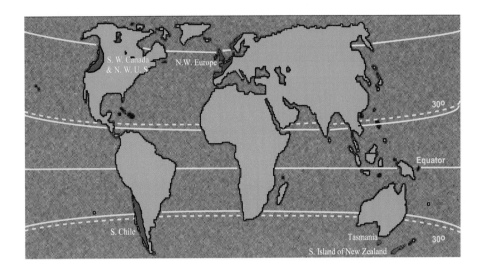

Fig 9.22
Areas of Cool Temperate Western Margin Climate

Climate

Sometimes called a North West European climate, this climate is experienced on west coasts between 40° and 60° north and south of the equator.

Apart from north west Europe, other areas which experience this climate include the North Western United States and British Columbia in Canada in the northern hemisphere and Southern Chile, Tasmania and the south island of New Zealand in the southern hemisphere.

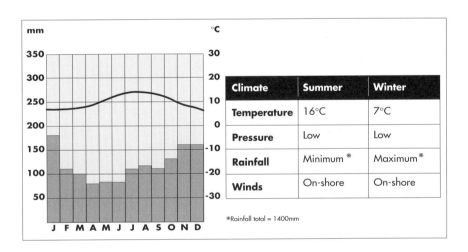

Fig 9.23
Temperature and Rainfall Chart for Cool Temperate Western Margin Climate

Climate	Summer	Winter
Temperature	16°C	7°C
Pressure	Low	Low
Rainfall	Minimum*	Maximum*
Winds	On-shore	On-shore

*Rainfall total = 1400mm

Summer temperatures are cool with the warmest months reaching 15°C to 17°C. This is the result of the low angle of the sun together with frequent cloud cover and the cooling influence of the sea. The same on-shore winds together with warm off-shore currents ensure that winters are mild. Temperature ranges, both diurnal and seasonal, are low and average winter temperature remains above freezing. Precipitation often exceeding 2,000 millimetres annually falls throughout the year. There is a winter maximum when depressions are more common.

Vegetation

Most of the original broad leaved deciduous forest has now been cleared for farming and urban development. Unlike the rainforest there are relatively few species with oak, ash, beech and chestnut being the most common. Leaf falls occurs in autumn which has the effect of reducing transpiration. The taller trees can reach heights of up to 40 metres with a lower scrub layer of holly, hazel and hawthorn up to five metres high. Moving northwards the deciduous forest gives way to coniferous trees of spruce and pine.

< Fig 9.24
Deciduous Trees

3. Cool Temperate Continental

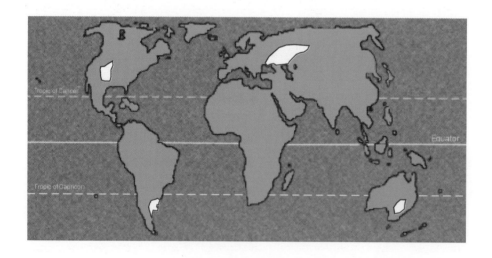

< Fig 9.25
Areas of Cool Temperate Continental Climate

Climate

The cool temperate continental climate occurs in the same latitude as the Maritime climate but there is a much higher annual range of temperature. This is especially true of the northern hemisphere.

These areas are situated in the middle of large land masses and so are removed from the moderating influence of the sea. Summer temperature then can rise to 18°C while winter can fall to -19°C giving an annual range of about 37°C.

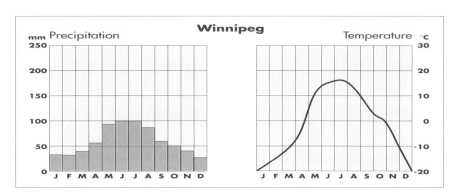

Fig 9.26
Temperature and Rainfall Chart for Cool Temperate Continental Climate

Precipitation shows a summer maximum when convectional uplift produces torrential showers. In winter, snow is common with the annual total rarely exceeding 520 millimetres.

Vegetation

Climatic conditions here favour grassland rather than forest. The main areas include the steppes of Russia, the prairies in North America, the pampas in Argentina and the Veldt in South Africa.

Little of the original natural grasslands exists today. The better watered areas have been ploughed to form the world's major wheatlands while the drier areas support cattle as in the ranches of America and sheep as in Australia and New Zealand.

4. Cool Temperate Eastern Margin (Laurentian)

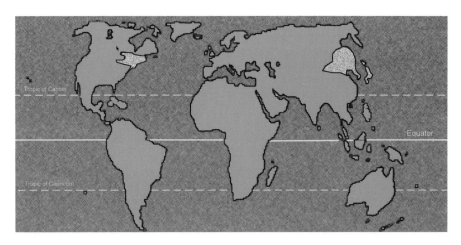

Fig 9.27
Areas of Laurentian Climate

Climate

The Laurentian climate occurs on the eastern side of North America and Asia between 35°N and 50°N. The Maritime Provinces of Eastern Canada are typical.

The cold winds which blow from the interiors of North America and Asia in winter together with cold offshore currents (Labrador and Kuril) result in cold winters when temperature falls to -9°C.

Summer temperatures can rise to 24°C (New York 22°C) resulting in a high annual temperature range of 33°C. Precipitation which falls as both rain and

Climate and Natural Vegetation

snow is fairly evenly spread throughout the year and can vary from 530 mm. to over 1,000 mm.

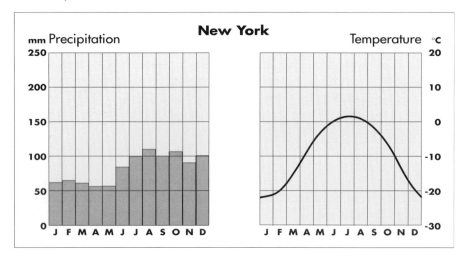

Fig 9.28
Temperature and Rainfall Chart for Cool Temperate Eastern Margin (Laurentian) Climate

Vegetation

Little now remains of the original natural vegetation of these regions. The cool temperate moist climate probably supported deciduous forests, broadly similar to that which exists in parts of Korea and Northern Japan. There tends to be a small variety of trees while stands of the same species are common, such as oak, elm, sycamore and chestnut. The deciduous forests of North America tends to be varied and luxuriant and include, together with those already mentioned, walnut, maple and hickory and numerous conifers such as cedars and spruces.

Fig 9.29
New England Trees

COLD CLIMATES

Above 60°N, the cool climates merge into cold climates where increasing distance from the equator means that in areas above 60°N, mean annual temperature falls to less than 6°C. Because plant growth ceases at 5.6°C cold climates are generally classified according to vegetation. These climates are absent from the southern hemisphere since land masses do not extend far enough polewards.

Climatic Divisions Based on Temperature

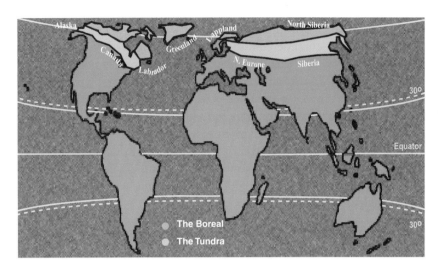

Fig 9.30 > Areas of Boreal and Tundra Climate

1. Taiga

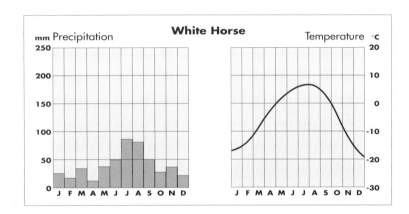

Fig 9.31 > Temperature and Rainfall Chart for Taiga Climate

Fig 9.32 > Coniferous Trees

Climate

Taiga is the Russian word for the coniferous forests of Siberia. This is the source region for the Pc air mass (see Chapter 8) with its cold, dry, stable air. High latitude, large land masses and winter darkness combine to reduce winter temperature to -50°C, while the summer may reach 17°C. Precipitation is generally low and frequently falls as snow.

Vegetation

Despite the fact that actual amounts of precipitation are small, the low temperature means little evaporation and so the 'effective precipitation'

Leaving Certificate Geography

is adequate to support tree growth. Coniferous forests (taiga) stretch right across the North American and Eurasian continents. The trees have to withstand the harsh climatic conditions of extremely cold winters and short cool summers. Growth is therefore slow with some species taking up to two centuries to mature. Transpiration is reduced to a minimum by needle-shaped leaves. The conical shape of the trees provides stability against strong winds while the downward sloping branches allow winter snow to slide off. Extensive stands of a single species are common. While the spruce, fir and pine are all coniferous, the cone-bearing larch is actually deciduous and sheds its leaves in winter. The undergrowth is thin and of poor quality which greatly facilitates the logging which is synonymous with the economy of countries such as Canada, Sweden and Finland.

2. Arctic – Tundra

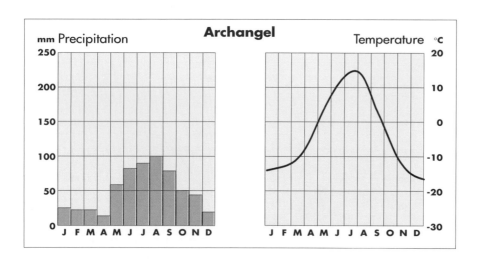

< Fig 9.33

Temperature and Rainfall Chart for Arctic Climate

Climate

This climate is experienced in Northern Alaska, Canada, Greenland and Russia in the Northern hemisphere and in Antarctica in the Southern hemisphere. In spite of the continuous hours of daylight, summer temperatures rarely manage to rise above freezing.

Winters are long dark and severe with icy winds resulting in temperatures of -30°C. Precipitation is light at around 250 millimetres often falling as snow.

VEGETATION

In Finnish, tundra actually means 'barren or treeless land'. This is the land of the permafrost (see Chapter 5) where the ground apart from the top 50 centimetres in summer, is permanently frozen. Plant life is rare and has to have a high degree of tolerance to the extreme cold. Lichens and mosses then are common and invariably low-growing to gain protection against the icy winds. In summer much of the region is waterlogged because the permafrost prevents infiltration. On south facing slopes where drainage is better, cushion plants such as arctic poppies and gentians provide a mass of colour.

Questions

ORDINARY LEVEL

1. Explain the following statement in some detail.
   Plants and animals of hot desert regions have adapted to their environment. *(20 marks)*

   Leaving Cert. 1996

2. Examine the following statement in some detail. The Sahara Desert is expanding southwards at over 50 km each year.
   (40 marks)

   Leaving Cert. 1995

3. Examine the following statement in some detail. Hot desert climates have a large daily temperature range. *(20 marks)*

   Leaving Cert. 1994

HIGHER LEVEL

1. (i) With reference to any **one** continent, illustrate and explain the relationship which exists between natural vegetation and climate.
   (75 marks)

   (ii) Examine how human activities interact with the relationship which you have referred to in (i) above. *(25 marks)*

   Leaving Cert. 1990

2. (i) With reference to any **one** continent, illustrate and explain the fact that major vegetation zones are related to major variations in climate. *(60 marks)*

   (ii) Explain, with reference to any **two** regions which you have studied, the ability of Humankind to change the plant and soil world.

   (40 marks)

   Leaving Cert. 1987

CHAPTER 10 — Soils

< Fig 10.1

Soil forms the thin surface layer which covers much of the earth's crust. It is one of three essential natural resources (together with air and water) as it forms the basis for plant and, in consequence, animal life on land. The ability of planet earth to feed its ever-growing population depends to a great extent on the efficient use of soil and its preservation from reckless exploitation and destruction.

How Soil is Formed

The formation of soil begins with the weathering process (see Chapter 2), which breaks up the earth's crust into small particles known as *regolith*. The nature of the soil therefore depends first on this 'parent' rock from which the regolith develops. It may be a result of the disintegration of the local bedrock or be transported by agents of erosion, e.g. boulder clay. When air and water enter the spaces between the particles in the regolith, chemical changes take place producing chemical substances. Bacteria and plant life now appear. When these plants die they decay and produce *humus*, an essential ingredient to soil fertility. Humus consists of the decayed remains of both plant and animal life and the role of bacteria is to assist in the decomposition of these remains. The end product of this complex chemical and biological process is soil.

Soil then is the result of:

(i) Parent material; (ii) Climate; (iii) Relief;

(iv) Living Organisms; (v) Time.

Parent Material

Where soil develops from the underlying rock, its supply of minerals is largely determined by the parent rock. The parent rock also helps to determine the depth, texture and colour of the soil.

Climate

Apart from its obvious effect on the rate of disintegration of the parent rock, the importance of climate can be gauged from the fact that the distribution and location of the world's soils correspond closely to patterns of climate and vegetation.

Precipitation affects the type of vegetation which grows in an area. Vegetation in turn provides humus. Generally speaking, the more rainfall, the greater the quantity of humus. Heavy rainfall also results in the downward movement of water which in turn transports mineral salts, a process known as *leaching*.

Temperature determines the length of the growing season (remember growth ceases below 5.6°C) which in turn affects the supply of humus. The growth and decay of vegetation is fastest in hot, wet conditions.

Relief

The relief or topography is important because it influences the local climate (micro-climate). As the height of the land increases, temperatures decrease but precipitation increases. Aspect also is important with south facing slopes enjoying higher temperatures in the northern hemisphere. The actual degree of slope is also important. Steep slopes encourage mass movement, thereby increasing the danger of soil erosion.

Living Organisms

The various types of plant and animal life all interact in the nutrient cycle. Plants take up minerals from the soil as they grow and return them when they die. This recycling is achieved by the activity of bacteria which assist in the decay and decomposition of the dead vegetation.

Time

The rate of rock disintegration and organic decay is obviously related to time. It has been estimated that it takes up to 400 years for 10 millimetres of soil to form.

While young soils exhibit many of the characteristics of the parent rock, soil continues to change over time and it can take between 3,000 to 10,000 years to produce sufficient depth of mature soil for farming.

Soils

THE SOIL PROFILE

The soil profile is a vertical section through the soil showing its various layers or horizons from the surface to the underlying bedrock.

Horizon A

This is where biological activity and humus content are mainly concentrated. It is also the zone which is mainly affected by leaching.

Horizon B

Beneath the upper layer is horizon B. It is a zone of accumulation where the minerals and other materials removed from the A horizon are re-deposited.

Horizon C

This consists mainly of recently weathered regolith which rests on the bedrock.

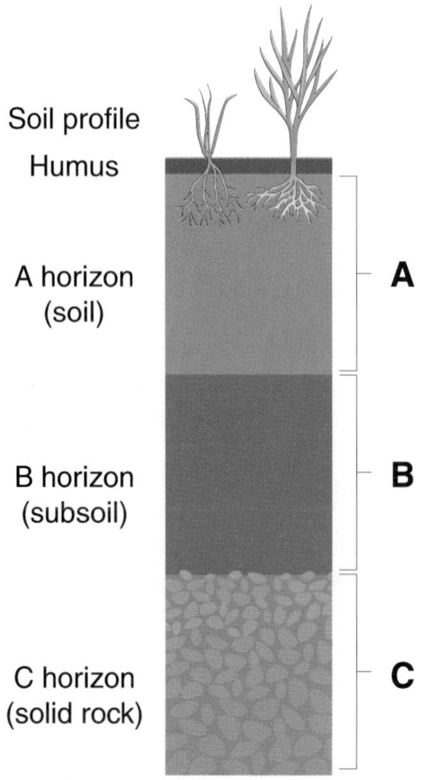

Fig 10.2 Soil Profile

SOIL CHARACTERISTICS OR PROPERTIES

It must be remembered that this threefold division of a soil profile is useful but it is also oversimplified. While some soils may have their humus content mixed throughout their depth, it tends to remain on the surface in cold, wet upland areas to produce an acidic and nutrient deficient soil as in peat moorland.

The inter-relationship between the four major components of soil, i.e. water, air, mineral and organic matter determines the properties or characteristics of any soil type. The most important being:

Soil Texture

This refers to the coarseness or fineness of the mineral matter in the soil. According to the ratio of sand, silt and clay present, soil can be classified into three groups according to texture.

1. Sandy Soils

Sandy soils occur when sand particles constitute 70 per cent of the soil by weight. Because of the large air spaces they are well drained, aerated and easy to cultivate. However, because nutrients and organic matter are easily leached they usually need considerable amounts of fertiliser.

2. Clay Soils

Clay soils occur when clay particles make up over 35 per cent of the soil by weight. With little space between the fine particles, water and air find it difficult to pass through. They are rich in nutrients and organic matter but are difficult to plough. After heavy rain they are prone to waterlogging and are often described as 'heavy'.

3. Loam Soils

Loam soils consist of a mixture of sand, silt and clay. They are well drained, well aerated and with a good mineral-holding capacity, they form the ideal soil from an agricultural viewpoint.

Soil Structure

Soil structure is concerned with the binding of soil particles to form lumps or *peds*. It is the shape of the peds which determine the size and number of pore spaces through which water, air, soil, organisms and roots can pass. Soil structure then influences the soil's suitability for cultivation. Soils with a 'crumb' structure are generally considered to be most suitable for cultivation.

Soil Humus

Humus is formed mainly from the decomposition of dead plants and animals, a process aided by the presence of bacteria. Humus gives the soil a dark brown or black colour. The highest amounts of humus are found in areas of temperate grasslands to form the chernozems or 'black earths' of the prairies of North America, the pampas of Argentina and the steppes of Russia. Humus is also a source of nutrients and acts as a cement, binding soil particles together and so reducing the risk of soil erosion.

Soil Acidity (pH)

The pH of a substance is its degree of acidity or alkalinity which can range from 0 (extreme acidity) to 14 (extreme alkalinity). It must be remembered that the scale is logarithmic (like the Richter Scale in Chapter 1). This means that a scale of 6 is ten times more acidic than a reading of 7. A soil with a pH of about 6.5 is most favourable for plant growth.

Generally the heavier the rainfall the more acidic the soils (as in upland areas). In areas where there is a balance between precipitation and evaporation, soils tend to be neutral (pH7). Acidic soils can be successfully farmed by the application of lime.

Soil Moisture

Water is retained in the soil by capillary attraction, i.e. it forms a thin film around the individual soil particles. Soil water is essential for plant growth. When a plant loses more water through transpiration, than it can take up through its roots, it is said to suffer *water stress* and so begins to wilt.

Soil Air

Air fills the spaces left unoccupied by soil water. Soil air is essential for successful seed germination, plant growth and animal life.

CLASSIFICATION OF SOILS

Soils are generally classified by using the zonal system. As a result soils can be divided into: (a) Zonal (b) Intrazonal (c) Azonal.

(A) Zonal

Tundra Soils

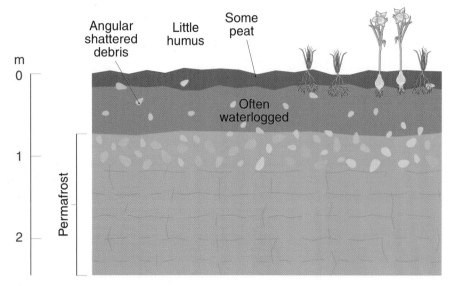

< Fig 10.3
Tundra Soil Profile

Tundra soils are found in Northern Canada, Scandinavia and Russia. The A-horizon is a mixture of organic and mineral matter with a dark grey-brown colour, to a depth of about 50 centimetres. This is the region of permafrost (see Chapter 5). Below this level, the ground is permanently frozen. Plant life is restricted to mosses and lichen. These soils are of little use for commercial agriculture but can support the caribou and reindeer of nomadic groups such as the Sami (Lapps) in Norway, Sweden and Finland.

Podzols

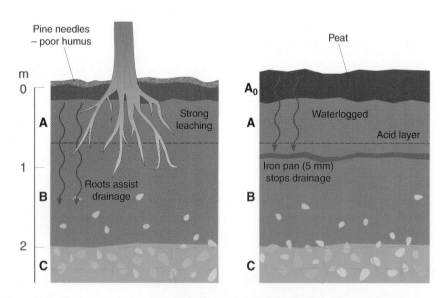

< Fig 10.4
Podzol Soil Profile

Podzols developed in areas where the natural vegetation was coniferous forest. They are heavily leached, especially when the snow melts and this can often lead to the development of a hard pan or iron pan from iron compounds. As this hard pan tends to be impermeable, the soil often becomes waterlogged. Generally, podzols are infertile and need large applications of fertilisers and lime to reduce acidity. They are common in the upland areas of counties such as Cork, Kerry, Donegal and Mayo.

Classification of Soils

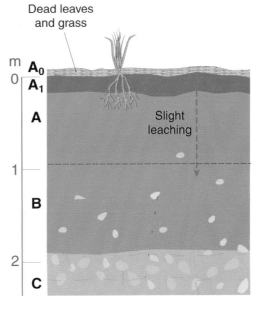

Fig 10.5 › Brown Earth Soil Profile

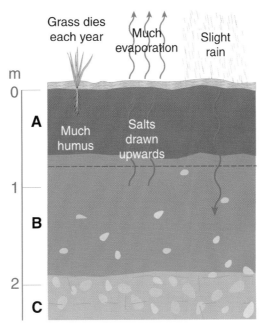

Fig 10.6 › Chernozem Soil Profile

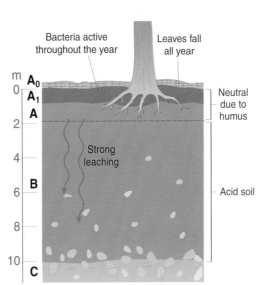

Fig 10.7 › Latosol Soil Profile

Brown Earths

Brown earths developed in the cool temperate areas of deciduous woodlands. They lack the distinctive horizons of podzols and have an almost uniform brownish colour as a result of the absence of leaching.

Chernozems

Chernozem is a Russian word meaning 'black earth' which refers to the humus rich thick black A-horizon.

It is largely associated with continental interiors where precipitation is low and grass is the natural vegetation. The soil is extremely fertile, has a neutral pH value and possesses a well developed crumb structure. As a result, it is associated with the world's 'bread baskets' (grain producing areas) of the interior of Canada, USA, the Ukraine and the pampas of Argentina.

Latosols

Latosols form where temperature and humidity is high and so are associated with the equatorial rainforest and tropical savannas. The A-Horizon is reddish in colour which is the result of the widespread occurrence of iron compounds in the soil. Leaching is common as a result of heavy rainfall. In spite of the dense forest cover, soil fertility is poor as the early colonists discovered. It is only the rapid decomposition and recycling of plant litter that enables the soil to support such vegetation. This is why the native people like the Boro tribe of the Amazon practice shifting or milpa agriculture.

Desert and Semi-Desert Soils

Desert and semi-desert soils are found in mid-latitude and tropical deserts and are red or grey in colour. The lack of vegetation results in the lack of humus which accounts for their light colour. Intense heat usually means the upward movement of ground water which evaporates to leave salt deposits near the surface.

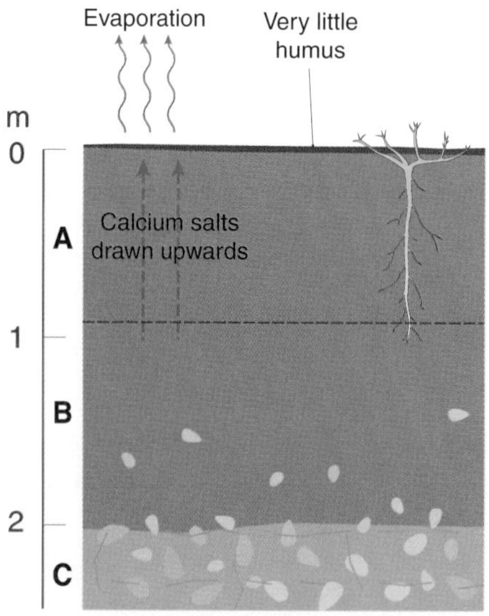

Fig 10.8 Desert Soil Profile

(B) Intrazonal Soils

Gleys

Gleys occur in saturated soils when the pore spaces are filled with water resulting in a lack of oxygen. This has the effect of reducing chemical weathering which gives the soil a grey-blue colour. Their excessive moisture content gives them a heavy texture which limits their utilisation. They are widespread throughout the drumlin belt (see Chapter 5) in the north of the country. Drainage is therefore essential before they can be utilised for intensive cultivation.

Peat

Layers of peat develop where a soil is waterlogged and the climate is too cold for organisms to break down the vegetation.

Essentially it occurs where the rate of accumulation of surface litter exceeds the rate of decomposition. Peat can be divided into two categories:

1. Blanket Peat, also called *climatic peat*: blanket peat is very acidic and develops in upland areas with heavy rainfall as found in the west of Ireland.
2. Raised Bog Peat: also called valley or basin peat, raised bog peat develops in lowlands with heavy rainfall. These are best developed in the Central Plain where they are harvested by Bord na Móna.

Rendzinas

Rendzinas form where limestone or chalk form the parent material and where the surface vegetation is composed of grasses. The A-horizon is usually black or dark brown. There is usually no B-horizon. The absence of regolith together with the permeable nature of the bedrock results in a thin soil covering. Rendzina-soils are common on the limestones of north-east Clare, east Galway and Roscommon.

Terra Rossa

As its name suggests, terra rossa is a red coloured soil and is sometimes called a 'red rendzina'. The removal of the calcium carbonate leaves residual deposits rich in iron giving it its red colour.

It occurs mainly in the Mediterranean where it is extensively used for viticulture.

(C) Azonal Soils

Regosols

Like other azonal soils, regosols are young soils which normally display characteristics of their origin, i.e. parent rock or agent of deposition. They have no distinct profile but can be extremely fertile as in the alluvial soils of the floodplains of the Nile, Tigris and Euphrates.

Lithosols

Lithosols are formed in upland areas. They are poorly developed and are sometimes referred to as 'scree soils'. Mass movement and agents of erosion remove the finer grains leaving only an immature stony covering behind with limited pastoral potential.

SOIL EROSION AND CONSERVATION

It usually takes thousands of years for a mature deep soil to develop. In spite of the fact that it is an essential natural resource, it is estimated that over 20 billion tonnes a year or seven per cent of the world's top soil is lost each year.

Causes of Soil Erosion

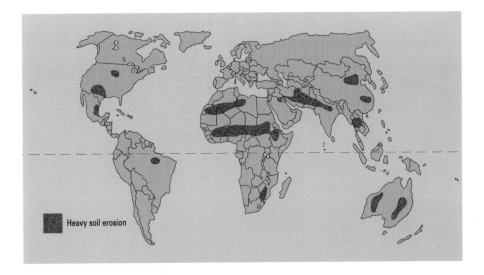

Fig 10.9
Areas of Heavy Soil Erosion

The major cause of soil erosion is the removal of natural vegetation cover which leaves the ground exposed to agents of erosion. The most serious is deforestation. In countries such as Ethiopia, the removal of trees for fuel results in the topsoil being loosened by *rainsplash*, which prepares it for removal by *sheetwash*. The lack of plant roots which help to bind the soil together encourage gulley erosion making the land useless for cultivation. The removal of trees also reduces the rate of transpiration which in turn limits the amount of moisture in the air. There are now serious fears that large-scale deforestation will turn areas at present under rainforest into deserts (See Desertification, Chapter 7).

Ploughing, coupled with drought, can be a major cause of soil destruction. In the 1930s, large quantities of top soil were blown away to create the American Dust Bowl. *Overgrazing* also reduces the grass cover which inevitably leads to an increase in exposed surface. *Monoculture* also leads to soil destruction: growing the same crop every year repeatedly uses up the same minerals.

Soil erosion is not confined to the developing world, as the American Dust Bowl of the 1930 testifies. Centuries of mismanagement in the Mediterranean has resulted in extensive soil erosion in countries such as Italy, Spain and Portugal.

Deforestation on the hillsides coupled with overgrazing by sheep and goats has denuded large areas of this vital resource.

What we must remember is that soil is dynamic; a 'system' where inputs must replace outputs if it is to be preserved.

Contour Ploughing

Ploughing up and down slopes creates a series of furrows which inevitably leads to gulley erosion but ploughing across the slope creates ridges which act as barriers to the downslope movement of water and in consequence, soil.

TERRACES

< Fig 10.10
Terracing in South-East Asia

Terracing involves the construction of a series of shelves or steps on slopes. Retaining walls are built which hold back soil and water.

It is common in South-East Asia where high population densities put pressure on agricultural land.

AFFORESTATION

Planting trees has the opposite effect to deforestation. Trees help to anchor soil, reduce the impact of torrential rain, help to maintain the water cycle and act as wind breaks.

It is important to remember that, as a natural resource, soil is dynamic. As a system its preservation depends on its outputs being replaced by inputs. Its function is not merely to provide people with food. It must also be viewed as the 'parent' of such resources as forests, as a source of much of the world's industrial raw material such as cotton and rubber – hardly a resource to be treated like dirt!

Questions

ORDINARY LEVEL

1. Examine the influence on soil formation of any **four** of the following: parent material; climate; relief; vegetation; animal organisms; people. *(80 marks)*

   Leaving Cert. 1992

2. Parent material, Climate, Relief, Organisms, Time.
   Above are listed some of the factors which influence the formation of soils.
   (i) In the case of **each** of any **three** of these factors, explain how it influences soil formation. *(60 marks)*
   (ii) 'Maintenance of soil fertility and the prevention of erosion are key aspects of soil management.' Briefly examine this statement. *(20 marks)*

   Leaving Cert. 1991

HIGHER LEVEL

1. The Earth System

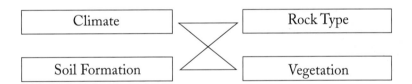

The arrows in this diagram illustrate relationships between different elements of the Earth System. Explain with the aid of diagrams, **four** of these relationships. *(100 marks)*

Leaving Cert. 1994

The World – Physical

The World – Physical

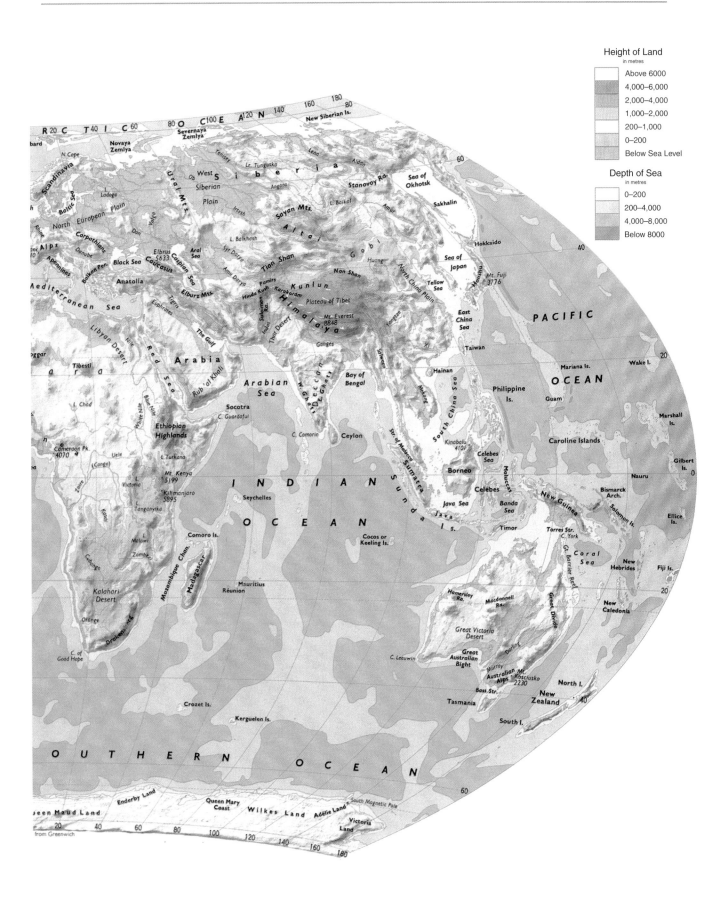

141

The World – Political

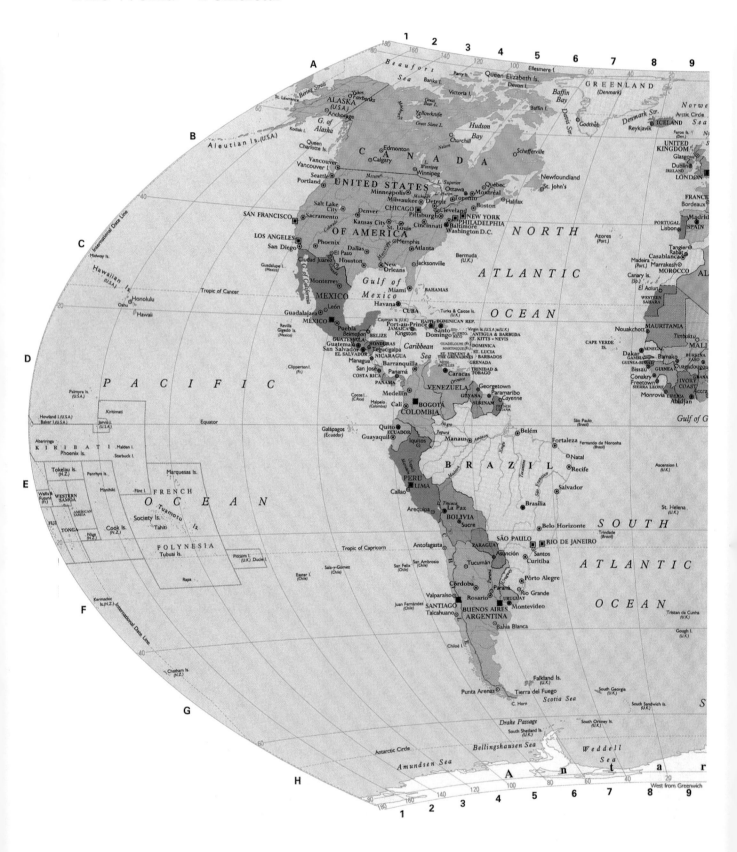

The World – Political

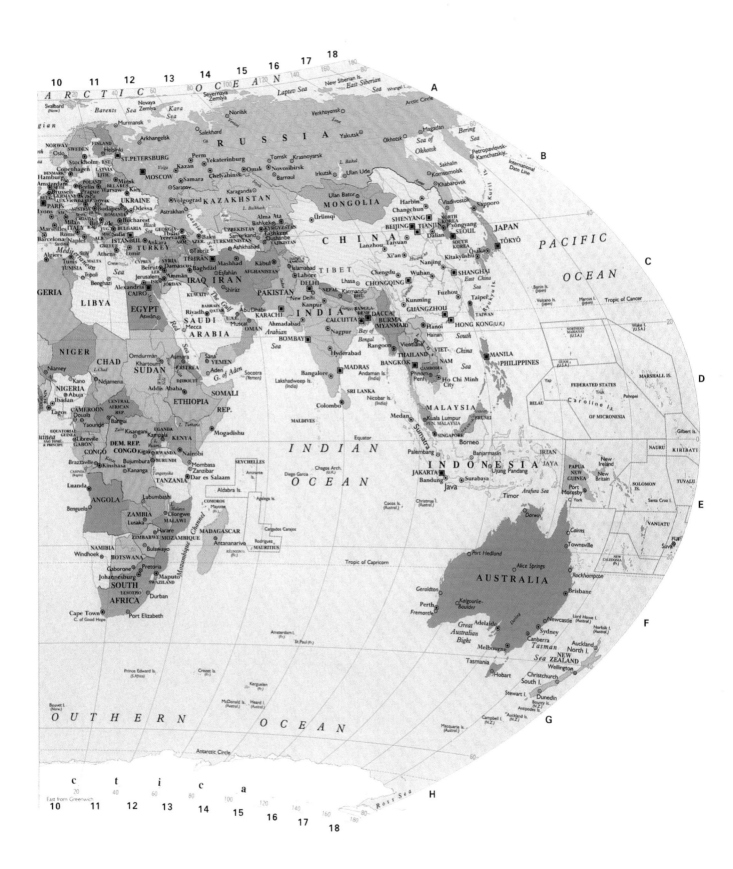

Leaving Certificate Geography

The World – Political

Questions

1. Examine the map of the world which accompanies this question and answer the following:
   (i) Identify, in the spaces provided, the features numbered as follows: River **1**; Mountain range **2**; Country **3**; Sea area **4**; City **5**; Island Group **6**. *(12 marks)*

   (ii) There are four regions marked **A, B, C,** and **D** on the map. The majority religions in these regions are Christianity, Islam, Buddhism and Hinduism. Link **EACH** region with its majority religion. *(12 marks)*

   (iii) Using spaces **W, X, Y, Z**, identify the following countries in the correct order:
   Brazil, Northern Ireland, South Africa, Rwanda. In the spaces provided, link **EACH** country with one of the following issues with which it is associated:
   * Abolition of Apartheid,   * Racial Conflict,
   * Rain Forest Destruction,   * Peace Process   *(24 marks)*

   (iv) Select any **TWO** of the countries named in part (iii) and for **EACH** one you select, describe **AND** explain the issue involved. *(32 marks)*

   Leaving Cert. 1996

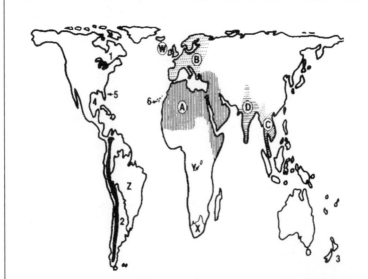

A Peters' Projection (as opposed to Mercator's Projection) is used for this map, which explains why the shapes of the countries are slightly different.

1		2		COUNTRY	ISSUE
3		4		W	
5		6		X	
A		B		Y	
C		D		Z	

CHAPTER 11

Population

Fig 11.1

Three babies are born and one person dies every second – there is a net increase of two people per second in the world's population. The present population of the earth, now heading for six billion, will double over the next 35 years. This population explosion has rightly been called the 'greatest problem of them all'.

Fig 11.2

The Growth in World Population from 1650

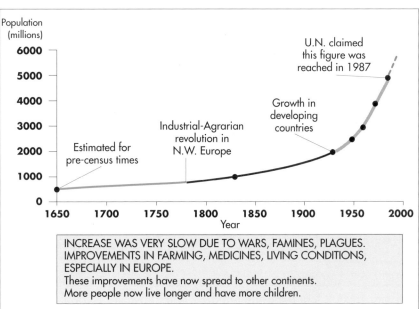

Leaving Certificate Geography

Population

CHANGES IN TIME – WORLD POPULATION GROWTH

At the end of the last Ice Age, the Pleistocene (see Chapter 5), the human population is estimated to have been less than eight million. People were nomads and survived by gathering wild fruit and hunting animals. With the neolithic revolution, the development of agriculture as we know it today began. As food supplies become more reliable, the need for constant movement disappeared and, rural settlement started. World population began to increase and by the year 1AD it is thought to have been about 260 million.

While population increase varied considerably from one area to another, depending on a region's capacity to sustain it, its growth was also irregular due to wars and natural disasters such as famines and plagues. It took one and a half thousand years for the population to double, so that by 1650 it is estimated to have reached 530 million.

One of the most notable checks on the European population during this period was the *Black Death* in the fourteenth century which is thought to have wiped out over one third of the entire population.

Dramatic increases in population are a relatively recent phenomenon and it is only in this century that the word 'explosion' can be applied with any degree of accuracy.

1650 – 1986

Since 1650 the population of the world and especially Europe began to show a steady increase in response to:

(i) The agricultural revolution which dramatically improved both the quality and quantity of food supplies.
(ii) The industrial revolution which improved clothing and living conditions (sanitation etc.).
(iii) The medical revolution which reduced the effects of diseases and so dramatically reduced the death rate.

People lived longer, more people married and had children and thanks to medical advances and better nutrition, fewer died in infancy. So began the population explosion.

Population per Continent, 1650–1986 (in millions)						
Date	Africa	Asia (ex. USSR)	N. America	Central, S. America	Europe (inc. USSR)	Oceania
1650	100	300	1	10	100	2
1750	95	480	2	13	140	2
1850	100	740	26	33	275	2
1950	208	1400	170	160	590	13
1986	586	2894	280	428	793	27

< Fig 11.3
Population Growth by Region, 1650-1986

Between 1650 and 1850 the world's population doubled to approximately 1.2 billion. It took less than one hundred years however to double again to 2.5 million by the year 1930. Less than 50 years later it doubled again to five billion. If present growth goes unchecked, it is estimated that the world's population will double again in 35 years. It is more worrying however that this increase is more dramatic in countries which can least afford it; while countries like Austria and Sweden have reached zero population growth (ZPG) countries in the developing world are set to explode. Nigeria's present population of around 100 million is expected to reach 500 million by the year 2085.

Birth Rates, Death Rates and Natural Increase

The increase in the world's population is due to the differences between birth rates and death rates.

- Birth Rate is the number of live births per thousand per year.
- Death Rate is the number of deaths per thousand per year.
- Natural Increase is obtained by subtracting the death rate from the birth rate.

Birth Rates (Fertility Rates)

Births per thousand in Developed Countries		Births per thousand in Developing Countries	
USA	16	Pakistan	51
Italy	16	Nigeria	50

Fig 11.4
Birth Rates in the Developing and Developed Worlds

As the table shows there are considerable variations between the birth rates in developed and developing countries.

Economic, social and educational improvements in Europe have dramatically reduced the birth rate. Many families now deliberately limit their family size to ensure a higher standard of living. As a result an increasing number of countries are now approaching or have attained zero population growth (ZPG).

In the developing world however, it is the lack of these economic, social and educational opportunities which combine to keep birth rates high. Children are seen as a source of cheap labour and as a social security for their parents. Other factors like religious beliefs and lack of education all militate against family planning.

Death Rates (Mortality Rates)

The essential factor in explaining the dramatic increase in population has been the reduction in death rates. The improvements in food supplies, living conditions and medical services have dramatically reduced the mortality rates.

THE DEMOGRAPHIC TRANSITION MODEL

From the study of birth and death rates in several industrialised countries a model has been developed suggesting that all countries pass through similar demographic transition stages or population cycles.

Population

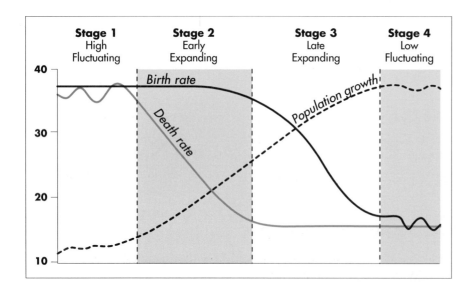

< Fig 11.5
The Demographic Transition Model

Present World Examples			
Stage 1	**Stage 2**	**Stage 3**	**Stage 4**
Tribes of the Rainforest	Peru	China	Canada
	Kenya	Cuba	USA

< Fig 11.6

Stage 1 – High Fluctuating Stage

The stage known as the high fluctuating stage occurred in the UK before 1750. Both birth and death rates are high (about 35 per 1,000) resulting in a small population growth.

The high birth rate is due to the following factors:

- Lack of birth control or family planning.
- Parents tend to produce more children because of infant mortality in the hope that some will live.
- Religious beliefs encourage large families.
- Children are needed to work on the land.
- Children are seen as a security against old age.

The high death rate is due to:

- Famine, uncertain food supply, poor diet.
- Few medical facilities, e.g. drugs, hospitals.
- Disease and plagues, e.g. cholera.

Stage 2 – Early Expanding

While birth rates remain high, the death rate reduces rapidly resulting in a dramatic increase in population.

The fall in death rate is due to:

- Improved medical care, e.g. vaccinations, doctors, etc.
- Improved water supply and sanitation.
- Improvement in food supply.

148 *Leaving Certificate Geography*

Stage 3 – Late Expanding Stage

Birth rates now fall rapidly while death rates decline more slowly. The overall result is a decline in population growth.

The fall in birth rates is due to the following factors:

- Family planning.
- With increasing mechanisation, fewer labourers are needed.
- A fall in infant mortality puts less pressure on parents to have more children.
- Improved employment and educational opportunities are available for women.

Stage 4 – Low Fluctuating Stage

Here both birth rates – approximately 16 per 1,000 – and death rates – 12 per 1,000 are low, resulting in little overall increase.

Difficulties with the Model

1. In some countries birth rates have actually fallen below death rates resulting in a negative population growth (NPG). As a result some demographers believe that the model should incorporate this with a fifth stage.
2. It seems increasingly unlikely that many of the less developed African countries will ever become industrialised.

Changes in Space – Migration

Migration does not affect world population figures but it can have a significant influence in areas where immigrants are more numerous than emigrants (population increase) or where emigrants are more numerous than immigrants (population decrease).

Migration is the movement of people from one place to another and may be either permanent or temporary. Migration may also be either internal or external, voluntary or forced.

Internal and External (International)

Internal migration refers to the movement of population within a country, e.g. from the west of Ireland to the Dublin region.

External migration on the other hand refers to the movement across national boundaries, e.g. from Ireland to the USA. Unlike internal migration, external migration affects the total population of a country. The term *migration balance* is applied to the difference between the number of emigrants (people who leave a country) and immigrants (people who arrive in a country).

Voluntary and Forced

Migration is said to be voluntary when people move of their own free will because they are looking for an improved quality of life. This movement is usually influenced by both 'push and pull' factors.

Leaving Certificate Geography

Push factors are those which make people unhappy with their home area (source area). Pull factors are those which attract people to another area (destination area).

When people have virtually no choice but have to move from an area due to wars or natural disasters, e.g. an earthquake, migration is said to be forced.

Voluntary Migration

The great age of Exploration and Discovery, between 1410–1520, resulted in a steady if small movement of people from Europe to America. This number dramatically increased throughout the nineteenth century as greater economic opportunities presented themselves.

Between 1890–1913 over 15 million Europeans entered America, many of whom were Irish. Today up to 40 million Americans lay some claim to Irish ancestry. The most recent example of voluntary migration on a large scale is the rural to urban movement which is presently taking place in the Third World.

This movement is partly accounted for by rural 'push' and partly by urban 'pull' factors.

Push Factors

Most families in the Third World do not own the land they work and so are easy prey to greedy landlords. Plots also are often too small to support a family and with increasing pressure on the land, over-cropping and over-grazing, crop failure and soil erosion quickly follow. Cash cropping too (see Chapter 20) often forces families to less attractive marginal land where even subsistence farming is difficult.

Pull Factors

Many people then have a perception of city life (usually far removed from reality) as a more attractive alternative. People move to the city in the hope of better housing, better job prospects, better services and a more reliable food supply. Dazzled by the 'bright lights' the reality is often a makeshift hut of tin and cardboard as they become the latest inhabitants of the shanty towns which are synonymous with Third World cities.

People who live in Shanty Towns (% population)	
Rio de Janeiro	30%
Sao Paulo	25%
Bombay	40%
Calcutta	60%

Fig 11.7

Forced Migration

The development of plantation agriculture in the southern states of North America witnessed the mass movement of over 20 million Africans to that region. Though the slave trade petered out in the nineteenth century, black slaves actually outnumbered white settlers until the mass European migration of the nineteenth century.

The expulsion of millions of Jews in Nazi Germany and the civil war between the Tutsi and Hutu tribes in Rwanda provide more recent examples.

Refugees

> The United Nation's definition of a refugee is 'a person who cannot return to his or her own country because of a well-founded fear of persecution for reasons of race, religion, nationality, political association or social grouping'.

Before World War II most refugees tended to become assimilated in their new host country. In the last 50 years however the number of permanent refugees has risen rapidly. According to the UN, the number had risen to 14 million in 1989. The first major post-World War II problem arose over the Palestinian Arabs in the Middle East. Since then, the problem has intensified with the conflict in Vietnam (boat people) and Rwanda. In 1996 an estimated one million Hutu refugees had set up camp in nearby Zaire. About four-fifths of the world's refugees are in the Third World.

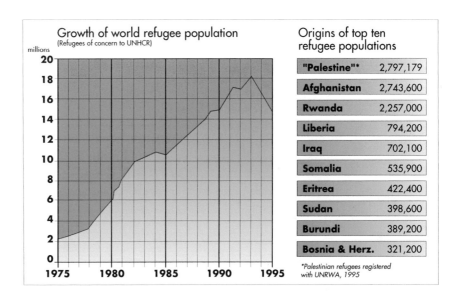

Fig 11.8 > **World Refugee Population**

As a result, the host country has few resources to deal with the problem. Refugees usually live in extreme poverty and lack food, shelter and clothing, educational and medical care. Generally devoid of citizenship they have few if any basic human rights and over time tend to lose their sense of dignity and purpose.

Effects of Migration On Source Area

Demographic Effects

The immediate effect on the source area is a decrease in population. Because migration tends to be age-selective, i.e. young adults form the majority of migrants, the source area tends to have very young and very old age groups – which distorts the population pyramid. Most migrants tend to be female, especially in developed economies as the farm is usually inherited by the 'male heir'. Thus migration tends to be sex-selective as well as age-selective which over time reduces the marriage rate and in consequence the birth rate.

Socio-Economic Effects

Outward migration inevitably leads to rural depopulation. Because this migration is both age and sex selective, many social activities are affected, e.g. discos, clubs, cinemas, etc. which may eventually close down. A decline in marriage rates inevitably results in a decline in birth rates which has a knock-on effect on social services such as health and education. The lack of a young, skilled workforce does little to improve farming techniques or to attract new industry. The problems of these peripheral regions are dealt with in more detail in Chapter 21.

Effects of Migration On Destination Area

Demographic Effects

Apart from the immediate distortion to the 'natural' population pyramid that an influx of migrants causes, the effects of inward migration on the destination area are both short and long term.

The immediate effect is a distortion of the natural population pyramid. The more long-term effects however include a more rapid increase in population as many migrants tend to be both young and female (of child bearing age).

Socio Economic Effects

The influx of a young, skilled, vibrant workforce can have many positive socio-economic effects. New industries are attracted by a skilled workforce, social activities are stimulated while the whole service sector enjoys benefits from the increase in numbers. Where the influx is too rapid however, the socio-economic effects can be catastrophic. Housing shortages, overcrowding, unemployment and overstretched medical and education services are only some of the problems which confront urban areas in both developed and developing economies.

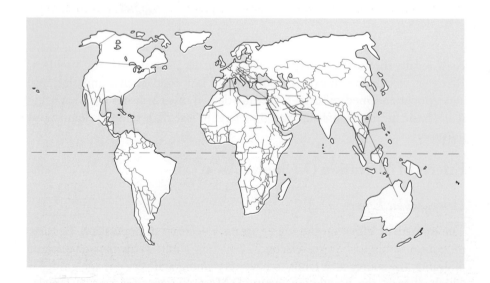

‹ Fig 11.9

Pattern of Major Recent Migrations

Population Distribution and Density

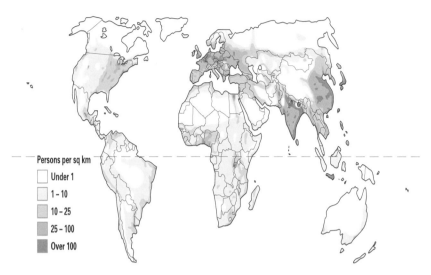

Fig 11.10
World Population Density

As the above map illustrates, the distribution of population over the world's surface is very uneven and there are considerable variations in density. While the factors which influence this distribution are both physical and human, one useful generalisation that can be made is that on a global scale this distribution is affected mainly by physical influences, while on a local or regional level the influences tend to be more human-related (political, social and economic factors).

FACTORS INFLUENCING WORLD POPULATION DISTRIBUTION

Physical Factors

Climate

People generally avoid areas with climatic extremes, i.e. too hot, too cold, too wet or too dry.

Hot Deserts

Fig 11.11

Leaving Certificate Geography

Population

Apart from the daily extremes of temperature (see Chapter 7) the greatest obstacle to human settlement is the lack of sufficient rainfall. It is this shortage that accounts for the virtual absence of two of the three essential resources namely, water and soil. Hence the extremely low population densities in areas like the Sahara, Atacama and Australian deserts. When a water supply is available from *exogenous* rivers (rivers which rise in mountains beyond the deserts) such as the Nile, their flood plains are often densely populated as they offer obvious attractions for the development of agriculture. On the other hand, population is attracted to areas with evenly distributed rainfall, with no temperature extremes and a lengthy growing season. Hence, the high population densities in North-Western Europe and Eastern and Southern Asia.

Relief

Mountainous regions as in the Andes, Rockies and Himalayas tend to have low densities of population. Here a combination of factors discourage settlement.

- Increase in altitude tends to result in climatic extremes, i.e. temperature decreases, while precipitation increases.
- Soils are often absent or thin and infertile.
- Rugged relief poses obvious transport difficulties.

Elevated areas within the tropics, as in Brazil and Venezuela, are often quite densely populated. This apparent contradiction is explained by the fact that in these regions the decrease in temperature due to altitude helps to offset the oppressive heat and in consequence makes them more attractive to human settlements.

⟨ Fig 11.12

Low-lying, well-drained regions like the North European Plain on the other hand are densely populated. The more fertile soils produce a greater food supply while the low-lying topography facilitates transport and human settlement.

Soils

Climatic and topographical extremes generally produce soils which are unsuitable for cultivation. The frozen soils of the permafrost regions of Northern Eurasia, the thin soils of mountainous regions and the leached soils of the tropical rainforest (Amazon Basin) help to explain the low densities of population in these regions. In contrast, the humus-rich soils of the Paris

Basin and the fertile river-deposited alluvium of rivers such as the Nile helps to explain their high densities of population.

Other physical factors such as vegetation and resources (coal, iron, etc.), while important, are less significant from a global viewpoint.

Fig 11.13

Human Factors

Climate, relief and soils play a major role in explaining the distribution of the world's population while the capacity of many areas to support population can be dramatically increased by human inputs. Some of the major concentrations of population occur where there are, or were large mineral deposits or energy supplies as in the Ruhr in Germany. This in turn gave rise to large scale industry which further increased the region's capacity. In other areas agricultural inputs such as fertilisers and machinery have dramatically increased farming output.

Reclamation in areas like the Netherlands has allowed that country to support one of the highest densities of population in the world. In contrast, it is the artificial application of water (irrigation) which has allowed semi-arid regions to increase their carrying capacity.

Population Structure

The study of migration and natural increase (or decrease) are only two aspects of the study of population in any region. A third is to look at the population structure of the area. This is important because the make-up of the population by age and sex, together with its life expectancy has major implications for future growth and economic development.

Population

POPULATION PYRAMIDS

The population structure of any society is best illustrated by a population pyramid. It shows both the age and sex of the population and tapers in towards the top as death rates increase with age. The basic technique normally divides the population into five year age groups, e.g. 5-9; 10-14, on the vertical scale, and into males and females on the horizontal scale. The number in each age group is given as a percentage (sometimes numbers in thousands are shown) of the total population and is shown by horizontal bars with males located to the left and females to the right.

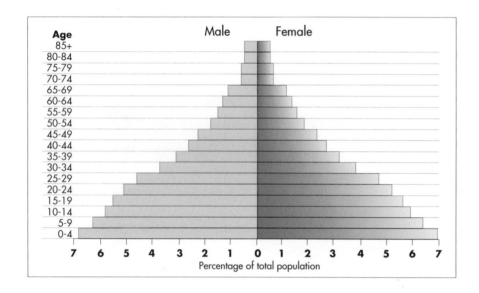

◁ **Fig 11.14**
Population Pyramid

Advantages of Population Pyramids

While the demographic transition model shows only the natural increase or decrease resulting from the balance between births and deaths, the population pyramid shows:

- The results of migration and its age/sex structure.
- The effects of large-scale wars and diseases.
- By the use of colour or shading the active or productive sector (15-64) can be highlighted. The under 15 and over 64 age groups then are referred to as the dependant sector.

The dependency ratio then can be expressed as:

$$\frac{\text{Children (0-14) and Elderly (65 and over)} \times 100}{\text{Those of Working Age}}$$

N.B. The dependency ratio does not take account of those who are unemployed.

Population pyramids are particularly useful for highlighting the demographic differences between developing and developed economies.

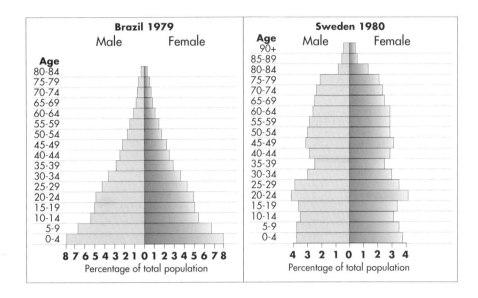

Fig 11.15
Population Pyramids of a Developing Country (Brazil) and a Developed Country

DESCRIPTION OF DIAGRAMS

Brazil

1. Very high birth rate – almost 16 per cent of population under five years.
2. Life expectancy is low with less than four per cent of the population over 65 years.
3. High dependency ratio especially of children (0-14 years).

Sweden

Examine the diagram for Sweden.

1. How does this diagram show low birth rates and low deaths?
2. If birth rates are low why is there a relatively high dependency ratio?
3. Sweden has entered Stage 4 in the demographic transition model. What advantages has the pyramid over this demographic transition model as a source of demographic information?

Population – The Future

Although the United Nations set aside 11 July 1987 as the date on which the five billionth human being arrived on earth nobody knows precisely how many people are living on the earth at a given moment. Recent evidence however shows that birth rates in the Third World may have begun to fall. In 1985 the United Nations Fund for Population Activities (UNFPA) claimed that the annual growth of the world's population had fallen from 2.1 per cent in 1965 to 1.6 per cent due in part to China's one child per family policy. What these figures fail to show however is the significant variations between developed and developing countries. At present the growth rate for developed countries averages 0.64 per cent per year compared to 2.02 per cent in developing economies. To achieve population stability, today's average family worldwide would have to consist of an average of 2.3 children.

The present figures of under two in Western Europe and North America stand in stark contrast to over four in Asia and over six in Africa.

POPULATION PROBLEMS

1. The World's Ageing Population

The improvements in medical facilities (hospitals, doctors, vaccines) has resulted in a considerable increase in life expectancy. In 1970 there were 291 million people over 60 years of age; that figure is expected to rise to 600 million by the year 2000. In the over 80 age bracket, the 26 million in 1970 is expected to be 58 million by the end of the millenium. The increased dependency of this sector will put a much greater demand on both families and governments to provide the relevant services.

The increasing demand for old age pensions, medical services, etc., will ultimately have to be borne by the active or productive sector. Also, the increase in life expectancy will initially result in a rapid increase in population with an associated strain on already stretched resources.

2. Zero Population Growth

Throughout history, on a global scale, birth rates have exceeded death rates, hence the growth in the world's population.

Recently however the birth rate has been falling in several European countries to the point where some countries are actually approaching zero population growth. In time, this will reduce their workforce and their competitive advantage in science and technology. While the problems of skilled labour shortages may be alleviated in the short term by migration, this will do little to guarantee long-term economic stability.

OPTIMUM POPULATION

In theory the optimum population of an area is the number of people which when working with all the available resources will produce the highest per capita income, i.e. the highest standard of living and quality of life. If at that point the population was to increase or decrease, the standard of living would fall. This concept is obviously true of a dynamic situation which changes with technology and the availability of resources, etc.

OVER-POPULATION

Over-population is where there are too many people relative to the resources and technology available to maintain an adequate standard of living. Countries like Bangladesh and India are often said to be over-populated as they have insufficient food and resources and seem to constantly live on the brink of starvation.

UNDER-POPULATION

This is where the population is too low to make the best possible use of the resources in an area. Countries like Canada and Australia are often said to be under-populated as their present populations are well below their 'carrying' capacities.

POPULATION AND FOOD SUPPLY

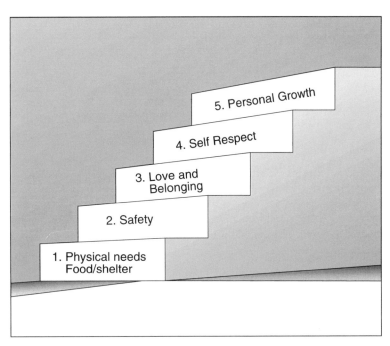

Fig 11.16 >
Hierarchy of Human Needs

Human development involves much more than feeding the world's population, but the provision of adequate food supplies is the first step on the human ladder to provide for people's needs. It is now 200 years since the British demographer, Thomas Malthus expressed his fears about world population outstripping food supply but in spite of recent famines, there is enough food for everyone in the world. There are however two major problems as is evident in Figure 11.17.

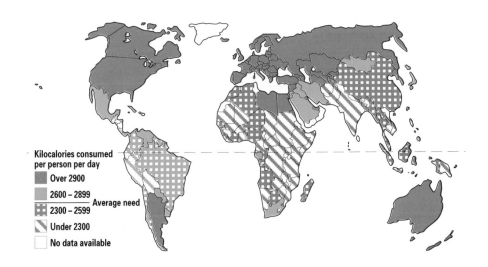

Fig 11.17 >
World Food Consumption

Leaving Certificate Geography

Population

(i) There are massive food surpluses in North America and the European Union (EU).

(ii) There are food shortages in many Third World countries.

It has been estimated that the average person needs an intake of about 2,400 calories a day. Developed economies enjoy over 3,300 calories, but developing economies consume less than 2,200 per person.

There is a difference however, between actual quantity and quality of food. The Food and Agricultural Organisation (FAO) estimates that while 25 per cent of the population in the Third World is suffering from insufficient food *(hunger)* over 50 per cent suffer from an inadequate diet *(malnutrition)*. At present, 45 million people die from hunger and hunger related diseases each year – almost half of whom are children. Even when malnutrition does not directly cause death, it reduces the capacity to work and increases susceptibility to diseases.

A more equitable distribution of food however is not simply a matter of transport. The fact remains that food is distributed, not on grounds of need but on one's ability to pay for it.

It is obvious then that while the various emergency aid programmes from developed economies are to be welcomed in alleviating immediate crises they do little to solve the longer-term problems.

Developing economies must increase their own food output. Methods to increase food output might include:

(a) An extension of cultivated land.

(b) Increasing agricultural yields from existing land.

(c) The development of new sources of food.

(a) An Extension of Cultivated Land

The world's food producing areas could be significantly increased by bringing more land into cultivation. This could be done by:

1. Terracing slopes in mountainous regions (see Chapter 13).
2. Extending and improving irrigation facilities in arid and semi-arid regions.
3. Clearing some forested areas in tropical areas.

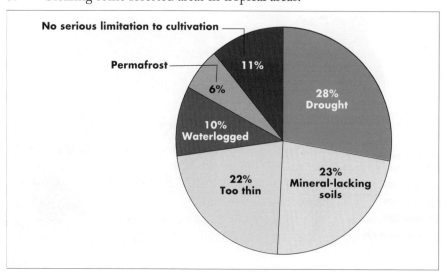

‹ Fig 11.18

Limitations on Cultivation of the World's Soils

While this would undoubtedly increase the food supply, it would be extremely costly in both time and money, none of which developing economies can afford. A more prudent approach then would be to 'make two blades of grass grow where one grew before' i.e. increase yields from existing land.

(b) Increasing agricultural yields from existing land

Higher yields from land already under cultivation can be achieved by:

(i) Increasing the use of fertilisers.

(ii) Using better seeds and improving animal strains.

(iii) Introducing and improving crop rotation.

(iv) Extending the use of insecticides to control pests.

(v) Using more efficient tools and farm machinery.

(vi) Controlling soil erosion (see soil conservation, Chapter 10).

(vii) Developing educational training for improved farming.

(c) New Sources of Food

While more food from improved farming techniques would make a significant difference, it should be remembered that the main product of agriculture is carbohydrates. What most people need however is protein food, e.g. milk, cheese, eggs, fish etc.

(i) Intensive fishing and 'fish farming' could significantly add to the protein requirements of developing countries.

(ii) Domesticating the wild animals of the Savannas, e.g. zebra would provide milk, cheese and meat.

(iii) Processing vegetable materials, e.g. sunflower, ground nut and soya plants.

All of these changes have one thing in common – they all require capital input – money which these countries simply do not have. Financial aid schemes from overseas may create problems as the recipient country is likely to fall into debt while the donor expects crops to be grown for export rather than for consumption by the local population. Corrupt administrations, political instability and civil wars exacerbate these difficulties in many developing countries.

Questions

ORDINARY LEVEL

1. (i) Rapid population growth has created problems in cities of the developing world. Describe **three** of these problems, referring to examples you have studied. *(30 marks)*

   (ii) The average number of children in a family in the developing world is 5.6, while in the developing world it is 2.3. Explain any **three** reasons for this difference in family size.

   (30 marks)

Population

    (iii) Rural areas in Ireland have problems that result from out-migration. Describe any **two** of these problems. *(20 marks)*

Leaving Cert. 1996

2. At the recent Cairo conference, the growth of world population was discussed.

    (i) Using the graph below, describe the trend in world population since 1600. *(12 marks)*

    (ii) Describe **two** reasons for this trend. *(24 marks)*

    (iii) Describe and explain any two effects of rapid population growth. *(24 marks)*

    (iv) In the graph below, there are two forecasts for the year 2100. Which one, in your opinion, is likely to occur? Give **two** reasons to justify your answer. *(20 marks)*

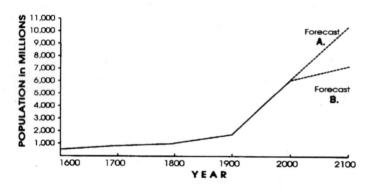

Leaving Cert. 1995

3. Study these population pyramids:

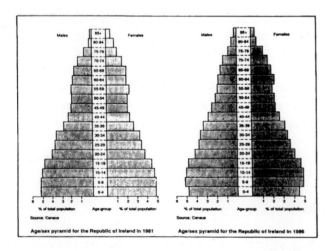

    (i) Estimate the proportion of the Irish population aged between 0 and 4 years in both the years shown.
Describe the apparent trend indicated by the figures and suggest an explanation for it. *(30 marks)*

    (i) If the trend indicated part (i) continued, describe one likely positive effect and one likely negative effect on Irish society in the future. *(30 marks)*

(ii) Suggest one major difference which you would expect to find between the Irish population pyramid for 1986 and a pyramid for a developing country.
Briefly describe how that difference affects developing countries.
(20 marks)
Leaving Cert. 1993

4.

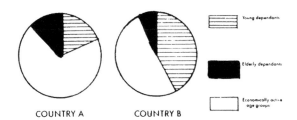

Examine these pie-charts, which illustrate the population structures of a developing and of a developed country.

(i) State which chart represents the developed country. Explain your answer. *(10 marks)*

(ii) Explain what you understand by the terms dependant and economically active. *(20 marks)*

(iii) Identify and explain some of the problems likely to face (a) developing and (b) developed countries in the future, due to the situation shown in these pie-charts. *(50 marks)*

Leaving Cert. 1992

HIGHER LEVEL

1. 'Europe has seen the development of several forms of international migration .... These include East-West migration, refugee movements and international retirement migration.'

Geographic Viewpoint 1993.

(i) Examine **two** of these forms of international migration.
(50 marks)

(ii) 'Rural-to-urban migration continues as a major trend across Europe.'
Examine some of the consequences of this movement for **both** urban and rural regions. *(50 marks)*

Leaving Cert. 1996

(iii) Different types of migration can be identified and these have implications – positive and negative – for both the donor and receiver areas. Assess the accuracy of this statement with reference to examples of migration you have studied. *(100 marks)*

Leaving Cert. 1993

(iv) With references to both causes and consequences, explain what you understand by the term 'over-population', using examples from regions which you have studied. *(100 marks)*

Leaving Cert. 1992

CHAPTER 12 *Urbanisation*

< Fig 12.1

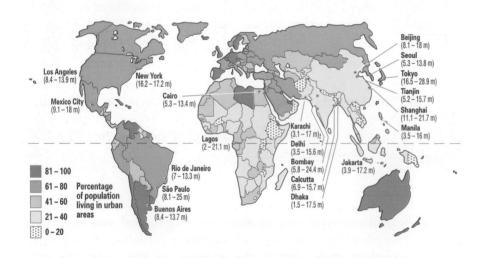

< Fig 12.2

Urbanisation. The Figures in Brackets show the Population in 1970 followed by the Projected Population for 2010

> Urbanisation is defined as the process 'whereby an increasing proportion of a country's population live in urban areas'.

While there is no universal agreement as to what constitutes an 'urban area' the growth in the number and size of large cities has been unprecedented in the last century in the developed world and over the past 30 years in Third

World countries. After the development of agriculture, urbanisation is arguably the greatest social change to have affected the human race.

Growth of Towns and Cities

It was the development in farming during the 'Neolithic Revolution' which gradually led to food surpluses and changed a nomadic society into a village community. This evolution in farming appears to have taken place independently but at about the same time in three river basins: The Tigris-Euphrates in Mesopotamia, the Nile and the Indus. By the year 3000 BC large towns had developed in these areas although recent excavations in the Jordan valley indicate that the city of Jericho may date from the seventh millennium BC. As a result, people were now needed to organise the collection and distribution of food supplies. Traders exchanged surplus goods with other centres. Others specialised in making farm equipment, pottery and other household articles – the world's first craftsmen made their appearance. While some towns developed for a specific reason, e.g. defence, the growth of business, trade and exchange was probably the main factor in urban development.

Fig 12.3
The Coliseum, Rome

The growth and expansion of both the Greek and Roman civilisations witnessed a further increase in both the number and size of towns and cities with Rome at the centre of Western 'civilisation'. The decline of the Roman Empire was paralleled with a decline in trade and so in the growth of towns. At the end of the 'Dark Ages' (c.1000 AD) the feudal system brought a certain stability and with it a renewed growth in commerce. Existing towns grew larger and new towns developed such as Venice, Genoa, Paris etc. Many medieval towns had castles and strong walls, testimony to the fragility of the political stability of the feudal period. In spite of this growth, human society remained predominantly rural so that by the year 1800 less than three per cent of the world's population could be classified as urban. Since then

Urbanisation

however there has been a dramatic increase. By 1900 the figure had risen to 14 per cent. At present it stands at over 46 per cent and by the year 2025 the United Nations estimate that five billion or over 61 per cent of the population at that time will live in urban areas. This urban explosion is perhaps best illustrated by the fact that the world's urban population grew by 320 million people – equivalent to 19 cities the size of New York in the last five years.

On a world scale, rapid urbanisation has occurred twice in time and space:

1. In the nineteenth century in developed countries when industrialisation led to a demand for labour in mining and manufacturing centres.
2. Since 1950 in Third World countries caused both by inward migration from rural areas and the high natural increase in population (high birth rate, falling death rate – see Chapter 11).

Percentage of world population living in urban areas				
Area	1950	1970	1985	2000 (est)
World	29.2	37.1	41.0	46.6
More developed regions	53.8	66.6	71.5	74.4
Less developed regions	17.0	25.4	31.2	39.3
Africa	15.7	22.5	29.7	39.1
Latin America	41.0	57.4	68.9	76.8
North America	63.9	73.8	74.1	74.9
Eastern Asia	16.8	26.9	28.6	32.8
South Asia	16.1	21.3	27.7	36.5
Europe	56.3	66.7	71.6	75.1
Oceania	61.3	70.8	71.0	71.3
Note: the UN prediction for the world in 2025 is 60%				

< Fig 12.4

WHY CITIES GROW

< Fig. 12.5
New York City

It is difficult to identify the exact reasons why people concentrate in clusters to the point where the environment cannot support them harmoniously. The United States for example has an average density of 26 people per square kilometre, yet New York's density is over 55,000 per square kilometre. The average density for Montreal is 52,000 per square kilometre yet Canada's average density is only a little over two people per square kilometre.

Apart from the natural increase which occurs in existing urban populations there seems to be two separate sets of forces which help to explain urban growth. On the one hand there are push factors, i.e. why people leave the land. On the other hand and of equal importance are the pull factors, i.e. the attractions, real or imagined which cities seem to offer.

1. Push Factors

Throughout the nineteenth century the number and size of farm families increased as the population moved from Stage 2 to Stage 3 of the demographic cycle. In the beginning, the extra hands were welcome but soon the additional labour exceeded the farm's ability to support them. The law of diminishing returns had set in. In short, farms became over-populated. This problem was now aggravated by a second – technology. The Industrial Revolution brought mechanical progress and with it, the tractor, the combine harvester etc. So while farm productivity increased, the actual demand for labour decreased. However unlike industrial output, the demand for agricultural products has a saturation point so that once a certain standard of living has been attained the proportion of additional income in the family budget spent on basic food declines. In short, with some exceptions, farming, both in terms of income and time was and is not perceived as an attractive livelihood. In fact rural poverty is now a major problem in developed economies especially in peripheral regions (See Chapter 21). Cities however have not grown simply because life on the land was unattractive. If that were the only reason then cities, as one writer put it, 'would resemble gigantic refugee camps'. The fact is that there are positive factors which 'pull' people into cities.

2. Pull Factors

The major benefits gained for a high density of population are *agglomeration economies*. These are the savings that can be made by serving an increasingly large market spread over a small geographic area. Economies of scale make production costs for each unit low, while the short distance separating buyer and seller cuts back the cost of transporting goods. A large output generally results in lower unit costs because machinery and plants are used more fully and labour can become more specialised.

While urbanisation is still closely related to the growth of manufacturing industry, the huge expansion in service employment has made the tertiary sector a major urbanising force in the twentieth century. Tertiary employment has grown with economic development, rising living standards and increased government administration.

Apart from employment and higher income opportunities cities were also perceived to have social advantages. The 'buzz' of living in the bright lights, varied leisure pursuits, better shopping facilities and a higher quality of education and medical services all helped to attract people.

The role of rural migration in the growth of cities in developed economies is diminishing and where growth does exist, it generally stems from the natural increase of the urban population.

But the advantages for both the secondary (manufacturing industry) and tertiary (services) sectors do not accrue indefinitely. The simple fact is that urbanisation will continue to increase as long as the benefits to be gained from crowding exceed the costs.

As cities grow, so do the costs of travel and the disposal of waste. The quality of life decreases with air and noise pollution, traffic congestion and rising crime rates. Recent research now shows that many of the world's leading western cities have stopped growing and in fact went into reverse. Developing countries however continue to show a strong urban growth with the population of Mexico city expected to reach 30 million by the year 2000.

WHERE CITIES GROW

Site and Situation

The 'site' is the actual land on which the town originally developed and was of major importance in the early establishment and growth of towns.

The 'situation' describes where a place is in relation to its surroundings. The situation together with human factors (e.g. political) largely determined whether the original settlement remained small or grew into a larger town or city.

Important Factors in the Selection of a Site

1. Wet point

Water is an essential resource. It is needed for drinking, washing, cooking etc. Because of this need and the fact that water is heavy to carry, early settlements usually located near rivers or springs.

2. Dry point sites

While water is an essential resource it is also important that the site be well drained and out of the reach of flooding.

3. Relief

Flat low-lying land was far easier to build on than steep high ground. This was particularly important in earlier times before the invention of machinery, etc.

4. Defence

Protection against neighbouring tribes was often a major reason for the selection of a site. Jericho, believed to date from 8300 BC is known to have had walls. In Scotland the site for Edinburgh was selected because the hill had a commanding view of the surrounding countryside. In Ireland, the Vikings selected an island (now King's Island) in the Shannon River for similar defensive reasons. This original settlement later developed into the city of Limerick.

5. Nodal Point

A nodal point is a site which is accessible from several different directions and as such, becomes a natural route centre, e.g. Paris.

6. Bridging Point

Because rivers often formed major topographical divides, settlement was inevitably attracted to a place where a river was easily bridged or forded, e.g. Oxford. In Ireland the Normans selected the Ford of Luan or Áth Luain now Athlone for similar reasons.

Generally most sites possessed a number of the above advantages. Where a site possessed a single advantage however, e.g. natural resource, it often tended to decline once the resource became exhausted, e.g. 'ghost towns' of the western United States were deserted when minerals ran out or became uneconomical to work.

FUNCTION

As settlements grew they tended to specialise in certain tasks or functions, i.e. the main activities. While all towns are essentially multi-functional (serving more than one function), certain activities may nevertheless dominate all others and consequently allow for a classification of towns based on function.

Capital City:	Canberra (Australia)
Religious Cities:	Lourdes (France)
Manufacturing City:	Turin (Italy)
Educational City:	Uppsala (Sweden)

Fig 12.6
Classification of Cities

CENTRAL PLACE THEORY

A central place is a settlement which provides goods and services. The area which is serviced by that settlement is known as its *sphere of influence* or urban field. The theory of a Central Plain was first put forward in 1933 by Walter Christaller. Later researchers have revised Christaller's terminology to include two simple concepts.

The Range

This is the maximum distance that people are prepared to travel to use a particular good or service. It is determined by the value of the good, the length of the journey and how often the service is needed.

The Threshold

This is the minimum number of people needed to support a particular service. As a rule, the more specialised the service the greater the number of people needed.

Urbanisation

The ranges and thresholds of different services vary considerably depending on the nature of the services and the frequency of their use.

Low Order Services, e.g. post office, grocery shop. These generally have small thresholds and low ranges.

High Order Services, e.g. furniture shop. These tend to have large thresholds and extensive ranges.

Schools provide a good example of a central place organisation.

The local primary school provides a lower order centre which serves a small part of a city or a small rural community. Above the primary school comes the higher order service provided by a secondary school. As the number of these centres become smaller, the complementary regions (the regions which serves them) become larger. At the top of the ladder stands the university.

To explain his theory, Christaller made a number of assumptions:

- All activity occurs on a featureless plain in which transport costs are equal in all directions.
- Population is equally distributed across the plain.
- Resources are evenly distributed across the plain.
- All customers have the same purchasing power (income) and make similar demands for goods.
- Customers always travel to the nearest service centre.

The ideal shape for the sphere of influence of a central place is a circle. However, when circles are drawn they leave certain areas which are left unserved by any central place while others are served by more than one central place. If the boundary is drawn midway through the overlapping trade areas this gives a series of hexagons.

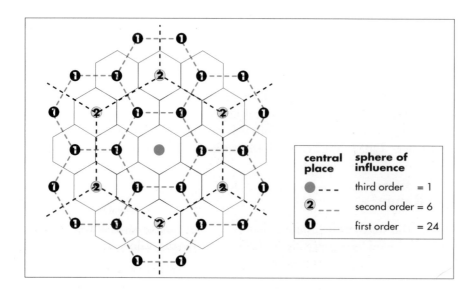

< Fig. 12.7

Christaller's Central Place Theory

Urban Hierarchy

Christaller graded settlements into separate categories according to their size and the number of functions which they served. He identified seven sizes of central places and arranged them into a hierarchy of central places. Higher level places offer more central services than lower members. Places at a given level in the hierarchy perform all the services of a lower level together with a group of services which distinguish them from and sets them above lower level places. This urban hierarchy is reflected in the names we generally apply to settlements of different sizes, e.g. hamlet, village, small and large towns, small and large cities, metropolis.

A number of factors combine to explain why Christaller's model cannot be found in the real world.

- Large areas of flat land rarely exist where transport costs are equal in all directions.
- Neither people nor resources are evenly distributed.
- People do not always have the same purchasing power.
- People do not always go to the nearest central place.

Models Of Urban Structures

As cities have grown in area and population, attempts have been made to identify and explain urban spatial patterns and variations in their structure.

1. Concentric Zone Theory – Burgess, 1924

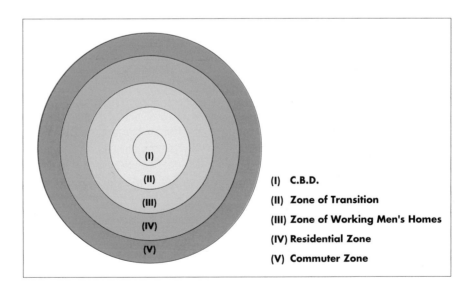

Fig 12.8
Burgess' Concentric Zone Theory

The morphology (shape) or structure of towns is an important aspect of urban geography. Most towns have distinctive areas, e.g. shopping centres, industrial areas, residential areas etc.

Urbanisation

EW Burgess began his land-use research in Chicago in the 1920s. In his Concentric Zone theory, it seems that Burgess made certain assumptions, including:

- The city was built on a flat surface.
- Transport facilities were of little significance as they were cheap and easily available in all directions.
- Land values were highest in the centre and declined in value away from the centre.
- There was no concentration of heavy industry.

As a result he identified five concentric zones:

Zone 1

The Central Business District (CBD) which contains the major shops and offices and is the centre of commerce and the focus of all transport routes.

Zone 2

This is a transition zone where the oldest housing is either deteriorating into slum dwelling or being 'invaded' by light industry from the CBD. This area generally tends to be inhabited by poor social groups.

Zone 3

This is a zone dominated by single family houses of working class people who have 'escaped' from Zone 2 and can afford to purchase their own houses. They are forced to live near their work to reduce travelling costs.

Zone 4

This zone is occupied by the middle and upper classes. They are single family residences but newer and better than those in Zone 3.

Zone 5

This is the zone of high class housing and dormitory towns occupied by people who can afford expensive properties and the high cost of commuting.

Apart from its simplicity, Burgess' model has little to recommend it. While his study was based on Chicago he suggested that this model of concentric zones was applicable to most 'Western' cities whether they were in North America or Europe. However he seems to have failed to take account of the fact that:

- Zones usually contain more than one land use.
- Cities are rarely built on flat land.
- The oldest housing is not always in or near the city centre.

2. The Sector Theory – Hoyt, 1939

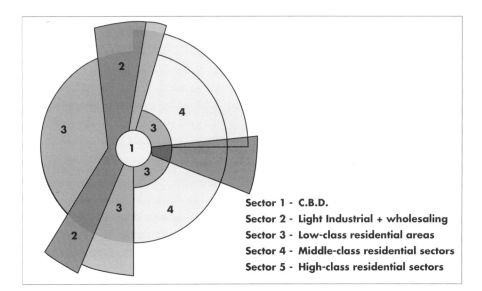

Fig. 12.9
Hoyt's Sector Theory

Sector 1 - C.B.D.
Sector 2 - Light Industrial + wholesaling
Sector 3 - Low-class residential areas
Sector 4 - Middle-class residential sectors
Sector 5 - High-class residential sectors

Hoyt's model was based on a study of residential patterns in the United States. Residential areas expanded outwards from the city centre in the form of wedges or sectors. He suggested that areas of highest rent tended to be alongside main lines of communication. Manufacturing would be attracted to railway lines or canals. Expensive housing on the other hand would tend to avoid these sectors and so be located on one side of the city.

Though many towns do exhibit a sector pattern Hoyt tended to base his model on housing, while neglecting other land uses. His study was based only on cities in the US and it was assumed that there were no planning laws or restrictions.

3. The Multiple Nuclei Theory – Harris and Ullman, 1945

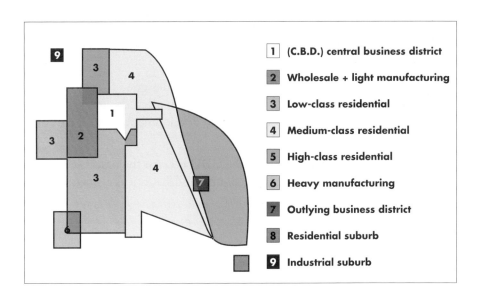

Fig. 12.10
Harris and Ullman's Multiple Nuclei Theory

1. (C.B.D.) central business district
2. Wholesale + light manufacturing
3. Low-class residential
4. Medium-class residential
5. High-class residential
6. Heavy manufacturing
7. Outlying business district
8. Residential suburb
9. Industrial suburb

Leaving Certificate Geography

The multiple nuclei theory suggested by Harris and Ullmann is a more realistic and consequently more complex model than that of Burgess or Hoyt.

They suggested that:

- Cities do not grow from one CBD but from several independent nuclei.
- Each nucleus acts as a growth point with a different function to other nuclei.
- In time, each nucleus grows outwards until it merges with the others to form one large urban centre. Harris and Ullman's theory can be applied to many large cities such as London which engulf and absorb nearby small towns as they grow, even though each may still continue to function as a minor focus within the larger urban agglomeration.

It is unlikely that any of the three theories which have been examined fits precisely any actual city. Most cities tend to have a development and morphology which is unique. Nonetheless, these urban concepts do help us to understand the structure of towns and to trace and analyse land use regions within the city. Whatever their precise morphology and development then, certain zones are recognisable in all towns.

Functional Zones Within a City

THE CENTRAL BUSINESS DISTRICT (CBD)

The CBD is regarded as the centre for retailing and service industries, e.g. banks, insurance offices. It generally contains the main public buildings and commercial streets and so forms the core of cultural, social and economic activities. Most CBDs tend to have well defined characteristics.

- It contains the major retailing outlets. Most of the leading department stores and specialist shops (jewellers) with the highest turnover and requiring large threshold populations are located here.
- Tall Buildings. The competition for space in these prime sites results in vertical growth – particularly true of North America, e.g. New York.
- Heavy Volume of Traffic. The CBD invariably enjoys a high degree of nodality. As a result traffic congestion is often synonymous with such regions.
- Constant change. Apart from traffic regulations designed to reduce congestion, taller office blocks, pedestrianisation, often declining retail outlets (due to competition from suburban developments) are all part of the dynamic nature of the CBD.

Apart from its economic contribution, the CBD is often the oldest sector of a city. This often makes it the focus of cultural and historical attractions, e.g. castles, museums, art galleries, churches etc.

MANUFACTURING INDUSTRY

Industry within urban areas has changed its location over time. While it is only a small use of land space, it is of major importance because of its provision of employment and its environmental impact. It is not easy to generalise about industry within cities but certain trends are noticeable. In the early nineteenth century it was usually sited within the city centre, e.g. textiles. With the Industrial Revolution however, factory units increased both in number and size. This resulted in a shift away from the centre towards the outer limits of the inner city where land was cheaper and where there was easy access to large quantities of cheap, unskilled labour. The twentieth century has witnessed a further shift to the suburbs. While this in part reflects the more 'footloose nature' of many of the industries, e.g. electronics, assembly and food processing, it is also in response to the lower cost of land, the development of road transport and an increase in car ownership.

RESIDENTIAL AREAS

Residential areas account for up to 50 per cent of the total land use in most urban areas. As cities grow outwards so do residential zones, so resulting in a series of concentric zones which usually reflect the social status and incomes of those who inhabit then. Though family size is important, income is the most important influence on residential choice.

Fig 12.11 > **Residential Zones**

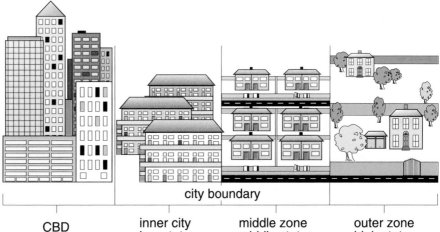

CBD	inner city low status	middle zone middle status	outer zone high status
Few residents, some homeless, some in penthouses	Lowest salaries, many unemployed, few skills, elderly, single parent and single member families, homeless, first time buyers	Average salaries, semi-skilled, second-time buyers	Highest salaries, many skills, professional/managerial groups, families with two or three children, perhaps both parents working, religious minorities

As with other urban functions, population densities are dynamic. Nevertheless certain generalisations can be made.

1. The Inner City

Because this is usually the oldest part of a town, buildings tend to be of poor quality. This area then is often inhabited by people with the lowest salaries, the unemployed, elderly and young migrants. It is often then an area of slums and ghettos. The word 'ghetto' originally meant the Jewish quarter of a city but has now come to mean any area occupied by a minority group or groups. Occupants of ghettos often tend to have similar racial, religious and economic characteristics. Today these inner city areas are often the focus of urban renewal and urban redevelopment.

> *Urban Renewal* is a programme aimed at eliminating slums and replacing them with improved residential ares.
>
> *Urban Redevelopment* is when old residential areas are cleared and the land is used for new office buildings, shops and car parks.

2. Middle Zone

Situated between the inner and outer zones, the middle zone tends to be an area of mixed housing, i.e. single family and multi-family residences.

3. Outer Zone – Suburbs

This zone tends to be occupied by those with the highest salaries. Increased mobility and higher incomes allow residences to be detached and more spread out. It is often the zone of new housing estates with shopping complexes and other facilities, e.g. schools, libraries, etc.

Problems of Urban Areas

ECONOMIC PROBLEMS

Inner cities have long suffered from a lack of investment. As traditional industries declined and new, more footloose industries spread to the suburbs, unemployment increased. In Britain for example unemployment rose from 33 per cent in 1951 to 51 per cent in 1981.

SOCIAL PROBLEMS

Most inner city housing tends to be of poor quality and high density. Accommodation is often rented which encourages neither landlord or tenant to improve conditions. Even where slum clearances occur, these have been replaced with poorly built high rise flats. Many inner city areas too have concentrations of ethnic minorities (ghettos) which often cause tension and riots.

It is often assumed that urban concentrations and crime go hand in hand. Derelict factories and houses, waste land, unemployment and the lack of adequate social amenities all combine to suggest reasons why crime rates tend to be high in city centre areas. However it is interesting to note that the most densely populated city in the world, Hong Kong, has one of the lowest murder rates of the world's largest hundred cities. In contrast, Rwanda, one of the world's least urbanised countries has the world's highest number of murders per hundred thousand population.

ENVIRONMENTAL PROBLEMS

Areas close to or part of the CBD inevitably suffer from noise and air pollution caused by heavy traffic and the few remaining factories. Visual pollution too in the form of derelict factories and houses together with waste land does little to foster community spirit.

Cities in the Third World

Fig 12.12
Shanty Towns

Percentage who live in Shanty Towns	
Rio de Janeiro	30%
São Paulo	25%
Bombay	40%
Calcutta	60%

CALCUTTA

Over 100,000 people in Calcutta live and sleep on the streets. For 60 per cent of its inhabitants, life is slightly more comfortable – a *bustee* with one room, no bigger than the average bathroom. These dwellings are made of wattle with tiled roofs and mud floors – materials which are unlikely to combat heavy monsoon rains. In Brazil these dwellings are known as *favellas* (wild flowers) so called as they have replaced these flowers on the hillsides.

Given such conditions, why then are cities in the Third World growing at twice the rate at which European cities grew in the nineteenth century?

Urbanisation

Nineteenth century urbanisation was 'migration-led', i.e. the vast majority of new urban dwellers were from rural areas. This is less true of today's developing countries. Although the popular image of the rural poor streaming into shanty towns on the edge of Third World cities are correct, the relative share of migration in the growth of urban populations is smaller in proportion to natural births.

Contrary to what we might imagine in much of the developing world today, the modern city is more healthy than were the cities of Europe and North America in the nineteenth century. The birth rate remains higher and the death rates lower. In short, most of the new urban dwellers are 'home-grown'.

PROBLEMS IN CITIES OF THE DEVELOPING WORLD

Housing

The main problem facing cities in the developing world is how to house the population. Inward migration and natural increase has far outpaced the ability of governments to provide adequate housing. In Addis Ababa, the capital of Ethiopia it is estimated that as many as 90 per cent of the city's population live in slums. In São Paulo (Brazil) the population grew by a staggering half million each year between 1970 and 1985. Urban dwellers then are faced with three choices: to sleep 'rough' on the pavement; to rent a single room (if they have the resources); or to build a shelter for themselves (favella or bustee). Many opt for the latter and build without permission, on land they do not own.

In time, some squatter settlements may develop into residential areas with basic services like a water supply, sewage systems and refuge disposal.

Services

Few cities in the developing world have adequate water supplies or sewerage mains. Rubbish is often dumped on the streets and rarely collected. Heavy rains block drains giving rise to obvious health hazards.

Drinking water is often contaminated with sewage which inevitably causes outbreaks of cholera, typhoid and dysentery. The uncollected rubbish is an ideal breeding ground for disease.

The atmosphere falls victim to uncontrolled emissions. Lahore, in Pakistan suffers smoke pollution ten times worse than New York. In Calcutta, it is estimated that three out of five people suffer from respiratory diseases caused by air pollution.

Crime

Crime is a fact of life in all the major cities of the Third World (and in many rural areas, e.g. Rwanda). As in the developed world, crime is closely associated with drugs. It is estimated that about 200 of the 500 favellas of Rio de Janeiro are dominated by drug gangs.

Children are always the most vulnerable and child labour, often more like slave labour, is more common than child education.

In Brazil alone there are millions of abandoned, homeless children most of whom live on the margins of the big cities.

Questions

ORDINARY LEVEL

1.

City	1980	2000 (Estimated)
	Population in millions	
New York	20.5	22.8
Tokyo	20.0	24.2
Mexico City	15.0	31.0
São Paulo	13.4	25.8
Shanghai	13.4	22.7
Los Angeles	11.7	14.2

   (i) Study this table and answer the following:
Name the city with the highest rate of growth.
Name the city with the lowest rate of growth.
Describe and explain one difference you notice between rates of growth of cities in developing countries and those in developed countries. *(30 marks)*

   (ii) 'Life ... in a large city can have its own problems – both social and economic.'
Examine this statement by referring to an example of examples you have studied. *(30 marks)*

   (iii) Explain why population density is relatively low in the centre of cities of the developed world. *(20 marks)*

Leaving Cert. 1994

2. (i) 'Residential areas in large towns and cities in developed countries vary greatly in age, design and accessibility to work and to amenities'. Examine this statement in detail. *(60 marks)*

   (ii) Describe and briefly explain **one** way in which the pattern of residential areas in cities in developing countries may be different from that described in part (i) above. *(20 marks)*

Leaving Cert. 1993

3. Between 1966 and 1986 the population of the Dublin area (city and county) increased from 795,000 to 1.02m. This rapid growth has produced a number of social problems.
   (i) Describe and explain some of the reasons for the growth of Dublin over the period. *(40 marks)*
   (ii) Examine some of the social problems which were caused by the rapid growth. *(40 marks)*

Leaving Cert. 1992

4. In the past 25 years, the population of Dublin has grown rapidly and continuously; it now stands at more than one million or 30 per cent of the total population of the country. This has led to social and economic problems, both for Dublin and for the rest of Ireland.

Leaving Certificate Geography

Urbanisation

(i) Explain **two** of the major reasons for the continual rapid growth of Dublin over the period. *(40 marks)*

(ii) Describe some of the problems caused for Dublin and for the rest of Ireland, referring briefly to ways in which planning authorities can try to lessen their effects. *(40 marks)*

Leaving Cert. 1991

HIGHER LEVEL

1. Urbanisation in the Developing World.
   (i) Identify and explain some of the causes of the rapid rate of urbanisation being experienced in the Developing World.
   (50 marks)

   (ii) With references to examples in the Developing World, outline some of the problems which have resulted from such rapid urban growth and some of the counter-measures which have been, or could be taken. *(50 marks)*

   Leaving Cert. 1996

2. The cities of the developed world have experienced enormous change in the processes active within them in recent years.
   Chief among these have been:
   - Inner city decline and renewal.
   - Urban sprawl

   (i) Explain the causes of these two **urban** phenomena, with reference to particular cities which you have studied, *(60 marks)*

   (ii) Discuss the strategies that planners and city authorities have used in order to deal with the problems caused by any **one** of the situations mentioned above. *(40 marks)*

   Leaving Cert. 1995

3. (i) Draw a simple diagram to show the model of the structure of a city you have studied. On your diagram, name the city and show the following:
   central business district; zone of light manufacturing; zone of high income housing; major transport facilities. *(40 marks)*

   (ii) Explain some of the major reasons why this city has developed in the way shown in your diagram. *(60 marks)*

   Leaving Cert. 1994

CHAPTER 13   *Agriculture*

Fig 13.1

The importance of agriculture as an economic activity can be gauged from the fact that no other activity has altered the 'natural face' of Planet Earth to the same extent. Recent estimates indicate that more than 10 per cent of the world's land surface is under cultivation while over 22 per cent is under permanent pasture.

Employment figures are equally impressive and while the actual percentage varies considerably between the developing and developed worlds it is estimated that on a global scale it employs in the region of 48 per cent of the world's working population.

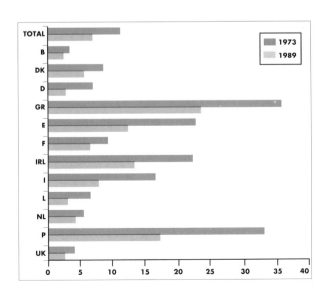

Fig 13.2

Percentage of Total Labour Force Employed in Agriculture

Agriculture

THE IMPORTANCE OF AGRICULTURE

The role of agriculture would appear to be quite obvious – to feed the world's population. A closer analysis however reveals that its role in the world's economy is more complex. Agriculture:

   (i)   Dominates the land use of the habitable world.
   (ii) Employs some 48 per cent of the world's labour force.
   (iii) Provides the world with the vast bulk of its food requirements.
   (iv) Is a major source of raw materials.
   (v)   Provides a market for a wide variety of industrial products.

1. Land Use

The amount of land devoted to farming varies considerably from one country to another and between regions of any one country. While the suitability of the physical environment (i.e. relief, climate, soils) is obviously a major controlling factor, these variations also reflect social, economic and political inputs.

From a global viewpoint however, over 10 per cent of the earth's land surface is under cultivation at present with a further 22 per cent under permanent pasture.

⟨ Fig 13.3

2. Employment

The proportion of the world's working population engaged in farming varies considerably from one country to another. In developed (more economically advanced) economies, this percentage often falls to single figures and at present the figure for the EU stands at 6.5 per cent varying from a high in Greece at 23 per cent to as low as 2.1 per cent in the UK. In developing (Third World) economies, on the other hand, the figure is often in excess of 60 per cent and in the case of Rwanda (Africa) it is over 90 per cent. Though it is impossible to calculate with any degree of accuracy, the 'optimum' percentage the figures do serve as a useful guide to a region's level of economic development. The high percentage employed in agriculture in

Ireland, Spain, Portugal (all around 12 per cent) and Greece (23 per cent) reflects their peripheral location and relatively under-developed economies.

3. Food Supply

In spite of the fact that only 29 per cent of the earth's surface is dry land and that the greater percentage of this land is unable to support agriculture of any kind, farming provides the vast bulk of the world's food requirements.

Cereals:	Wheat, barley, rice etc.
Fruit:	Oranges, lemons, apples etc.
Vegetables:	Carrots, lettuce etc.
Beverages:	Tea, Coffee
Meat:	Beef, pork, lamb
Milk:	Used fresh and for dairy products e.g. butter, cheese etc.

Fig 13.4

4. Major Source of Industrial Raw Materials

Apart from supplying the world with the vast bulk of its food requirements, agriculture also provides a wide variety of industrial raw materials such as cotton, rubber, flax, wool, hides, etc.

5. Market For Industrial Products

Apart from being a major producer of food and industrial raw materials, the agricultural sector is also a major consumer of industrial products. The use of fertilisers, insecticides, pesticides and farm machinery is now synonymous with agriculture where advanced farming techniques are practised and as a result it provides an enormous market for industries that specialise in the production of these products. In fact these capital inputs form one of the outstanding differences between farming in the developed and developing world.

THE ORIGINS OF AGRICULTURE

Towards the end of the last ice age, the world's population consisted of small bands of hunters and gatherers in subtropical lands. The search for food inevitably led to a migratory life style. Two major technological changes known as the 'Neolithic Revolution' turned a nomadic hunter-gatherer into a sedentary farmer. The first was the domestication of animals (sheep and cattle), the second was the introduction and cultivation of new strains of cereals (wheat, maize, rice). This evolution in agriculture appears to have taken place about the same time in the Tigris-Euphrates (Mesopotamia), Nile, and Indus river valleys. It is still unclear whether agriculture developed in each of these centres independently but it is generally agreed that its development in the New World (Mexico) around 5000 BC was totally independent to that in the old world.

Arguments of time and place aside, it is certain that the emergence of agriculture was a crucial step in the cultural development of the human race. With a reliable food supply, people became sedentary and rural settlement began.

Agriculture

With food surpluses, trade developed and as a result towns appeared (see Chapter 12). The eighteenth and nineteenth centuries witnessed the second 'agricultural revolution' although the rate at which improvements in crop rotation, animal breeding and the use of fertilisers were adopted was gradual, and it would probably be more accurate to think of an agricultural 'evolution' rather than 'revolution'.

Despite these developments in agriculture, modern farming techniques are far from universal as the primitive methods of 'slash and burn' practised by the Boro tribe in the Amazon basin today testifies.

The development in modern farming techniques then inevitably gives rise to two questions.

1. If modern farming techniques help to increase both the quantity and quality of agricultural production, why are these methods not adopted universally?
2. Since these modern techniques have resulted in an enormous increase in food production is there any actual limit to the amount of food any given area can produce and if so what are the controlling factors?
    The first question belongs to Chapter 20, but the second is relevant at this point.

Factors Controlling Agricultural Production

The presence of agriculture in any region, its level of development, the type and intensity of farm produce is due to an interplay between physical, cultural and economic factors, the most important of which are:

1. The physical environment.
2. Land tenure and farm size, i.e. cultural.
3. Capital.
4. Markets.
5. Political status.

1. The Physical Environment

In spite of the influences which technology and market forces exert on modern farming, the physical environment continues to play an important role in influencing the type, quality and level of agricultural production. From an agricultural viewpoint the relevant components of the physical environment are:

(a) Relief and drainage.

(b) Climate.

(c) Soils.

Factors Controlling Agricultural Production

(a) Relief and Drainage

The nature of the topography in any area and the extent to which it is well drained or otherwise plays a major role in dictating both the presence and type of agriculture practised there. High altitude generally results in a decrease in temperature but an increase in precipitation both of which affect crop growth. Slopes on the other hand have a major influence on soil depth, moisture content and its pH (acidity) and therefore the type of crop which can be grown. Slopes also encourage soil erosion and place obvious limitations on the use of agricultural machinery which, even with modern technology is limited to slopes of less than 11°. From an agricultural viewpoint, the relief of Western Europe is mixed. Most of its northern landmass forms part of the North European Plain which begins in France and sweeps north-eastwards into Germany.

Fig 13.5 > Terraces

In contrast, however, Southern Europe is less fortunate and the devastation caused by the Alpine Storm (folding) has considerably reduced the availability of suitable land in countries such as Portugal, Spain, Greece and Southern Italy. Where such land is available, maximum use is often made of aspect so that south facing slopes, *adrets* or *sonnenseite* with their increased sunshine are considerably more attractive than *ubacs* or *schattenseite* – (north facing slopes). We should take care however before condemning any region to a 'negative' category, even from a relief point of view. Certain types of farming are less demanding than others. Areas which might be unsuitable for large-scale arable farming may be quite adequate for certain types of pastoral farming. Necessity often demands that areas are utilised despite their drawbacks. The terraced farming (step farming) of South-East Asia provides an excellent example.

Necessity also dictates that the mountain slopes in countries such as Norway and Switzerland be used as summer pastures *(saeters)* for animal grazing. This has given rise to a distinctive, if declining, farm practice known as *transhumance* (see p. 189).

(b) Climate

< Fig 13.6

Climate is the most important physical input in farming and arable farming is more sensitive to climatic variations than pastoral farming. While each crop has its optimum conditions (i.e. the perfect conditions for its growth) it also has its *absolute limits* (i.e. extreme conditions under which the crop will not grow). As distance increases from the optimum, conditions tend to become less ideal. As a result, profit margins are reduced and the law of diminishing returns operates.

Because optimum conditions are restricted to a limited area, it is more realistic to consider the *economic limits* of production, i.e. where crops can be grown or animals reared on a profitable basis.

Temperatures

Temperature is critical for plant growth and while some crops are more tolerant to temperature variations than others – and scientific research continues to extend their absolute limits, each plant or crop nevertheless requires a minimum growing temperature and a minimum growing season. Certain crops have clearly-defined, absolute limits and as a result the vine in France is confined to areas south of the frost line. Generally speaking, in temperate latitudes the critical temperature is 5.6°C. Below this figure, most cereals like wheat and barley cannot grow, the exception being rye which can be grown in more northerly latitudes. The growing season is defined as the number of frost-free days which are required for plant growth. The actual amount of growth however is not solely related to the length of the growing season.

Below 0°C	Freezing
Below 6°C	(Soil temperature) Growth ceases
Above 6°C	Growth takes place and increases with temperature
Above 15°C	Vigorous growth

< Fig 13.7

Higher temperatures are also important: vigorous growth occurs above 15°C. While Ireland has milder climatic conditions than either the Netherlands or Denmark and, as a result, has a longer growing season, vigorous growth is rarely experienced.

Duration of Growing Season (in weeks)			
Temperature	Ireland	Netherlands	Denmark
6°C	40	35	30
10°C	24	25	20
over 15°C	5	13	10

Fig 13.8

Examine the figures in the chart above. The actual growing season for grass in Ireland is 40 weeks but only 30 in Denmark. On the other hand the number of weeks during which grass growth is vigorous is twice as long in Denmark as it is in Ireland. In the Plain of Lombardy in Italy, vigorous growth results in the production of up to nine crops of fodder in one year. Latitude is not the only influence on temperature however. Aspect and the increasing length of daylight in the summer months can have a considerable effect on the level of insolation. Because many crops have a growing period of less than six months, the absolute limits of crop production can be quite extensive.

Precipitation

Water is vital to all forms of life so its absence *(aridity)* imposes obvious limitations on all forms of agriculture. The effective rainfall (i.e. total rainfall minus evapotranspiration) of any region is of the utmost importance to farming. Few crops can grow successfully with less than 250 mm per year in temperate latitudes while 500 mm is required in the tropics. The distribution of that rainfall and whether it is seasonal, e.g. the Mediterranean climate of Southern Europe or perennial is equally important, (see Chapter 9). Most crops need water at the vegetative stage (i.e. growing season), but excessive rainfall at the reproductive (ripening) and harvesting stages can prove disastrous.

It is perhaps unfortunate that in Western Europe the areas which have the least attractive relief are also those which experience climatic extremes. The south of Europe, already 'marginal' by virtue of its upland topography is climatically a seasonal desert. Tropical high pressure moves north over the region in summer and, with descending air currents, ensures hot dry conditions. Summer drought is further aggravated by high evaporation. The climate is less severe further north and west however. In the west the maritime influences coupled with the warming influences of the North Atlantic Drift are brought on-shore by the prevailing south-westerlies and as well as moderating temperature, they also bring perennial rainfall.

Fig 13.9

These mild, damp conditions favour pastoral farming and help to explain its importance in areas such as Ireland, Belgium, Denmark and Western France. Generally speaking the determining factor for ending the grazing season is the state of the ground rather than the cessation of growth. Diseases such as liver fluke and roundworm thrive in mild, wet conditions. Bad weather can also also induce stress which can have a serious effect on milk yields.

(c) Soils

Soil is the third major physical input in agriculture. Farming depends upon depth, texture, water retention capacity, pH and mineral content of soil (see Chapter 10).

Clay soils then which tend to be heavy, acidic and poorly drained are more suited to permanent grass. Sandy soils on the other hand tend to be lighter, less acidic and well drained and so more suited to vegetables and fruit. The light texture of lime soils make them particularly suitable to cereals.

While soil types often reflect the topography, rock type and climatic conditions of a region, the Quaternary Ice Age has been particularly influential in the distribution of soils in Europe.

Glacial erosion stripped many of the upland regions of their soil cover. This helps to explain the barren rocky nature of areas like the Scandinavian Highlands, and it is estimated that up to 72 per cent of Norway is actually devoid of soil cover. Other areas however were more fortunate and while regions like the Golden Vale, Eastern Denmark and Southern Sweden (Scania) owe much of their fertility to boulder clay deposits, the limon, humus-rich deposits in the Paris Basin are loess in origin, having been blown from the edge of the Quaternary Ice Sheet in the German uplands. Not all lowlands were as fortunate however and the Geest or poor heathlands in areas such as the North German Plain are the legacy of fluvio-glacial outwash.

Components of the physical environment (relief, climate, soils) play a major role in determining both the type and level of agricultural production but they are not the sole arbiter. Human ability to adapt, if not overcome the physical environment is also crucial as the following studies show.

Adaption to Relief

From an agricultural viewpoint, people's greatest impact on relief is found in areas where land is at a premium. The high density of population in South-East Asia has necessitated the utilisation of 'marginal' if not altogether negative land and step farming, or terracing forms an integral part of the landscape. It was the high density of population, together with the danger of flooding in the Netherlands that gave birth to some of the most remarkable feats of modern engineering, i.e. the Polderlands of the Zuider Zee.

Adaption to conditions is no less significant and the practice of transhumance in countries such as Norway, Switzerland, Spain, France and Germany is in direct response to the same scarcity of agricultural land.

Transhumance

> The word transhumance comes from the French 'transhumer' meaning to migrate seasonally. It may be defined then as 'the vertical movement of livestock, on a seasonal basis to mountain pastures, in response to a shortage of both arable and pastoral land'.

A study of the agricultural land use of Norway and Switzerland provides excellent examples of this farming practice.

Norway

The physical environment (relief, climate, soils) has placed serious restrictions on farming in Norway. At present, arable land occupies a little over three per cent or one million hectares of the total land area.

From an economic viewpoint, pastoral farming is far more important than arable farming and therefore transhumance has an important place in the agricultural pattern. In the late spring and early summer, cattle are brought up to mountain slopes (often by truck) and left to graze on *saeters*.

Saeters are summer pastures (the result of the retreating snowline) usually found above the tree line (or among the forests where trees are widely spaced).

Accompanied by herdsmen (either hired or members of the farmer's own family) the cattle are left to graze while fodder crops are grown on the farm below for winter fodder. Milk from the dairy herd is sent down to the farm below through pipes *(pipelaits)* or by truck.

Switzerland

Pastoral farming, due to its greater adaptability to slopes, poor soils and harsher climatic conditions also dominates the agricultural land use of the Swiss mountain slopes. Stall-fed in winter, the cattle are moved in early spring or summer to the *moyenne* (middle level slopes) to graze on mountain pasture, vacated by the retreating snow-line, a process often hastened by the warm *Fohn* winds. Later the cattle are moved to *saeters* (high levels) or alpages. These are generally found on the *adret* or *sonnenseite* slopes which face south and so receive more sun than the *ubac* or *schattenesite* (north-facing slopes). During these months the milk is transported down the slopes by polythene pipes or pipelaits (long plastic tubes usually buried underground) to creameries in the valley. During the same period crops are intensively grown and harvested in the valleys and plateaux for winter fodder.

Saeters are not as rich in grazing as the valley farms but they provide a welcome addition to farmers in areas starved of agricultural land. Nevertheless, the number of saeters is declining and the 10,000 saeters in the Sogne fiord region of Oppland (Norway) is only half the number 50 years ago. The disruption in family life, the social isolation, the decrease in small holdings have all contributed to its decline. Better job opportunities in other sectors of the economy are also contributing factors.

Irrigation

The planting of drought-resistant crops in areas deficient in rainfall, the stall feeding or 'factory farming' in Denmark and the glass-house horticulture now synonymous with the Dutch have all developed in response to market demands and the climatic conditions in those regions.

As water is vital to all types of farming it is hardly surprising that the practice of irrigation is one of the oldest and perhaps, most basic ways from an agricultural viewpoint in which people have overcome the deficiencies in the physical environment.

> Irrigation is defined as 'The artificial application of water to land in order to overcome deficiencies in rainfall and thereby increase the growth of crops'.

Climatic Conditions Under Which Irrigation has Developed

1. In arid and semi-arid regions, e.g. Egypt and Pakistan, rainfall amounts are totally deficient to support plant growth. The importance of the Nile to Egypt is obvious in statements such as 'the Nile is Egypt'. The total population of Egypt is crowded into a productive area similar in size to that of Belgium.
2. Areas of seasonal aridity, e.g. the Mediterranean where the high pressure which dominates the climate of the Mediterranean basin in summer results in seasonal aridity.
3. In humid regions, e.g. South-East Asia, the application of water to crops with specific water requirements such as rice, is vital to survival.

Types of Irrigation
1. Basin irrigation.
2. Perennial irrigation.

Basin Irrigation

Basin irrigation or annual flooding is a relatively primitive form of irrigation. Fields along the flood plain of a river are enclosed by earth banks and the flood waters of the river spill into the 'basin' (formed by these embankments) – and the fields are flooded to a depth of about one metre. While this method has the advantage of enriching the fields with sediments, it is totally dependent on the river's regime and normally only permits the growing of one crop per year.

Perennial Irrigation

Perennial (throughout the year) irrigation has the obvious advantage of being able to supply water throughout the year. Despite its higher construction and maintenance costs, it is more commonly adapted in modern schemes. The water is supplied from wells or from reservoirs and applied either to the surface or distributed underground through porous pipes.

The method of application of surface water must take account of the gradient, type of crop, nature of the water supply and the ability of the soil to hold and absorb water. Float-irrigation, which involves covering the surface with sheets of water is more suited to land which is level, while corrugation irrigation which supplies water to small shallow furrows is used where the slope is more severe.

Problems Associated with Dam Construction

1. High Cost

The enormous cost associated with irrigation dams, like those of the Grand Coulee (US) and the Aswan (Nile) requires that they be multi-functional. Besides their contribution to agriculture, they are often used to (a) supply hydro-electric power; (b) regulate river regimes, and (c) provide social amenities.

2. Loss of Water

Water stored behind dams is subject to evaporation and seepage. Considerable amounts of water are also lost in the distribution of water to the fields.

3. Loss of Soil Fertility

In the case of the Nile Project, the soil suffered severe loss of nutrients which previously resulted from basin irrigation.

4. Effects on Coastal Fishing

The retention of water behind the dam has increased the salinity in the Nile Delta and along the Mediterranean coast which has resulted in a decline in sardine fishing in that area.

5. Spread of Disease

Reservoirs like Lake Nasser (behind the Aswan Dam) provide ideal breeding grounds for diseases like bilharzia which now severely affects the population of that area.

6. Displacement of People

The construction of dams and the flooding of land to fill reservoirs often entails the displacement of large numbers of people and the inundation of villages and even towns. Historic buildings and archaeological sites may also be lost or damaged.

Displacement of People by Dams	
Location	Number displaced
Three Gorges (U/C China)	1,200,000
Mahweli (U/C Sri Lanka)	1,000,000
Almatti (U/C India)	240,000
Aswan (Egypt)	120,000
U/C = Under construction	

Fig 13.10

2. Land Tenure and Farm Size

< Fig 13.11

Land tenure can have a vital bearing on the level of agricultural production. Farmers who own their land have a greater incentive to invest capital, increase output and exercise greater care in the use of their farms.

On the other hand, large estates in areas such as Southern Italy and South America provide no such incentive to those who work them. They are often owned by absentee landlords and the land is worked by *braccianti* (farm labourers) who have neither the incentive nor the interest in producing beyond subsistence level.

Farm size can also exercise a considerable influence on the level of agricultural production. Small farms and farm fragmentation severely handicap mechanisation as it is uneconomic to use expensive machinery on small farms.

3. Capital

The availability of capital to invest in agriculture is one of the outstanding differences between farming in developed and developing economies. Farm machinery, buildings, fertilisers, insecticides, antibiotics, improved seeds and animal breeding have all been responsible for the dramatic increase in food production in the developed world. On the other hand, in many areas of the developing world, the absence of capital has resulted in farming techniques changing little over thousands of years and in consequence the level of food production has failed to keep pace with the increasing population.

4. MARKETS

The influence of markets, together with the availability of capital form the two most fundamental differences between agriculture in the developed and developing worlds. In developing economies most of the produce is consumed directly by those who produce it, so farming is said to be *subsistence*. In developed economies however the bulk of the produce is sold so that farming is said to be commercial. It is the quantity (actual numbers) and quality (buying power) of the market which dictates both the type and level of agricultural production.

Because distance from the market place serves to increase transport costs and because much farm produce is of a perishable nature, the location of any farm in relation to the market place has a vital bearing on the type and level of agricultural produce. Market gardening (horticulture) is usually located near large urban centres due to the demand for its produce and its perishable nature. The fact that it occupies land with a high value also helps to explain the high costs of its production. In this respect Irish agriculture is removed from the Central Market Axis which has developed within the EU and suffers from both its maritime and peripheral location and in consequence, the *friction of distance*.

5. POLITICAL STATUS

In developed economies, political factors are assuming an ever increasing role in deciding both the type and level of agricultural production. The value of self-sufficiency in foodstuffs and agricultural raw materials is obvious for any country. Farming communities for any country comprise a significant if declining percentage of the voting population, a fact that few governments can afford to ignore. The importance of politics in both the type and level of agricultural production is probably best illustrated by the CAP. Agriculture in most Western European countries is governed by the Common Agricultural Policy (CAP).

> The original aims of the CAP were:
> 1. To increase agricultural productivity.
> 2. To ensure a fair standard of living for people in the agricultural sector.
> 3. To stabilise markets.
> 4. To guarantee a steady and permanent supply of food at reasonable prices.

The Common Agricultural Policy is now over 30 years in existence and in terms of the objectives laid down in the Rome Treaty can, broadly speaking, be considered successful.

Spectacular increases have been achieved in production and, with the aid of increased trading, consumers have a more reliable and varied food supply. In real terms, food has become less expensive while agricultural incomes have, on the whole improved in most member states.

This success however has commanded a high price. Increased productivity has been achieved by protecting farmers against over-production and falling prices. The European Agricultural Guidance and Guarantee Fund (EAGGF) bought up surplus production from farmers and stored it in warehouses and freezers. In the case of Irish agriculture this guaranteed market introduced a complacency which did little to encourage the exploitation of new markets. A policy generally referred to as *intervention* led to the so-called grain, beef and butter *'mountains'* and wine and milk *'lakes'* which by the late 1980s were costing in excess of IR£500 million pounds a year for storage alone.

In an effort to reduce these mountains and lakes, a quota system was introduced which set up strict limits on the levels of agricultural production for these commodities. Farmers were also encouraged to set land aside and are now actually being paid not to produce foodstuffs. Whatever arguments may be advanced to justify such policies within the EU, they are more difficult to support against a wider background of world hunger and malnutrition.

Increased productivity had other effects also. Modernisation and mechanisation placed greater emphasis on larger farming units so that in the ten-year period between 1977–1987 almost 550,000 holdings ceased work and the agricultural labour force fell by almost 2.4 million. This had dramatic socio-economic effects on rural areas and particularly on the more peripheral regions.

Against this background, it is not difficult to understand the reluctance of countries like Norway to joining the EU. The physical environment in Norway is marginal if not altogether negative in promoting commercial agriculture. Farming is only possible through government intervention in providing high subsidies and grants. Entry to the EU would reduce these subsidies to the point where farming would simply not be a viable economic prospect. This would lead to rural migration on a vast scale (in a Norwegian context) and destroy the whole fabric of the rural community.

The Farming System

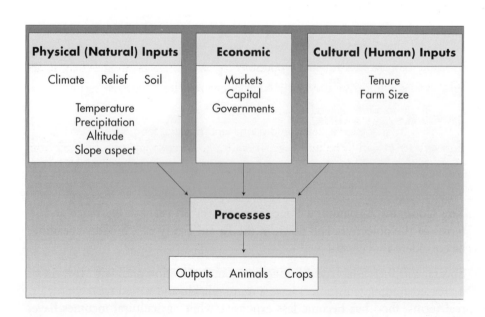

< Fig. 13.12
The Farming System

Types of Farming

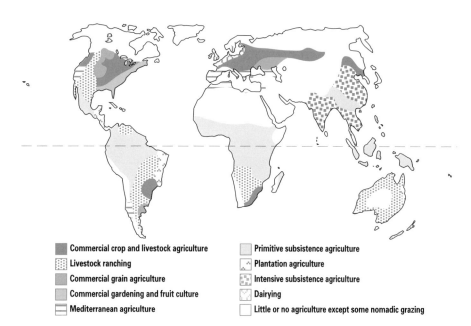

Fig 13.13
Types of Farming Worldwide

Varying physical environments in different parts of the world have inevitably led to a variety of farming systems. Add to this variations in land tenure, farm size, economic stimuli and political influences and any effort to identify farming systems might well seem futile. Nevertheless, generalisations can be made as the following terms indicate.

Subsistence Agriculture

Subsistence agriculture is a term applied to farming where the produce is largely consumed by the cultivator and the cultivator's family. In short, there is little if any surplus left to sell.

Commercial Agriculture

Commercial agriculture is the opposite to subsistence farming. Most, if not all, the produce is sold in the market place.

Intensive Agriculture

Agriculture is said to be intensive where the maximum yield is obtained from the minimum holding.

Extensive Agriculture

This is the opposite to intensive farming and so yields per hectare are low in contrast to the high yields of intensive farming.

Based on the above terms it is possible to identify four main farming systems.

(i) Intensive Subsistence   (iii) Intensive Commercial

(ii) Extensive Subsistence   (iv) Extensive Commercial

Agriculture

INTENSIVE SUBSISTENCE AGRICULTURE – AN UNFAIR RETURN?

Fig 13.14
Rice farming in South-East Asia

Few regions of the world provide a better example of intensive subsistence farming than the farmlands of South-East Asia. With population densities of over 800 per square kilometre there is a huge demand for food from a limited land area – hence the intensive use of the land. In spite of high output, high density of population means that there is rarely sufficient food, let alone a surplus for sale. At present it is estimated that almost two-thirds of the world's farmers can be classified as intensive subsistence cultivators and they support almost half of the world's population from approximately one quarter of the world's cultivated land. It is hardly surprising then that multicropping and intercultivation are commonly practised.

Multicropping

Multicropping is the production of two or more crops from the same land in the same year, e.g. wheat, beans, peas, etc., are often grown in winter when the rice has been harvested.

Intercultivation

The production of two or more crops at the same time on the same piece of land is known as intercultivation.

As farm holdings are small (rarely exceeding two hectares), huts and indeed whole villages are usually sited on the less productive land. Wet padi farming dominates the land use of much of South East Asia. The word padi derives from a Malay word meaning rice and wet padi rice (as distinct from dry upland rice) refers to a unique system of farming based on the provision, (from monsoon rains) and precise regulation of water supplies to flights of artificially embanked terraces or pond fields which are planted with rice and other crops.

Characteristics of Intensive Subsistence Agriculture

1. Land use is intensive and land holdings are small.
2. Arable rather than pastoral farming predominates.
3. It is labour intensive and work is often done by hand.
4. There is little if any capital investment, e.g. machinery.
5. The produce is consumed by the cultivator and dependants with little, if any, left for sale.

EXTENSIVE SUBSISTENCE AGRICULTURE

The essential difference between extensive subsistence and intensive subsistence agriculture is that the agricultural land is plentiful but rarely of good quality. It is largely confined to tropical areas where infertile soils are quickly exhausted. Consequently, farming is rarely sedentary as tribes move from one area to another in search of richer pastures.

Nomadic Herding

Fig 13.15
Nomadic Herding

Agriculture

Nomadic herding, as practised in the northern fringes of the Eurasian land mass (the Sami (Lapps) with their reindeer), the Asian steppes (cattle, sheep, horses) and the Savanna lands of Africa (the Masai with their cattle) represents one of the earliest forms of agriculture developed by people. Due to an unfavourable physical environment, livestock is moved continuously from one unmanaged pasture to another following the seasonal variations in food supplies. The actual movement however is well organised and families or tribes often have well defined territories. The animals reared are those most suited to the particular environment. So while goats are more adapted to the drier regions and the reindeer to the colder areas, e.g. Northern Scandinavia, cattle and sheep are more common in the warmer and wetter savanna lands of Africa. In most cases, the animals provide the entire necessities of life: meat, milk and cheese for foods, animal fibre for clothing, hides for tents, manure for fuel and bones for tools. As the quality, if not the quantity of pastures is often unreliable, food supplies can vary considerably from season to season and year to year which sets natural limits to the density of population which is invariably low, often less than one person per square kilometre.

COMMERCIAL INTENSIVE AGRICULTURE – A LOT FROM A LITTLE

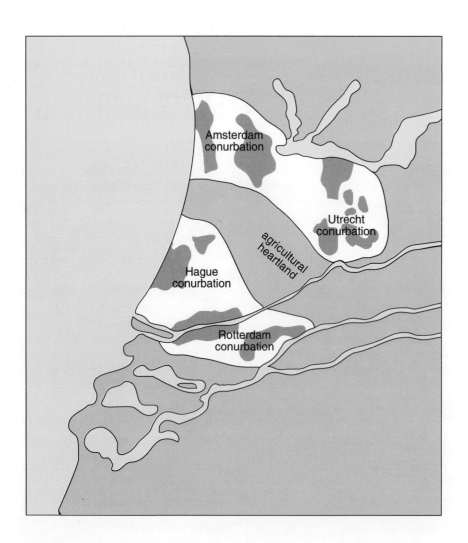

Fig 13.16

Horticulture

Fig 13.17
Flower Fields in Holland

'The more expensive the land the more intensive its use' is an economic principle amply illustrated by horticulture in the Netherlands which is one of the best examples of commercial intensive farming anywhere in the world. The high population of the Netherlands necessitates that the land be used intensively. Proximity to one of the world's greatest concentrations of urbanisation – the Central Market axis of the EU allows Dutch farmers to exploit the constant demand for fresh fruit, vegetables and flowers. There is a wide variety of products grown including cucumbers, lettuce, tomatoes, but the production of cut flowers and plants dominates. This *floriculture* as it is called is particularly important in the area known as the 'Greenheart of Holland' – South-West of Amsterdam.

The small rectangular holdings are capital intensive with glasshouses artificially heated by gas or oil. The careful use of sprinklers allows for precise regulation of water supply and the rate of growth is controlled by the use of black plastic sheet coverings. Intensive farming methods such as these invariably lead to soil exhaustion and so the soil is heavily fertilised by the use of superphosphates. Much of the work of weeding and harvesting has to be done by hand so this type of farming is both labour and capital intensive.

The perishable nature of the produce often results in much waste and in consequence necessitates quick and costly methods of transport.

Agriculture

COMMERCIAL EXTENSIVE AGRICULTURE

Fig 13.18
Wheat Ranch

The presence of commercial grain farming on a large scale in areas such as the USA, Canada, Argentina and the USSR represents one of the most recent developments in agriculture. It has developed primarily due to:

1. The market demands for wheat and other cereals in developed economies.
2. Improved transport facilities in the late nineteenth century, e.g. steamship.
3. Improved agricultural machinery, e.g. combine harvester.
4. The presence of land space where crops would be grown within economic limits.

The commercial grain farming regions of the USA, Canada, Australia, Argentina, the Ukraine, Russia and France contribute half the world's wheat output. Farms are generally large and highly mechanised so that yields per person are very high. In spite of the high return, commercial grain farming is monocultural and in consequence is exposed to its associated dangers.

Agriculture – The Future

As a hungry world keeps crowding onto worn out soils, few would argue against the need to increase food supplies both in quantity and quality. The various ways in which this can be accomplished are dealt with in Chapter 20.

Obviously, conservation of the planet's soil as a healthy and productive resource is vital, but soil is being lost or contaminated on a vast scale. According to the latest report from the World Resources Institute, quoting estimates by the world's leading soil scientists, more than 1.2 billion hectares of vegetated land – an area as large as India and China put together have been significantly degraded since World War II. The UN estimates that six million hectares of land are still being turned to desert each year and that a further 21 million are being so badly degraded that crop production cannot be supported. In Europe, most of these problems arise from the industrialisation of agriculture. The rich countries have applied mass production techniques to farming as well as to manufacturing. Now the problems are beginning to mount up. In southern Sweden, soil has been compacted by heavy machinery. In eastern England, nitrate contamination of drinking water has reached alarming levels. At present there are over 35,000 different commercial products to control insects, weeds, fungi, etc., in use in the USA alone. It is difficult to see how the problem of soil degradation can be solved in the near future. As long as profit remains the yardstick for successful farming, soil has little chance. In developed countries, many farms have become locked into a cycle of use which requires them to make up for deteriorating soil quality by ever larger applications of fertiliser. In Europe the application of nitrogenous fertilisers grew from 509 pounds per hectare in 1980 to 579 pounds in 1990. The excessive use of pesticides also contaminates and depletes the soil which then needs more fertiliser, and so it continues in a vicious circle of decline.

Source: *The World Bank*	Fertiliser consumption (kilos per hectare)		Index of food production (per head 1979-81 = 100)
	1970/71	1989/90	1988/90
Bangladesh	15.7	99.3	96
China	41.0	261.9	133
Egypt	131.2	404.3	118
Germany	384.4	370.5	112
India	13.7	68.7	119
Indonesia	13.3	116.6	123
Japan	354.7	417.9	101
Malawi	5.2	22.7	83
Netherlands	749.3	642.4	111
Nigeria	0.2	12.1	106
Pakistan	14.6	89.0	101
Philippines	28.7	67.4	84
UK	263.1	350.2	105
USA	81.6	98.5	92

Fig 13.19

Agriculture

Questions

ORDINARY LEVEL

1. (i) Any farm **or** any manufacturing industry may be viewed as a system with inputs, processes and outputs. Examine this statement with reference to an example you have studied.
    (60 marks)

    (ii) Describe briefly the possible negative effects of **either** agriculture **or** manufacturing industry on the environment.
    (20 marks)

    Leaving Cert. 1994

HIGHER LEVEL

1. Farming and manufacturing can both be viewed as economic systems, with inputs, processes and output. Discuss this statement with reference to **either** a farming **or** a manufacturing business which you have studied. *(100 marks)*

    Leaving Cert. 1996

2. Farms and manufacturing industries may be viewed as systems with inputs, processes and outputs, all of which are inter-related.

    Select **either** farming **or** manufacturing industry to examine the accuracy of this statement. *(100 marks)*

    Leaving Cert. 1993

3. 'Reform of the European Community's Common Agricultural Policy is inevitable. This complex system of supports for the agricultural sector has led to production outstripping demand for a range of commodities and has created an unacceptable burden on the Community budget.'
    (i) Briefly explain the main aims of the Common Agricultural Policy. *(30 marks)*
    (ii) Describe examples of 'production outstripping demand".
    (30 marks)
    (iii) The EC is reluctant radically to reform the C.A.P. but other countries, including USA, demand just that.
    Examine the arguments of both of these groups.

    Leaving Cert. 1992

4. Examine some of the factors which have influenced the development of agriculture in **any** country which you have studied.

    (100 marks)

    Leaving Cert. 1991

CHAPTER 14 Fishing

Fig 14.1

It has been estimated that there are about 12.5 million fishermen worldwide. If their families were to be included the number increases to about 50 million with a further 150 million dependent on incomes from related activities.

The Importance of Fishing

1. Human Diet

Fish is an important part of the human diet in many parts of the world. It is a reliable source of protein, vitamins and fats and is one of the most nutritious of all foods. Statistically, humans eat an average of 13 kilogrammes of fish per person every year.

2. Direct Employment

While fishing is an important industry for most coastal states worldwide and employs about 12.5 million fishermen, 80 per cent of the global catch is accounted for by the top 20 nations including Japan, China and Peru. At present it employs about 260,000 people in the EU.

3. Indirect Employment

The Food and Agricultural Organisation (FAO) of the United Nations have estimated that as many as 150 million people are dependent on incomes from on-shore processing, distribution, fleet servicing etc. Its role becomes all the more important when we realise that fishing is often associated with difficult physical environments so that it helps to maintain settlement in areas which already suffer from out-migration. In consequence, fishing has a stabilising influence on the social fabric of rural societies.

In the EU at present the ratio of direct to indirect employment in fishing is estimated at 1:5.

4. Exports

Fish are a renewable resource and form an important part of many countries' exports. More than 90 per cent of Norway's catch is exported to countries as far away as the USA and the Mediterranean, and accounts for over 15 per cent of all exports, so earning valuable foreign currency. In Iceland, cod catch alone accounts for 30 per cent of total exports.

CHANGES IN WORLD FISHING

In 1939 the global catch of sea fish was less than 20 million tonnes. After World War II it began to rise at about six per cent a year. By the 1980s, the total reported catch had reached 82 million tonnes. Despite a big increase in the number of fishing operatives and the development of highly sophisticated techniques for locating and catching fish, the total harvest seems to have reached a plateau. The FAO has reported that stocks in four of the world's major regions are seriously depleted while catches in nine other areas are declining. Many now believe that total tonnage can only be sustained by 'fishing down the food chain', i.e. by catching smaller, younger or less palatable fish.

Fishing Methods

GILL NETS AND DRIFT NETS

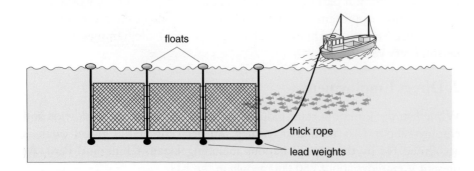

‹ **Fig 14.2**
Drift Netting

Gill nets may be suspended in the water from floating buoys forming a vertical curtain held in position by weights. Left to float free they are known as drift nets. The fish swim into the net (whose mesh is practically invisible) and get caught by their gills. The mesh size determines the size of the fish that are caught. Drift netting produces high catches but also results in *by-catch* and death to a huge number of dolphins, whales, sharks and porpoises. In 1991 the 'incidental' catch of dolphins was over 20,000.

SEINE PURSES

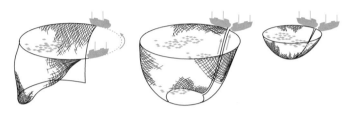

Fig 14.3
Fishing with Seine Purses

A seine purse is a circular net which works like a shopping bag with a drawstring. A shoal of pelagic or demersal fish, e.g. herring or mackerel is surrounded by the net. The lines are then pulled trapping the fish inside.

TRAWL NETS

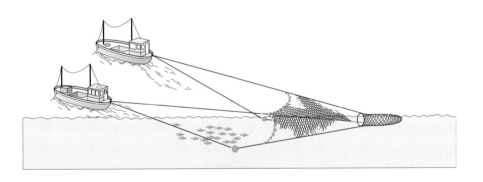

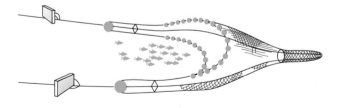

Fig 14.4
Trawling

Trawl nets are generally used to catch demersal fish. They are funnel-shaped nets up to 80 metres wide and 300 metres long.

Since World War II and particularly since the 1970s, there has been an increase in the number of larger and more modern vessels engaged in fishing.

Leaving Certificate Geography

Fishing

A fully equipped super-trawler today may cost between £60 to £100 million. These factory ships with on-board freezing and processing can operate far from home and stay at sea for months at a time. The introduction of synthetic materials for ropes and netting together with the mechanisation of hauling gear and the use of powerful engines has enabled fleets to use much bigger and larger nets. Electronic aids which help in fish detection and navigation has enabled trawlers to return to the richest fishing grounds with pinpoint accuracy.

Nowadays, fishing is increasingly subject to government regulation covering everything from limitations on catches and fishing grounds to the types of gear that can be used. In 1970, the then members of the EC agreed to an extension of territorial limits to 320 kilometres. These Exclusive Economic Zones (EEZs) gave the countries involved control of both fish and other resources (oil, gas etc.) around their respective coastlines. In 1983, after much debate the Common Fisheries Policy finally came into existence.

> **Common Fisheries Policy**
>
> 1. Conservation and Management of Stocks
>   (i) Agreed on a 320 kilometre limit for EC member states while each state would have an exclusive 19 kilometre limit.
>   (ii) TACs (Total Annual Catches) were agreed upon which gives each nation its annual quota.
>   (iii) Setting limits on minimum net sizes etc.
> 2. Organisation of the Markets
>   By agreeing on minimum standards in produce e.g. quality, size, packing etc.
> 3. Structural Changes
>   Loans and grants made available for infrastructural investment in harbours processing plants, boats etc.
> 4. International Agreements

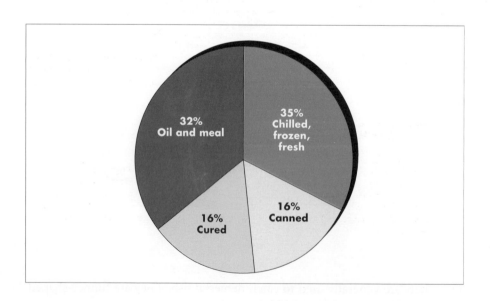

< Fig 14.5
How the Catch is Used

By agreeing with other countries such as the US, Canada and developing countries, a more global approach to conservation can be undertaken. The Convention of the Law of the Sea for example recognises that the sea is part of the heritage of all people.

Fishing – The Future

FAO reports conclude that the production of fish has reached a plateau in most of the major fishing areas and is in decline in some of them. In some cases production levels are being maintained by catches of younger fish – the very breeding stock on which fish supplies depend. Demersal fish (such as cod, plaice and haddock) are already fully exploited. The outlook for increasing catches of pelagic fish (herring and mackerel) seems pessimistic. Depletion of stocks has been in part due to the technique of large-scale drift netting. Successive UN resolutions against this culminated in an international moratorium on their use with effect from 1 January 1993. The second threat to the world's fishing stocks comes from pollution.

The apparently limitless extent of the world's oceans has long tempted people to dump their waste in the sea. Only 12 per cent of marine pollution is due to oil spills. The 44 per cent which comes from the run-off and discharge from land and the 33 per cent from airborne emissions is more serious. Approximately 83 per cent of all marine pollution is derived from land-based activities. The Baltic Sea alone gets more than 900,000 tonnes of nitrogen and 50,000 tonnes of phosphorus each year. While these run-offs initially enrich the sea they can eventually cause *eutrophication* and the death of all marine life through lack of oxygen. Residues of insecticides, pesticides and heavy metals can eventually find their way into the food chain.

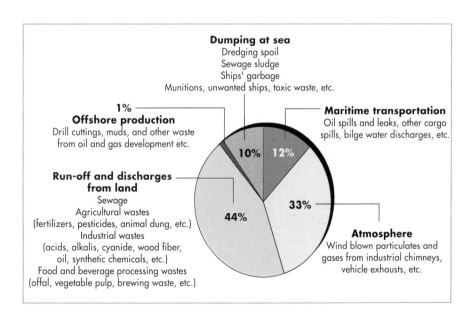

Fig 14.6 › Marine Pollution – Poisoning the Sea

Fishing

In the Newfoundland region of Canada, the fishing grounds are almost completely empty and more than 30,000 people are out of work. Before 1950 the annual catch of cod in this region was about 250,000 tonnes. By the 1980s it had soared to 800,000 tonnes. At present, the total catch is less than 200,000 tonnes. The Canadian Government's response was to impose a two-year moratorium (suspension) on fishing, which has now been extended.

CONSERVATION

The need to conserve fish stocks has long been recognised but agreeing and implementing the necessary measures has proved extremely difficult and complicated by the migratory habits of many species. The most common approach has been to set quotas or total allowable catch (TAC) for different species which appear under threat. In the EU, the mesh size of the net is controlled to help protect younger fish.

The FAO has suggested that as far as large-scale industrial fisheries are concerned, structural change should involve three basic steps:

1. *Removal of government subsidies.* Subsidies can encourage intensive or undesirable fishing methods.
2. *Introduction of a system of use rights.* The present system of open areas encourages over-fishing whereas owners of inclusive use rights would have a long-term interest in preserving fish stocks.
3. *Increase in the real prices of fish.* This would enable fishermen to make a decent living without having to resort to excessive catches.

In 1993, the FAO stated:

> *'There will be a significant global shortage of supply of fish in the future. Although the severity of the shortage will differ among countries, the overall effect will be a major rise in the real price of fish which will have critically important consequences in several regards.'*

A Solution – Aquaculture?

The increasing demand for fish together with a decline in the fish catch has encouraged the commercial development of fish farming or aquaculture. This industry has grown rapidly and has become an important source of employment and income for many coastal communities. Within the EU it produces 925,000 tonnes or 13 per cent of the total catch of the EU countries. Trout and salmon are the main species. France (36,000), Denmark (32,000) and Italy (31,000) produce the bulk of the trout. Fish farms now produce more salmon than are caught by fishermen, the bulk of which comes from the United Kingdom. Norway, however, with over 120,000 tonnes, is way ahead of other EU countries in this type of production. In the southern countries of Spain and Italy, the emphasis is on molluscs, mainly mussels and oysters in which France dominates producing 130,000 out of a total of 137,000 tonnes.

A Solution – Aquaculture?

Fig 14.7
Fish Farming Output (1,000 Tonne Live Weight)

Country	Fish	Molluscs and shellfish	Total
Belgium	2	0	2
Denmark	32	0	32
France	44	185	229
Germany	49	0	49
Greece	5	2	7
Ireland	7	16	23
Italy	42	102	150
Netherlands	1	109	110
Portugal	2	7	9
Spain	18	247	265
United Kingdom	44	5	49

While the long-term future of both world and EU fishing may well lie with aquaculture (fish farming) its large scale development is already raising problems in relation to the environment. The chemicals used in such operations can have negative effects on the environment of other fish species, e.g. shellfish. Setting aside large inlets can cause conflict with other water-related recreation activities, e.g. boating. The decrease in sea trout in the West of Ireland is alarming. In the late 1980s, up to 12,000 catches of this species were recorded each year. By 1992 the catch had fallen to less than 500. Many attribute this decline to the growth of fish-farming in the area, which they claim increases the incidence of fish lice.

CASE STUDY I – NORWAY

The relevant components of the physical environment have combined to make Norway the most successful fish-producing country in Western Europe.

1. Relief

The Norwegian coastline is over 12,000 miles long and as the inland topography is unattractive for human habitation, the sea is accessible to most of the country's settlement. The numerous fiords, protected from rough seas by the *Skerries*, provide excellent, safe harbours for the country's fishing fleet. The shallow water of the Continental Shelf enables sunlight to penetrate, encouraging the growth of plankton – the food of fish.

2. Climate

The arctic waters off the Norwegian Coast are warmed by the North Atlantic Drift which helps to keep the coast ice-free as far north as Hammerfest and allows for year-round fishing. As they mix with the cooler arctic waters, these warm waters provide a variety of plankton and in consequence a variety of fish. Before the advent of modern technology, the cool temperatures helped fishermen preserve their catch.

Fishing

3. Human Inputs

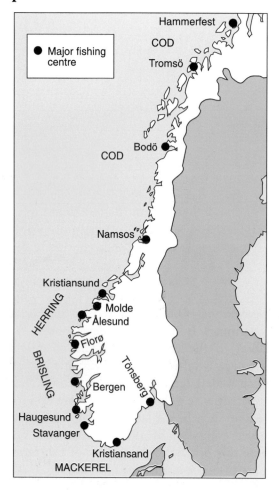

< Fig. 14.8

The unattractiveness of the physical environment from an agricultural viewpoint meant that the sea was always viewed as a source of food. This helps to explain the unlikely combination of fishing and farming as part-time occupations. That the physical environment deteriorates moving northwards from an agricultural viewpoint can be gauged by the increasing dependence on fishing in these regions. In the northern provinces such as Finnmark, for example, up to 50 per cent of the entire workforce is involved with the fishing industry. Coastal settlement provides other advantages for the fishing industry also. It creates a domestic market for fresh fish. The fish processing plants are close at hand making costly 'break of bulk' unnecessary.

Coastal settlements also provide a ready labour supply and the other requirements of a modern fishing fleet. The Norwegian government has played its part, offering grants and low interest loans for boats and equipment. Norway's rejection of EC membership in both 1973 and 1995 guarantees it an exclusive 320 kilometre fishing limit, vital to the survival of the industry.

4. Fishing Areas and Processing

While there is a wide variation in the percentages of the different fish which make up the annual catch of over two and a half million tonnes, herring and brisling (small herring) account for 70 per cent of the total. Cod make up the bulk of the remainder.

A Solution – Aquaculture?

Fig 14.9

Fish Types and Ports	
Cod	Lofoten Islands, Tromsø, Hammerfest.
Herring	Trondheim; Stavanger
Brisling	Bergen
Mackerel & Haddock	Stavanger
Whales & Seals	Hammerfest

Though Norwegian fishermen travel as far as Greenland, 80 per cent of the catch comes from their own coastal waters. In spring the area around the Lofoten Islands is important for cod-fishing, making Hammerfest and Tromsø busy ports. Further south, along the fiorded coast of the south-west, herring, brisling and mackerel are the main catch with Trondheim, Stavanger and Bergen being the main ports. Because of the small population, about 95 per cent of the entire catch is processed to produce a wide variety of products such as fishmeal, margarine and liver oils (cod). More than 90 per cent is exported to countries in the Mediterranean and to the USA.

As a resource, fish are a valuable commodity to the Norwegian economy accounting for over 15 per cent of all exports. The recent decline in fish catch, then, due to over-fishing, competition and pollution must be a cause of some concern.

CASE STUDY 2 – IRELAND

Like Norway, Ireland possesses many physical advantages for the development of a fishing industry.

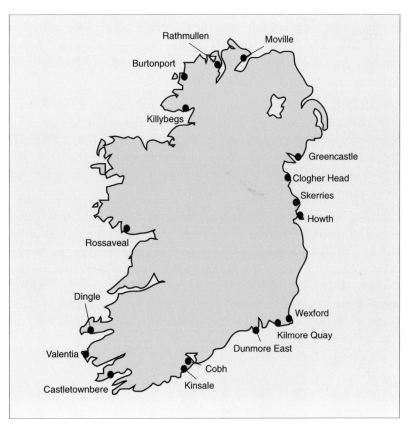

Fig 14.10
Irish Fishing Ports.

Relief

The Irish coast is over 6,500 kilometres long, of which 5,800 kilometres falls within the jurisdiction of the Irish Republic. Like Norway, it is indented with numerous inlets which offer safe harbours for fishing trawlers (see map). The relatively shallow waters of the continental shelf allow sunlight to penetrate, encouraging the growth of plankton. It has been estimated that Ireland can lay claim to almost 0.9 million square kilometres of the Atlantic.

Ireland's continental shelf area is 14 times the area of the country itself and 14.6 per cent of the EU total. In relation to population size, Ireland has the largest sea-bed territory in the EU.

Member State	% Seabed Area
Portugal	27.36
Spain	16.61
UK	14.29
Ireland	14.64
Italy	9.33
Greece	7.75
France	6.16
Denmark	1.83
Germany	1.00
Netherlands	0.95
Belgium	0.08
EU Total	100.00

◁ Fig 14.11

Ownership of EU Sea-bed Territory (before the Accession of Austria, Finland and Sweden)

Climate

The warm waters of the North Atlantic Drift (NAD) mix with the cooler waters of the temperate latitudes to ensure a variety of plankton and in turn a variety of fish. The stocks on which the fishing industry depend are distributed over the continental shelf usually at a depth of between 100–150 metres.

Under the 1983 Common Fisheries Policy (CFP), most of the fisheries in which Ireland participates are regulated by Total Allowable Catches (TACs) divided into national quotas according to a fixed allocation key. These TACs and quotas however have posed serious problems for the Irish fishing industry. The system obviously favours those countries which had highly developed fishing industries at the time the allocations were made in 1983.

The physical environment in Ireland, however was particularly suited to farming. Traditionally, Irish people tended to prefer a 'beef steak to a cod steak'. The Irish fishing industry was underdeveloped in the early 1980s and Ireland was allocated a mere 4.4 per cent of the total fish allocated under the TAC system despite having more than 14 per cent of the community waters. Apart from quotas, the industry is governed by regulations concerning minimum mesh size, closed seasons, closed areas and individual boat quotas in order to conserve fish stocks.

The Irish Fishing Fleet

Fig 14.12

At present the Irish fishing fleet comprises about 1,400 vessels. Of these, 30 are full-time pelagic boats, and 100 are part-time pelagic boats. About 240 vessels concentrate on demersal fishing while the remainder are involved in shellfish and salmon fishing.

Total Irish landings by the Irish fleet range between 21,300 to 24,700 tonnes with an annual value of £97 million or about 0.5 per cent of GDP. At present the most important species is mackerel which accounts for 35 per cent of landings by volume. Under EU guidelines, the present fleet must be reduced by five per cent. However, the future aim is to reconstruct and modernise the fleet increasing the volume and quality of landings while meeting the EU guidelines.

Over 50 per cent of landings in actual value terms takes place in the top five ports in the country, namely Killybegs, Howth, Rossaveal, Dunmore East and Castletownbere. Fish processing comprises 220 enterprises employing about 3,500 people mainly in coastal areas. About 30 large firms account for 70 per cent of processed output.

Processing has now reached a production value of £200 million.

The sector is still relatively underdeveloped with less than 20 per cent of fish being processed to secondary level.

The Irish fishing industry is concentrated mainly along the south west, west and north west coasts. These regions are remote and peripheral with poor socio-economic conditions relative to the EU core. This isolation, both in an absolute and relative sense, coupled with a poor infrastructure and lack of raw materials, does little to attract secondary industrial development. Dependence

Fishing

on primary sector employment, together with out-migration characterises many of these regions. Consequently in many areas fishing is the main source of full-time or part-time employment. In some areas, fisheries can account for up to 25 per cent of local employment.

< Fig 14.13

Development of Irish Landings 1983-1990				
Year	Demersal	Pelagic	Shellfish	Total
1983	36,011	118,159	16,063	170,233
1986	36,451	162,331	24,709	223,491
1988	44,464	180,057	22,782	247,303
1989	42,346	148,656	25,164	216,166

< Fig 14.14

Development of employment in the Irish Fleet 1983-1989			
Year	Full-time	Part-time	Total
1983	3,431	5,141	8,572
1986	3,561	3,809	7,370
1989	3,380	4,520	7,900

< Fig 14.15

Total employment in the sea-fish industry, including fishing vessels and related on-shore industries is about 15,000.

Trade

Exports of fishery products grew between 1983–1990 from a value of £85 million to £155 million. Exports of fish are mainly in fresh, commodity frozen or semi-processed form to importers, processors and distributors in the country of destination.

Aquaculture in Ireland

The main aquaculture products produced in Ireland are as follows:

1. Mussels

Cultured by dredging and relaying in areas of plentiful food produced 7,000 tonnes in 1990. Cultured by rope culture in western coastal areas produced 3,000 tonnes in 1990.

2. Oysters

At present about 1,500 tonnes are produced each year.

3. Rainbow Trout

Cultured in fresh and salt water – up to 600 tonnes per annum.

4. Atlantic Salmon

Production has grown rapidly in the last 20 years. In 1991 about 8,000 tonnes were produced.

Ireland's production represents 2.5 per cent of world production valued at £40 million compared to Norway 60 per cent and Scotland's 14 per cent.

About 6,000 people depend on mussel, salmon and oyster fisheries for their livelihood. It is particularly important in the peripheral regions of the west of Ireland.

Questions

ORDINARY LEVEL

1. (i) Name four countries which have important fishing industries.
   (20 marks)

   (ii) Name and explain the physical and human inputs which have contributed to the development of the fishing industry in any one of the countries named in part (i) *(60 marks)*

2. Write an account of the Irish fishing industry under the following headings:
   (a) Physical advantages.
   (b) Contribution to the economy.
   (c) Problems which confront its development.

Fishing

HIGHER LEVEL

1. 'In nature all the fishing grounds of the world's oceans are typified by shallow, cool, mixed, waters, by nearness to continental run-off indented shorelines and forests and by periodic exposure to storms, fog and ice-bergs'.
   (i) Explain how some of the factors mentioned in the above passage are conducive to a thriving fishing industry while others do not facilitate that industry.
   (ii) The successful harvesting of fish may also be affected adversely by ocean shipping, overfishing and pollution. Explain briefly how each of these three factors could affect the weight of fish landed.

2. 'In spite of an attractive marine environment the Irish fishing industry is poorly developed'.
   (i) Name and explain some of the reasons for the low level of fish-catch in Ireland.
   (ii) Examine some of the effects which the present Common Fisheries Policy (CFP) is having on the Irish fishing industry.

CHAPTER 15 *Forestry*

Fig 15.1

Forests appeared on earth about 225 million years ago and at one time covered almost 50 per cent of the earth's land surface, and almost the whole of Western Europe with the exception of areas of climatic and topographical extremes. That figure has now been reduced to less than 30 per cent in response to the growth in world population.

The principal reasons for deforestation are the following:

- Land cleared for farming
- Trees used for building materials
- Trees used for fuel
- Fire damage

THE VALUE OF FORESTS

Forests are a multi-purpose resource and the destruction brought by acid rain to the forests of Germany – *Waldsterben* (forest death) is understandably viewed with growing concern.

But forests are more than just a resource. Some scientists now believe that their role in the ecosystem is far more complex, linking as they do the three essential resources of air, water and soil. It is all the more difficult then to understand the almost frenzied annihilation of these silent giants at the staggering rate of *one acre per second*.

Forestry

Physical Advantages of Forests
- They help prevent soil erosion in upland regions.
- They help to prevent flooding and avalanches.
- They provide oxygen and prevent the build-up of carbon dioxide.
- They regulate climatic conditions especially rainfall in tropical regions.

Social Advantages of Forests
- They provide habitats and sustenance for wild life.
- They offer recreational facilities – forest walks, nature trails etc.
- They aesthetically add to the landscape.

Economic Advantages of Forests
- They provide raw material for building.
- They provide fuel – especially important in developing countries.
- They provide raw materials for furniture, paper, etc.
- They give direct and indirect employment.

	The Great Destroyers (acres lost each year)				Fastest Devastation (per cent destroyed each year)		
			of total				acres
1	Brazil	8.5m	(0.7%)	1	Uganda	6.38%	(211,000)
2	Indonesia	3.0m	(1.0%)	2	Burundi	3.75%	(18,000)
3	Bolivia	2.5m	(2.1%)	3	El Salvador	3.74%	(11,000)
4	Mexico	2.3m	(1.9%)	4	Kenya	3.30%	(90,000)
5	Venezuela	2.1m	(1.5%)	5	Rwanda	3.21%	(33,000)
6	Zaire	1.8m	(0.6%)	6	Costa Rica	2.72%	(112,000)
7	Peru	1.4m	(0.9%)	7	Burkina Faso	2.63%	(232,000)
8	Burma	1.2m	(1.6%)	8	Philippines	2.58%	(585,000)
9	Sudan	1.1m	(1.0%)	9	Guatemala	2.54%	(299,000)
10	Malaysia	0.9m	(2.0%)	10	Bangladesh	2.49%	(60,000)

< Fig 15.2 Forest Loss

Distribution of Forests

The factors which influence the growth and distribution of forests can broadly be divided into two groups: physical and human.

Physical

While relief, drainage and soils all influence the growth of forests, the most important physical input is climate.

Climate

Two elements of climate are crucial in accounting for the presence and type of forest cover, i.e. temperature and rainfall. Most trees need a growing season of at least three months and a minimum temperature of 6°C. In this respect it is interesting to compare of the forest cover of Sweden and Ireland. Sweden has a total forested area of over 60 per cent while Ireland's has seven per cent – one of the lowest in Western Europe. Yet in terms of actual temperature requirements growth in the North of Sweden (where most forest cover exists) is confined to about four months while the maritime conditions in Ireland allow for almost all year round growth.

Rainfall is also essential and while 350 millimetres is sufficient for growth in temperate regions, the warmer temperatures of the Mediterranean with greater evaporation and increased rates of transpiration require up to 750 millimetres.

Human

Human activities play a significant role in accounting for both the presence and type of forest cover. Irish forests which once covered most of the country were devastated for both economic and political reasons so that by the 1920s only 1.5 per cent of the land area was covered. That figure now stands at seven per cent as a major drive in reafforestation is being undertaken.

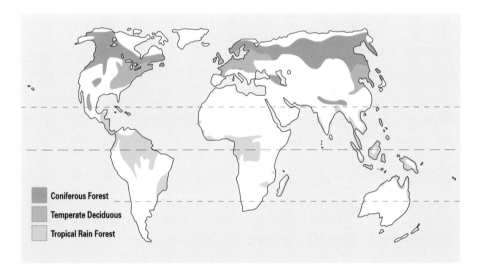

Fig 15.3
Distribution of Forests

AFFORESTATION AND REAFFORESTATION

> Afforestation is planting trees in areas where they were not grown before. In reafforested areas the types of trees planted may not be similar to the original species.

> Reafforestation is the planting of trees in areas from which they have been cleared.

In Ireland at present the native oak, ash and elm have largely been replaced by the coniferous sitka spruce and contorta pine.

The advantages of coniferous over deciduous trees are essentially commercial because:

- Coniferous trees grow faster (around 50 years) – and so produce a greater economic return.
- They need less space than deciduous trees. Up to 2,400 coniferous trees can be planted per hectare.
- They are more adaptable to harsher climatic and soil conditions, and so can thrive on marginal land, e.g. mountain slopes.
- There is a greater demand for softwoods.

TROPICAL FOREST – BROADLEAF EVERGREEN

Location of Tropical Forests	
Amazon Basin	South America
Congo Basin	Africa
East Indies	Asia

Climatic Conditions

(See Chapter 9)

Abundant rainfall, throughout the year – over 1,500 millimetres. Temperatures are constantly over 25°C. The high temperatures and rainfall result in growth throughout the year.

Types of Trees

While there is a wide variety of trees, *stands* (large numbers of the same species) are rare. The principal commercial species are hardwood, e.g. mahogany, ebony, rosewood, and teak.

Lumbering

Tropical forested areas are located among the most underdeveloped economic regions of the world and the actual lumbering (felling) and transport of the timber faces many difficulties.

Because extensive stands are rare, a wide area has to be 'harvested' to obtain sufficient quantities of a required species. Most of the trees are hardwoods which are difficult to cut and transport. Machinery therefore, is both costly and difficult to operate in poor infrastructural conditions.

Transport

Dense undergrowth, swampy conditions, a poor road network and distance from the consumer market place makes transport difficult and costly.

Working Conditions

High temperature and rainfall results in humidity rising to 80 per cent making working conditions oppressive. Hot, wet conditions like these provide ideal breeding grounds for diseases like malaria which are endemic in these areas.

The exploitation of tropical forests is labour intensive and provides valuable employment in both the primary and secondary stages of production. Also, the export of hardwoods to the developed economies of the North provides valuable foreign currency. The timber itself is a valuable source of fuel for heating and cooking. In Ghana, over 75 per cent of the timber harvest is used for fuel while on a global scale it has been estimated that up to 40 per cent of all timber is burnt.

Fig. 15.4 > Coniferous Forest

FORESTS IN MIDDLE AND HIGH LATITUDES

The deciduous forests (generally hardwoods) in the mid-latitudes and boreal forests (coniferous soft woods) in high latitudes are the main types of forest cover in these regions. Because mid-latitude regions are more attractive to human settlement, the coniferous forests are more extensive and are the most commercially developed forests in the world. The principal forest areas are Northern Canada, and Northern Eurasia (Norway, Sweden, Finland, Russia).

Climatic Conditions

(See Chapter 9)

The climate of these regions is characterised by long cold winters and short cool summers. The mean temperature rises above 6°C (the minimum required for plant growth) for only four or five months of the year. Precipitation amounts are low, varying from 250 millimetres to 1,000 millimetres with a summer maximum. This is adequate however as evaporation and transpiration rates are low.

Types of Trees

The major species include Scots Pine and Norway Spruce. They have adapted to their harsh environment in a number of ways:

- They are grouped close together and have conical shapes so reducing the effect of the strong winds.
- Shallow roots allow them to tap the surface moisture and nutrients.
- Branch movement allows snow to slide off and so prevent breakage.

Exploitation

Most species occur in almost pure stands reducing the need to harvest unwanted species. Because the wood is 'soft' it is easier to cut. This was particularly important before mechanised cutting. The lack of undergrowth facilitates access and harvesting.

Fig 15.5
Wood Processing Plant

Transport

Traditionally timber was cut in winter when the logs could be easily moved over the frozen ground. They were then placed on the frozen rivers which thawed in spring and the timber floated downstream to the sawmills. Today, however rail and road are more common as they are more reliable and allow for year-round delivery.

Case Study 1 – Sweden

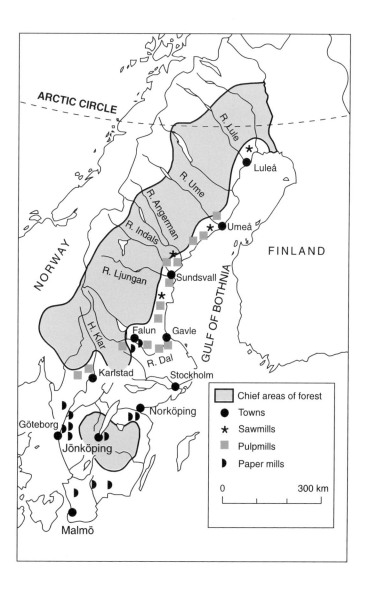

Fig 15.6
The Swedish Forestry Industry

Forests are one of Sweden's greatest natural resources with over 60 per cent of the land area devoted to their production. Physical and human inputs combine to account for the importance of forestry.

Relief and Drainage

Over 80 per cent of Sweden's forest are in the Norrland region. Here, gentle slopes facilitate access for both planting and harvesting. The numerous parallel-flowing rivers such as the Ume, Lule and Skellefte provide more than 32,000 km of natural transport and together with artificial channels reduce haulage distance to an average of three kilometres. Many of these rivers are also ideal for the development of hydro-electric power. While the rivers no longer play a role in transport due to competition from road and rail, the contribution of these waterways to the development of the forestry industry has been considerable. Today, the role of these rivers is no less significant as they provide a cheap form of energy (HEP) which has brought the industry into the twentieth century.

Climate

The contribution of climate to the development of the forests of Northern Sweden is difficult to assess as it has both positive and negative inputs. The low temperatures throughout most of the year mean that trees can take up to 150 years to mature, over twice as long as those in the south of the country. Traditionally, however the same cold winters had positive inputs.

The low sap content made the trees easier to cut and the frozen ground facilitated the haulage of the felled trees to the nearby waterways. As the rivers thawed, the logs floated downstream.

Markets

Since market forces control the levels of economic production in developed economies, Sweden is particularly fortunate in being close to a region which has a noted deficiency in timber. Ever since the Industrial Revolution created a demand for raw materials, Sweden has been able to exploit the growing demand for timber products in Western Europe through careful forest management. Today, the Swedish saw-milling industry is the largest in Europe and accounts for over eight per cent of the world's exports. Britain (21 per cent) and Germany (15 per cent) are the main market outlets. Sweden is the world's largest exporter of paper and also exports safety matches (invented by a Swede – Lundström in the nineteenth century), plywood, furniture and cellulose, etc.

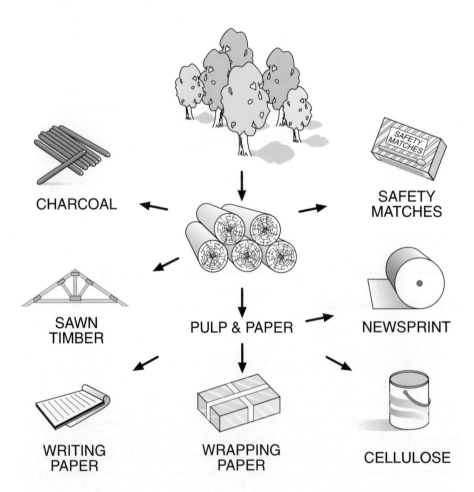

⟨ **Fig. 15.7**
The uses of timber

Forest Management

The clearing of over 200,000 hectares of forest each year (half a million acres) to meet the constant demand for timber and its by-products would in itself justify careful management to conserve this national asset. To the Swedish government however, the role of forests extend beyond the economic, to the social and physical needs of its people.

In the peripheral, less developed regions like Norrland, forests make use of marginal land and together with an income from farming help to make small holdings viable and thus stem the flow of rural depopulation. Forests and woodlands are also seen to have a recreational value and as in other Nordic countries, Sweden has a 'common access prerogative' which allows anyone to walk freely in woods and fields even though 50 per cent of forests are privately owned.

	Private Ownership	State forests	Other forms of public ownership
Total	60.4	17.5	22.1
Belgium	51.8	13.1	35.1
Denmark	65.5	30.4	4.1
France	74.1	9.9	16.0
Germany	43.8	31.1	25.1
Greece	14.7	73.2	12.1
Ireland	19.6	79.9	0.5
Italy	60.0	7.2	32.8
Luxembourg	56.1	7.9	36.0
Netherlands	51.9	30.9	17.2
Portugal	84.7	2.5	12.8
Spain	66.1	5.4	28.5
UK	59.5	40.5	0.0

Fig 15.8 Ownership of Forests

The Plan

Understandably, forest conservation is a high priority with the Swedish government and its people. The basic plan is two-fold – reafforestation and protection. Once a stand of mature trees has been felled, it must be replaced immediately by new trees through planting and natural generation. Quota systems also ensure that cutting does not outstrip the growth of new trees. Special forest schools are funded by the government where training and research are carried out. Aircraft surveillance and the use of the media help to combat the danger of forest fires. In spite of this strict conservation policy, however, the forestry sector faces major problems. The clearing of forests in the south for agriculture places great pressure on the more Northern regions (where growth is slower) to conserve the nation's supply. In spite of vigilance, forest fires are a major hazard and at present number 1,000 a year, but the cause of the problems which result from acid rain are beyond Swedish

Government control and so must be seen as a far more serious threat to Sweden's 'Green Gold'.

Forest Cover for Selected EU States	
Country	Percentage
Ireland	7 per cent
France	27 per cent
Italy	21 per cent
Portugal	32 per cent
Greece	44 per cent

Fig. 15.9 Forest cover

Case Study 2 – Ireland

At the turn of the century, only 1.5 per cent of the land area of Ireland was under forests. Since then the figure has risen to seven per cent which is still extraordinarily low when compared to our EU counterparts (See Fig. 15.9).

Most of the initiatives in afforestation and reafforestation were taken by the state with very little investment taken by farmers and other private individuals.

This is understandable when we realise:

- the high capital investment required
- the 40–50 years needed for rotation
- little generation of income for the first 15–20 years
- membership of the EC in 1973.

On joining the EC in 1973, the incentives that farmers received for their produce was a further disincentive to allocate land to forestry.

Changing Times

Over-production in agriculture in the 1980s led to a change in the Common Agricultural Policy (CAP) and ultimately to the introduction of farm quotas and this, together with the need to increase EU wood supplies (the EU imports over 50 per cent of its needs) made forestry an attractive option which led to EC support for Ireland's Forestry Operational Programme.

This programme, along with related measures, includes a package of grants for planting, woodland improvements, forest roads, forest harvesting machinery, forest nurseries, forest training and back-up measures such as studies, pilot projects and an annual premium for farmers who afforest their land.

To assist with the high capital costs involved in establishing a forest, the State pays grants of up to 70 per cent of the cost to non-farmers. A grant of up to 85 per cent of the costs and the forest premium scheme which provide them with an income for up to 20 years, makes forestry a particularly attractive option for farmers today.

The Aims of the Irish Forestry Operational Programme are as follows:

- To provide the raw material base for an expanded and improved forestry based industrial sector.
- To stimulate rural development.
- To make a positive contribution to the environment.
- To generate employment particularly in rural areas.
- To expand and diversify the rural economy.
- To contribute to the reform of agricultural structures.

These incentives, together with the limited possibilities for expansion in traditional farming output, have made forestry both a viable and attractive alternative.

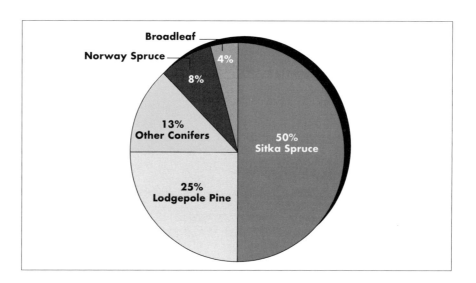

Fig 15.10 >
Types of Trees in Irish Forests

Under 10 per cent of the land will be under forests by 2000 AD and 14 per cent by the year 2020 at the present rate of planting.

The annual planting target including reafforestation is 30,000 hectares per year.

Ireland is one of the few areas within the EC where large-scale forest expansion can occur. This together with the fact that Ireland is a net importer of wood and wood products to the value of over £200 million annually would suggest that forestry is an attractive proposition. When we realise however that the EC imports 50 per cent of its timber requirements – a staggering £15 billion worth and that this figure is increasing, the opportunities appear limitless.

Forestry

Questions

ORDINARY LEVEL

1. Forests once covered 50 per cent of the earth's land surface; that figure has now been reduced to 30 per cent and, in Ireland's case, it stands at seven per cent – the lowest in the EU.
   (i) Briefly explain the reasons for this deforestation. *(20 marks)*
   (ii) Examine the physical, social and economic role of forestry in any country which you have studied. *(60 marks)*

2. Compare and contrast the forests in the tropics with those in temperate latitudes under the following headings:
   (i) location; (ii) types of trees; (iii) methods of harvesting;
   (iv) uses of timber *(80 marks)*

3. Examine the table which shows forest cover in selected European Union (EU) States (1993)

Country	Forest area in '000s hectares	Percentage of land cover
France	14,765	27 per cent
Greece	5,755	44 per cent
Ireland	460	7 per cent
Italy	6,403	22 per cent
Portugal	2,967	32 per cent
Total for E.U States	53,764	24 per cent

   (i) Name the country which
   (a) has the lowest percentage of forest cover;
   (b) the greatest area of forest cover;
   (c) is closest to the European Union average percentage;
   (18 marks)
   (ii) The present forest planting target in Ireland is 30,000 hectares per annum. Explain any **two** advantages which the increase in area under forests will bring to the Irish economy. *(32 marks)*
   (iii) The amount of forest wildlife will have declined by nearly 50 per cent during the twentieth century. Describe and explain any **one** cause of this decline **and** any **one** concern which results from it. *(30 marks)*

   Leaving Cert. 1995

4. Since the beginning of the twentieth century, the total area of forest in the world has declined by 40 per cent.
   Environmentalists and other authorities are becoming more and more alarmed at the long-term results of this rapid decline.
   (i) Describe and explain some of the reasons for the rapid forest decline. *(50 marks)*

(ii) Examine some of the causes of the alarm referred to above and briefly outline possible steps which could be taken to ease the long-term results of forest decline. *(30 marks)*

Leaving Cert. 1992

HIGHER LEVEL

1. Apart from topographical extremes the Irish landscape was almost completely covered in forests. By 1922 the actual cover had been reduced to 1.5 per cent; it now stands at seven per cent and is increasing rapidly.
   (i) Examine some of the causes of deforestation up to the turn of the century (1900). *(30 marks)*
   (ii) Account for the extraordinary renewed interest in afforestation and reforestation since the 1980s. *(70 marks)*

2. Examine the contribution of the forestry industry in any European country with which you are familiar. *(80 marks)*
   Comment briefly on some methods of conservation employed in that country. *(20 marks)*

3. In 1950, 15 per cent of the earth's land surface was covered by tropical forest. By 1975 this had declined to 12 per cent and by 2000 it is likely that the figure will stand at only seven per cent.
   (i) Describe and explain **two** causes of this rapid rate of forest clearance. *(50 marks)*
   (ii) Concern over this forest clearance arises for a number of reasons, both localised and global.
   Examine this statement. *(50 marks)*

Leaving Cert. 1990

CHAPTER 16 — Minerals and Energy

Fig 16.1
Strip Mining of Copper, Utah, USA

While mining is an activity which has taken place for thousands of years, its spectacular growth is tied to the Industrial Revolution. Unlike other primary activities, however, mining is a 'robber economy' i.e. once the resource is exhausted it cannot be replenished. The presence of human settlement in the most inhospitable environments, e.g. oil mining in the Sahara is testament to the value and importance of certain minerals and it has been the reason for the existence and development of many cities, e.g. Johannesburg (gold), Dallas (oil), etc.

THE IMPORTANCE OF MINERAL WEALTH

1. Raw materials for industry

Examples of this include iron, copper, lead, tin etc.

2. Source of Energy

Examples of this include coal, oil gas.

3. Employment

Employment in the minerals sector can be both direct or indirect. Even though only about one per cent of the world's labour force is employed in mining, its presence can be of major economic importance to specific regions, e.g. coal mining in South Wales.

4. World Trade

The demand for oil in the developed economies of the 'North' has made oil the 'black gold' of the twentieth century and a major 'money spinner' for countries in the Middle East, (e.g. Saudi Arabia) which have few other resources.

The world distribution of mineral resources is very uneven, a consequence of the complex geological forces which have led to their formation. The actual mining of a mineral however depends on a number of factors.

The size of the deposit is a major consideration. In general, mining is costly so small deposits may not be worth mining. The grade (the proportion of mineral contained in the ore body) is also relevant. The minimum grade to justify mining iron ore can be up to 30 per cent while for gold it can be as little as 0.001 per cent.

The depth of the deposit has a major influence on the cost of extraction. Open cast or surface mining is usually the cheapest but not the most environmentally friendly method. Thin layers of mineral also increase mining difficulties and costs particularly where the mineral occurs underground. Ideally then the ore body should be continuous over a large area with thick seams (layers) and uninterrupted by faulting or folding.

Markets and Transport

Since most minerals are used in the production of other goods, i.e. manufacturing industry, developed economies form the main markets. Therefore, world consumption is dominated by North America, Europe, Russia and the Far East (e.g. Japan). The price of minerals, which reflects their rarity and demand, is also relevant. Over-production of a mineral may lead to the closure of the less economic mines.

Transport has always been a major factor in the location of the mining industry because most minerals tend to be high-bulk, low value commodities. Access to water and rail transport is important; hence the building of canals and railways at the beginning of the Industrial Revolution. While transport remains an essential element of mineral production, its actual cost has decreased with the development of bulk carriers and the primary processing of ore, which reduce the amount of *gangue* (waste) before shipment.

Political

The importance of mining as an economic activity can be gauged from the influence which governments exert on its location. Governments may encourage, subsidise or themselves engage in the exploitation of mineral resources. The nature of ownership and the size of the tax and royalty payments can have significant effects on the distribution of mining.

Mining Methods and Environmental Effects

Surface Mining

Surface mining techniques are used, (i.e. open-cast, open pit or strip mining) where the ore occurs at or near the surface and the overburden (the material above the ore) can be economically removed.

Surface mining contributes over half the world's mineral output. Once the overburden is removed, the ore may be extracted by power shovels, explosives or scrapers and then transported out of the pit. While it is particularly sensitive to weather conditions especially in high latitudes (where mining may cease in winter) it is the most economic method of ore extraction. The detrimental effects on the landscape however can be severe. The surface is often scarred by huge unsightly pits while mining in lowland areas can often result in the loss of agricultural land. The landscape can however be restored by careful landscaping (planting grass and trees, flooding quarries etc.) but this can add significantly to the overall cost of mining.

Fig 16.2
Open Cast Mining

Underground Mining

Underground mining takes place when the mineral deposits occur deep beneath the ground surface. There are three types of mining:

1. Shaft Mining – where the miners approach the ore body by a lift.
2. Slope Mining – where a tunnel is dug into the ore body at an angle.
3. Drift Mining – where the ore occurs in horizontal layers in valleys, drift mining methods are used.

Apart from the increased costs, working conditions are unpleasant and often dangerous. The main effect of underground mining on the landscape is usually spoil heaps of waste material.

Coal tips and slag heaps often blight the landscape and can be a landslip hazard as in the Aberfan Disaster in 1966 which killed 116 children and 28 adults in South Wales. Once the mine is abandoned, subsidence may occur affecting roads and houses.

Apart from the negative effects on land, mining can also poison air and water (waste from lead, copper etc.). While this can lead to the destruction of the local flora and fauna, air and water are 'flow resources', the pollution of which can have effects far beyond the mining environment.

Mining Methods and Environmental Effects

Fig 16.3 > Underground Mining

CASE STUDY – IRON-ORE MINING IN SWEDEN

In terms of employment and exports, mining makes a major contribution to the Swedish economy. While the iron-ore deposits of the Bergslagen field, mined at Dannemora supply the domestic steel industry, it is from the north and in particular the Kiruna and Gallivare fields that the vast bulk of Sweden's ore comes. Mining began in the Gallivare Region as early as 1890 and the Kiruna field began large scale production in 1900. In spite of the harsh physical environment which inevitably resulted from an Arctic location, the state owned mining company – the Arctic Mining Company – was attracted by both the massive reserves estimated at around 3,000 million tonnes and its high grade magnetite ore – up to 70 per cent iron. Apart from the obvious mining difficulties which result from an Arctic environment, the greatest problem was one of transport. The town of Luleå acted as the main port during the summer months but the Gulf of Bothnia is frozen in winter. The problem was finally overcome by the construction of a railway – the Baltic Railway Line – allowing access to the ice-free port of Narvik on the Norwegian coast. Apart from transport, mining here faces two other major difficulties.

(a) Climate

Lying deep inside the Arctic Circle temperatures can fall to −40°C posing obvious difficulties for both miners and machinery.

The northerly latitude also results in 24 hours of darkness for almost two months in winter.

(b) Labour

The difficulties of attracting a skilled workforce to a remote and harsh environment are obvious. Higher wages (25 per cent above the national average) and other benefits (housing etc.) have to be offered as 'carrots' to attract a suitable labour force.

Because most of the ore mined at Kiruna is exported, the ore used in Sweden's domestic industry comes from the Bergslagen ore field in the north-east of the Central Lowlands. It was here that the world's first steel was produced in 1858 at Sandvicken (north of Stockholm) using the Bessemer process.

Today it forms the basis for the Swedish-produced ball-bearings, saws and razor blades which are known throughout the world. It has also played a major role in the development of ship-building in Sweden and in the manufacturing of cars at the Volvo and Saab-Scania plants.

Energy

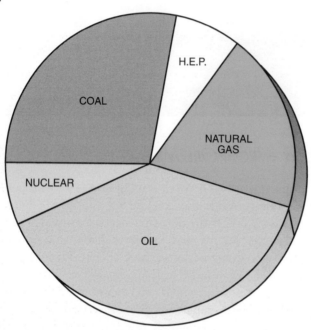

< Fig 16.4
Sources of World Energy

Fossil fuels have provided a cheap and convenient energy supply since the Industrial Revolution but they have also contributed significantly to atmospheric pollution, acid rain and global warming.

COAL – ITS RISE AND FALL

Coal is derived from vegetation which only partially decayed under swamp conditions and which was compacted by subsequent deposition of sedimentary material on top of it. Three grades of coal, based on carbon content, are usually recognised: lignite, bituminous coal and anthracite.

Because coal is bulky, heavy and expensive to transport, coal-burning electricity power stations are located where fuel can be brought to them relatively cheaply. As a result they are often located on coalfields, on inland waterways or on coastal sites to which the coal can be shipped. The major exporters are the USA, Poland, Germany and Russia.

Its Rise

The Industrial Revolution of the late eighteenth and early nineteenth centuries placed enormous demands on coal, especially coking coal for the smelting industry. Because of the huge quantities of coal needed it was cheaper to bring the raw materials to the coal fields and export the finished products. Hence the coal fields of Europe became synonymous with the great industrial regions.

Coal also provided the raw material for the chemical industry, was used as a domestic fuel and made a major contribution to transport (steam engines). By the late nineteenth century coal was 'king' and people working in the coal-mining regions could boast of higher wages if not the most salutary working conditions. By 1950, coal accounted for 90 per cent of Europe's energy needs.

Its Fall

Fig 16.5 > Oil Tanker

Since then, however, coal has been replaced as Europe's major energy source although it still retains a vital role in the regional economy of some countries.

The decline in coal-mining is related to a number of factors including the following:

Increased Costs
- As the more accessible seams become exhausted the cost of mining increases.

Competition
- Oil and gas are easier to transport and are perceived as being more environmentally friendly.
- Government policies of closing down uneconomic mines have been implemented.

The decline in coal mining in certain areas in Europe however involved more than just the loss of an industry. For regions like South Wales, North-East England, the Nord and the Sambre-Meuse Valley in Belgium it represented the loss of a way of life.

Minerals and Energy

CASE STUDY – THE SAMBRE MEUSE VALLEY – ITS RISE AND FALL

Cause of the Problems

Because coal exercised a considerable pull effect on industry in the nineteenth century, coal-mining regions developed into centres of heavy industry. *Geographical inertia* (inactivity) further boosted their importance and together with industrial agglomeration, their economic future seemed assured. Throughout the nineteenth century, the Sambre-Meuse region enjoyed high levels of employment and could boast of up to 250 coal-producing pits operating at the same time. Coal, however, is a finite resource and as the more accessible seams became exhausted the remaining thin and discontinuous seams became uneconomic to mine.

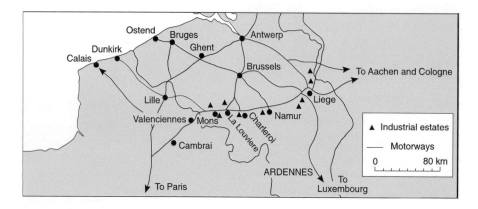

◁ Fig 16.6
Sambre Meuse Valley

Apart from the rising cost of coal mining, coal was already beginning to lose its magnetic effect on industry. The increasing use of oil and gas, the development of power grids and the improvements in transport all helped to release the grip that these areas exerted on industrial location. With the increasing cost of coal and the use of imported iron-ore, modern steel-plants moved to waterside locations. The decline in coal mining also spelt a decline in the associated chemical and engineering industries. As employment fell so did the demand for housing which in turn sent the construction industry into decline. It was a classic case of industrial agglomeration – in reverse.

The Problems

One of the most depressing aspects of the problems confronting the Sambre-Meuse region is the visual scars it has left on the landscape. Abandoned collieries (coal mines), overgrown slag heaps, derelict factories and industrial slums – ghosts of a once thriving economic region are now the tell-tale signs of a 'has been' industrial area. While the decline in coal mining affected the whole region, it was particularly severe in the west where the last colliery closed in 1986.

These closures had negative spin-off effects on associated industries such as chemical and engineering. As a result the region was devastated by unemployment especially among the male population. Lack of job opportunities resulted in out-migration leaving a middle-aged labour force whose skills have outlived their usefulness. Today with unemployment exceeding 23 per cent, it poses major challenges for the Belgian government. The need to find solutions to the problems of the south become all the more pressing when the cultural divisions within the country are taken into account.

Solutions

- Vast amounts of Belgian and EU funding for the region to re-train workers and provide alternative employment.
- Infrastructural improvements especially to roads and railways. The region is now linked to Brussels and the surrounding cities by new motorways and an electric rail network.
- Industrial estates have been developed which have had moderate success to date in attracting light industries.

OIL

Oil was first drilled in the USA in the mid-nineteenth century. One hundred years later it had replaced coal as the leading source of energy. This spectacular growth can be explained by the following factors:

1. Government Policy

In the 1960s most governments especially in Western Europe decided on a policy of cheap energy. Little thought was given to the strategic risk of depending upon source areas which could be regarded as politically unstable, e.g. the Middle East.

2. Developments in Transport

The construction of supertankers often over 1,000 feet in length and with capacities of 250,000 tonnes guaranteed oil supplies to Western Europe. The construction of oil pipelines from the 1960s resulted in a comparatively cheap method of transporting oil throughout Europe from ports such as Rotterdam – the 'oil jugular' of Europe.

3. Multi-purpose Resource

    (i) Crude oil can be separated into different fuel products e.g. petrol (automobiles), paraffin, etc.

    (ii) Generation of electricity.

    (iii) Raw materials for the chemical industry e.g. chemicals, synthetic fibres (nylon), fertilisers etc.

4. The Increased Demand for Energy

As population and living standards increased so too did the demand for energy in domestic, industrial and transport uses.

The spectacular growth in oil consumption was checked by the energy crisis in 1973–74. Up to the early 1970s, most oil-producing countries received very little for their oil. The 13 oil-producing countries (including Saudi Arabia, Algeria, Nigeria and Venezuela) formed the Organisation of Petroleum Exporting Countries (OPEC). They decided to:

    (i) Increase their share of oil revenue which resulted in a four-fold increase in price within two years.

    (ii) Control extraction rates which would help to maintain prices and extend oil reserves.

In response, the countries of Europe reduced their dependency on imported oil – between 1973–88, imports fell from 614 to 303 million tonnes. This was achieved by:

- a greater reliance on their own energy resources.
- the use of natural gas, e.g. Norway, Ireland.
- research into alternative energy sources, e.g. solar, wind.
- increased taxation of oil products to reduce demand.
- improvements in energy conservation, e.g. insulation.

Though fossil fuels are a relatively cheap and convenient form of energy their use has resulted in environmental damage. They are also a finite resource which will become exhausted sooner or later. Though natural gas reserves are now known to be higher than once thought and new discoveries have made predictions that the world would soon run out of oil seem premature, the finite nature of fossil fuels is indisputable.

Even the vast oil fields of Saudi Arabia, if pumped at the current rate of about 8.1 million barrels a day, will be emptied in about 75 years.

The greatest problem with the use of fossil fuels to generate energy is that they give off harmful emissions when burnt. While technology continues to reduce emissions of sulphur dioxide and nitrous oxides the emission of carbon dioxide continues to cause concern to those worried about global warming. Advocates of nuclear energy also have to admit the dangers associated with its production including the disposal of nuclear waste.

It is against this background then that the search for clean and renewable forms of energy has been intensified in recent years.

SOLAR ENERGY

Every year the sun delivers free energy to the earth equivalent, in theory, to 10,000 times mankind's energy needs. The main difficulty is harnessing this resource. Solar energy only contributes 0.01 per cent of energy consumption worldwide. Even in Israel where 80 per cent of homes have solar hot water systems, total solar power contributes only two per cent of total energy consumption. Solar radiation can be used to warm or cool buildings, to provide hot water and to generate steam for turbines producing electricity. Solar light can also be turned directly into electricity by *photovoltaics* (PV), a branch of solar science which is developing rapidly and which one day could lead to a 'pollution free' economy. Broadly speaking, solar energy can be divided into three categories.

**‹ Fig 16.7
Solar Building**

1. *Passive Solar* – the basic design of buildings to take maximum advantage of the sun's radiation.

2. *Solar Thermal* – using the sun's radiation to heat water systems, swimming pools and living space or for the generation of electricity.

3. *Photovoltaics (PV)* – using solar cells to generate electricity directly from sunlight.

Annual Hours of Sunshine	
Cairo (Egypt)	3,717
Bilma (Niger)	3,681
Giles (Australia)	3,370
Almeria (Spain)	3,053
Tunis (Tunisia)	2,907
Rome (Italy)	2,537
Stockholm (Sweden)	1,973
Lagos (Nigeria)	1,862
Syowa (Antarctica)	1,849
Berlin (Germany)	1,818

Fig 16.8 >

Seasonal and regional variations in climate complicate the task of harnessing the sun's energy.

Countries with year-round sunshine have an obvious advantage. But even the sunniest climes have to cope with the cut-off of energy flow that accompanies the daily sunset.

While countries which bask in sunshine for most of the year have an obvious advantage in using solar power, the idea that solar power is not relevant to temperate regions is misguided. Northern Europe may sit under cloud cover for much of the year but solar power could still make a big contribution. The breakthrough in photovoltaics has changed conventional thinking. Photovoltaics is based on solar cells – which operate simply from light – in the same way as hand-held calculators have done since the 1970s. The biggest problem with photovoltaics is cost. However desirable solar power may be in environmental terms, people will not buy it if it is too expensive.

In Norway, most lighthouses along the coast and over 50,000 holiday homes get electricity from solar panels. Even in Finland, 20,000 holiday homes are PV powered. However, the big breakthrough in cost has not yet happened. There is widespread international research into improving solar cells but they remain much more costly than conventional energy sources. Solar cells however are reliable, silent and long lasting (20–30 years if properly protected). The only maintenance required is occasional washing to prevent dirt from blocking the sun's rays. They create no pollution in operation and relatively little during manufacture. Unlike solar thermal systems which require direct sunlight to be effective, solar cells work even in cloudy conditions, though they obviously produce more electricity in bright sunny conditions.

Millions of villages worldwide lack electricity for water pumps, refrigeration, lighting, communication equipment and cooking.

Minerals and Energy

Linking them to national grids is, financially, not feasible. The PV technology then offers a cost-effective and environment-friendly solution. Yet the capital cost continues to put solar installations out of reach.

Advantages	Disadvantages
No need to burn fossil fuels.	PV electricity is at present very costly.
Plentiful and cheap supply of basic cell material (silicon).	Major seasonal and daily variations in energy output.
Long operating life (20-30 years). Noiseless in operation.	Need for storage (e.g. batteries).
No carbon dioxide, sulphur or nitrous oxide emissions.	High initial capital cost.
Easy installation.	Low energy density, therefore large surface area of PV cells needed to generate high power.
Low maintenance cost.	

‹ **Fig 16.9**
Advantages and Disadvantages of Solar Power

WIND

‹ **Fig 16.10**

As a source of power, wind has been used for centuries from the early sailing ships to pumping water, to draining the land in the Netherlands. The development of steam power, and with it the Industrial Revolution, relegated windmills to the status of tourist attractions rather than working machines. The oil crisis in the 1970s, however, together with the increasing awareness of the environmental damage caused by the use of fossil fuels has resulted in a new interest in wind as a source of energy.

World Wind Energy Resources	
Region	per cent of land area with usable wind power*
North America	41
Latin America & Caribbean	18
Western Europe	42
Eastern Europe & CIS	29
Middle East & North Africa	32
Sub-Saharan Africa	30
Pacific	20
Central & South Asia	6
Total	27
*Area with annual mean wind speed between 5.1 m/s and 8.78 m/s, i.e., with good potential for electricity generation by wind power.	
Source: *World Energy Council*	

Fig 16.11

In theory, wind resources could provide twice the world's present electricity production. Many of the more suitable sites however are too remote to be useable.

The siting of large scale wind farms close to population centres or in areas of natural beauty may provoke much public opposition to be implemented. Nevertheless, wind power is likely to provide a growing proportion of world energy production. According to experts at the Royal Institute of International Affairs, wind power could eventually supply more than 20 per cent of global electricity demands.

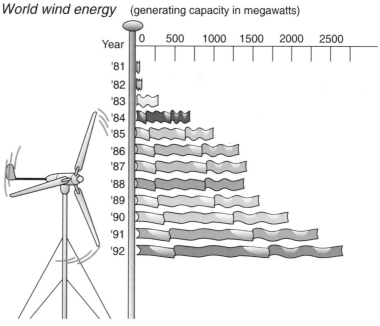

Fig 16.12

Most wind power installations have taken place in California and Denmark followed by Germany the Netherlands, the UK and Spain.

Minerals and Energy

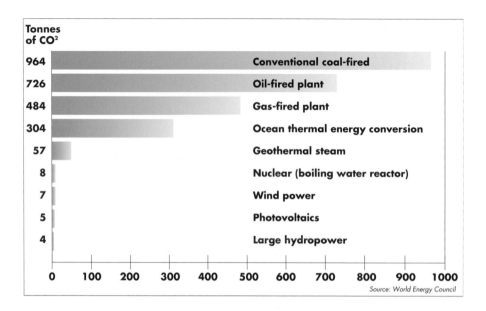

Fig 16.13
Carbon Dioxide Emissions from Power Stations

The most fundamental problem with wind power is the irregularity of wind speed. If the wind blows too strong, the turbines may be damaged. If wind speed is too slow power generation stops.

Because wind travels more slowly at ground level due to friction wind turbines are typically 25-35 metres off ground. While the increase in altitude however reduces the effect of friction it also reduces air density and so reduces the ability of the wind to turn the turbines. Turbulence is another important consideration in the siting of wind turbines. Where the topography is level, wind can flow across the surface smoothly.

Obstacles such as hills, trees, buildings, etc., set up patterns of turbulence in the air. Rotor blades are thus subjected to conflicting pressures which decrease the efficiency of power generation and shorten the life of the machinery. Coastal and sea-based wind farms then would seem to offer the ideal locations, but apart from the increased capital costs there are major problems associated with the corrosive effect of sea water.

Advantages	Disadvantages
Virtually no pollution.	High initial expense in setting up.
Wind fuel is free and widely available.	Negative visual impact on environment.
No transport or storage costs for fuel.	Noise pollution.
Running costs are low.	Irregularity of wind speed.
Safe form of power generation.	Possible danger to wild life e.g. migrating birds.

Fig 16.14
Advantages and Disadvantages Wind Power

At present Denmark leads the world in terms of reliance on wind power for electricity generation. Yet, despite a decade of government support the proportion generated by wind is only three per cent.

The European Union has set a target of 8000 MW capacity from wind power by 2005, enough to satisfy one per cent of its electricity needs. The portion of electricity generated by wind could reach 10 per cent by the year 2030.

NUCLEAR ENERGY – A SOLUTION OR PROBLEM?

As living standards and world population increases, world demand for electricity is expected to double within the next 30 years. The harmful emissions from the burning of fossil fuels, together with the unreliable, and as yet underdeveloped nature, of renewable sources has resulted in an increasing reliance on nuclear power in countries such as Japan.

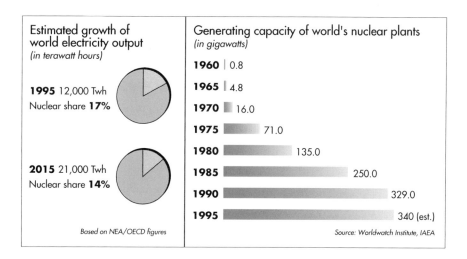

Fig 16.15

One of the strongest arguments in favour of nuclear energy is the lack of atmospheric emissions from nuclear power plants. Without the world's nuclear reactors, annual carbon dioxide emissions would be almost two billion tonnes higher, adding to the greenhouse effect.

Fig. 16.16 1 tonne of uranium = 16,000 tonnes of coal

Uranium is the raw material which provides the fuel for nuclear power stations but the quantity is small in comparison with the volumes of coal and oil needed for fossil fuel power plants. One tonne of natural uranium ore can produce the same quantity of electricity as 16,000 tonnes of coal or 80,000 barrels of oil. In terms of land-scarring and transportation, energy production which relies on coal mining or drilling for oil causes vastly more harm than nuclear power generation. Despite the fact that coal mining has killed thousands of workers over the last century and coal smoke has damaged the

Minerals and Energy

health of millions, the fear of radiation – the invisible hazard from nuclear power stations – strikes fear in the hearts of most people, a fear which the explosion in Chernobyl in 1986 has done little to abate. Numerous studies too have been carried out into the effects of discharges into the Irish Sea from the nuclear plant at Sellafield in North West England. Claims of links to leukaemia however remain inconclusive. It is indisputable, however, that some of the waste which is created remains highly dangerous for thousands of years.

Advantages	Disadvantages
Only small quantities of fuel are needed.	Extremely costly to set up.
Nuclear plants emit no sulphur dioxide – the main cause of acid rain.	Radioactive waste is costly and dangerous to dispose of.
Emits only small amounts of greenhouse gases.	High consumption of cooling water increasing the danger of 'thermal' pollution.
Nuclear power produces large amounts of electricity.	Danger of a reactor exploding and the associated effects on public health.
They provide a reliable and uninterrupted supply of electricity.	

‹ Fig 16.17
Advantages and Disadvantages of Nuclear Power

Questions

ORDINARY LEVEL

1. Examine the following statement in some detail. 'There are positive **and** negative aspects to nuclear power as a source of energy in Europe today.' *(40 marks)*

   Leaving Cert. 1996

2. Examine the following statement in some detail: The decline in coal production has led to social and economic problems. *(40 marks)*

   Leaving Cert. 1994

3. 'In the push to exploit the earth's natural resources, damage can be caused to the environment and to people'.
   (i) Examine this statement, referring to examples which you have studied. *(50 marks)*
   (ii) Describe briefly how such damage can be limited or repaired. *(30 marks)*

   Leaving Cert. 1993

4.

Energy Source	Percentage contribution to total use		
	1950	1975	2000 (est)
Wood, vegetation	21	13	5
Coal	44	27	30
Petroleum	25	40	39
Natural gas	8	15	15
Other sources (mainly hydroelectric & nuclear)	2	5	11

(i) Study the information given in this table. Describe the trends and changes indicated. *(20 marks)*

(ii) Conservation of energy is an important economic aim in most countries today.
Explain the reasons for this and describe some of the methods which are being used in order to achieve the aim. *(45 marks)*

(iii) Consider the figures given in the table for the year 2000. Describe what you think might be the environmental consequences of such a pattern of energy use. *(15 marks)*

Leaving Cert. 1992

HIGHER LEVEL

1. The table below gives details of the consumption and production of commercial energy, by source and region for 1988.

Commercial Energy Production and Consumption by Region and Fuel, 1988						
	Oil		Natural Gas		Coal	
Region	Production	Consumption	Production	Consumption	Production	Consumption
North America	22,857	36,171	21,203	21,211	23,866	21,542
Latin America	14,278	9,551	3,601	3,308	888	959
Western Europe	8,290	24,875	6,297	8,332	7,821	11,033
Middle East	30,954	5,669	2,734	2,278	29	105
Africa	10,991	3,609	2,227	1,264	4,187	3,048
Asia and Australia	6,816	19,507	4,539	4,091	9,877	12,779
Centrally Planned Economies						
USSR	26,127	18,385	29,045	22,982	16,409	12,984
China	5,699	4,216	528	561	24,251	24,331
Other	888	5,238	2,617	4,262	14,994	14,881
World Total	126,900	127,222	72,791	68,290	102,322	101,660

Source: British Petroleum (BP) BP Statistical Review of World Energy (BP, London, 1989), pp. 4, 22, a. Conversion factor: One million metric tons of oil equivalent = 41,87 petajoules.

Leaving Certificate Geography

Minerals and Energy

(i) Contrast the main production and consumption patterns of the listed commercial energy sources with reference to the above table. *(70 marks)*

(ii) Examine the current position regarding nuclear power as a source of commercial energy. *(30 marks)*

Leaving Cert. 1993

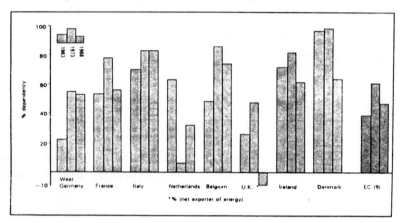

Percentage Dependency Upon Imported Energy in the EC(9), 1963–1988

2. Examine the bar-chart above and answer the following:
   (i) Explain, using the chart, why 1973 is seen as a turning-point in the pattern of energy supply in the European Community.
   Refer in your answer to:
   • The overall pattern of change
   • Variations between individual member states. *(25 marks)*

   (ii) Explain by referring to particular examples, the changes which have occurred in Europe in:
   • Diversification of energy sources
   • Development of internal energy sources
   • The environmental impact of these changes. *(75 marks)*

Leaving Cert. 1991

CHAPTER 17 *Manufacturing Industry*

Fig 17.1

No revolution has changed society and divided the world in an economic sense as has the Industrial Revolution. For thousands of years, articles were fashioned by hand and even when simple machines were invented, they too were worked by hand. Then in the late eighteenth century, some people began to see the practical use for something that they had known since the first lid rattled on a kettle of boiling water – steam power! The age of mass production – the Industrial Revolution had begun. From an economic viewpoint the Industrial Revolution divided the world in two: the industrialised or 'developed economies', on one side, the non-industrialised or developing economies on the other. In spite of the lapse of over 200 years, this broad division still exists and at present over 80 per cent of the world's manufacturing output (in terms of value) comes from a limited area known as the Power Belt – a region which stretches east from the Mississippi in the USA to the Ural Mountains in Eurasia.

While industrialisation has improved the living standards of these nations, material progress has commanded a high price which many believe far outweighs the benefits.

Manufacturing Industry

Definition of Manufacturing Industry

In its widest sense the word industry is used to cover all forms of economic activity – primary, secondary and tertiary, whether we talk about the agricultural industry (primary) or the tourist industry (tertiary). In this chapter we will confine the term to manufacturing industry.

> Manufacturing industry may be defined as 'Any process which involves changing raw materials into finished products'.

The changing of raw materials into 'finished' or consumer products often involves a number of stages and what one industry may regard as a finished product is simply a raw material to another.

A simple example illustrates this point:

WHEAT –> FLOUR –> BREAD

To the milling industry flour is a finished product, to a bakery, however it is essentially a raw material.

> A more accurate definition of manufacturing industry then might be:
>
> 'The processing and changing of materials whether in a raw or partly raw state to increase their usefulness'.

It is possible then to distinguish between two levels of production.

Primary Manufacture

The term 'primary manufacture' is used to refer to industries which process basic raw materials into partly raw materials to be used by other industries, e.g. wheat to flour.

Secondary Manufacture

Secondary manufacture refers to industries which use these partly raw materials and produce 'finished' or 'consumer' products.

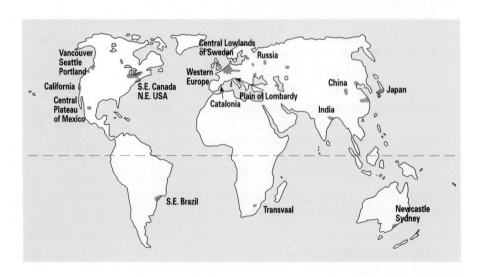

Fig 17.2

Major Industrial Regions

Factors Influencing Manufacturing Industry

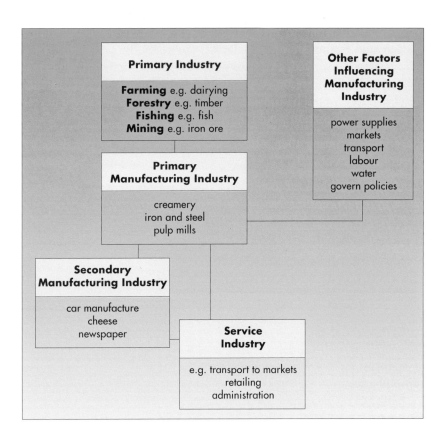

Fig 17.3
Types of Industry

Factors Influencing Manufacturing Industry

1. Raw Materials

Fig 17.4
Iron Smelting

In 1909, Alfred Weber a German economist, published his findings on a study made on industrial location. Though considerably modified since then, the basic principles still apply.

He showed that industries whose commodities involved a loss of weight in their production tended to be *resource-orientated*, i.e. located near the source of the raw material. The conversion of wood to pulp involves a 60 per cent weight loss, so paper mills tend to be located near forested areas. The actual amount of metal which is extracted from ore bodies shows a dramatic fall in weight and, as a result, some degree of concentration of the ore is usually done near the mine which reduces the amount of useless material, known as *gangue* and so lowers transport cost. Other examples include refined sugar which is only one-eighth the weight of cane and butter only one-fifth the weight of milk.

Leaving Certificate Geography

If industries which use the raw materials from forestry and mining tend to be resource-orientated due to their bulk or weight, then those depending on agriculture and fishing are often influenced more by the perishable nature of the raw material involved. Fruit canning, fish processing and wine-making then often tend to be resource-orientated as the transport cost of the raw material involved would be considerable (as refrigeration would be necessary) and a certain deterioration would inevitably follow.

Primary manufacturing then, either by reasons of bulk or perishability are often resource-orientated.

As a pull factor on the location of modern industry in general, resources have declined in influence. This decline can be attributed to the following:

(i) Improved transport facilities.
(ii) More use being made of what was heretofore regarded as waste.
(iii) Greater sophistication in industrial products which are multi-resource, i.e. use many raw materials in their production and as all these are rarely found in one area then raw materials exert less of a pull.

As a generalisation however, it is still true to say that primary manufacturing tends to be resource-orientated while secondary manufacture is less conscious of this pull effect.

2. Power

Power is essential to all manufacturing industries so it has played a decisive role in the history of industrial location. Before the Industrial Revolution, the use of hand-power allowed manufacturing to be decentralised. The use of water power in the eighteenth century gave a particular significance to areas which possessed waterfalls and rapids and much of the settlement and industrial development in the east of the USA is associated with the 'fall line' location of towns such as New York, Boston etc. The invention of the steam engine, however which was directly responsible for the Industrial Revolution dramatically changed the spatial face of industrial location.

‹ Fig 17.5
ESB Power Station at Poolbeg

As large quantities of coal were needed to generate steam power and because coal is bulky and heavy and so costly to move, coal-mining areas became synonymous with industrial location.

Factors Influencing Manufacturing Industry

The exhaustion of the more accessible seams, the increasing use of oil and gas, the development of power grids and the improvement in transport all combined to dethrone 'King coal' and release the 'iron grip' that these areas exerted on industrial location. Certain industries however are greedy users of electricity and as a result the location of the electro-metallurgical and electro-chemical and wood pulp industries in Norway and Sweden are closely tied to areas of cheap hydro-power production. The decline in coalmining in general has resulted in major problems for many of the regions involved. Once thriving centres of industry, they are now economically depressed and with high unemployment are major problem areas of a maladjusted nature (see Chapter 21).

> Geographical inertia is the tendency of an industry to remain in the same place even though those factors which originally attracted it there may have ceased to exist.
>
> Geographical momentum is the tendency of places with established industries to attract other industries to the same location.

Fig 17.6
Industry On The Move: Power Supplies and Changing Location of Iron and Steel works

Time	Sources of Power	Examples of Location
Early Iron Industry	Charcoal	Wooded areas – The Weald
Later Iron Industry	Waterwheels	Fast-flowing rivers Sheffield)
Early Steel Industry	Coal	Coalfields (South Wales)
Present Day	Electricity	Coastal – Port Talbot

3. Markets

The sole purpose of manufacturing industry is to produce commodities for sale. The ability to identify and satisfy market demands is central to industrial development. In identifying a particular market then an industrialist must take into account:

The Quantity of a Market

This refers to the actual numbers of people in a particular area.

The Quality of a Market

This refers to the purchasing power of the people.

The distinction is important. Third World countries offer poor market outlets. Even though many are heavily populated they have little purchasing power. Developed countries on the other hand provide excellent 'quality' markets as their standards of living are high and their purchasing power considerable. Once a market has been identified the choice facing an industrialist is whether to locate the industry within the market place or elsewhere, say at the source of the raw material.

Generally speaking, industries producing the types of product below are *market orientated*, i.e. they locate in or near the market place.

Leaving Certificate Geography

i) Perishable Commodities

There is no market for stale bread or 'old news' so industries such as bakeries and newspapers are located in or near the market place.

ii) Bulky and Weighty Products

When a finished product occupies a lot more space in comparison to the sum of its components it makes economic sense to set up the industry in or near the market place. The automobile industry whose finished product (i.e. cars) occupies up to four times the space as that occupied by its parts, is located in the market area – hence the Japanese assembly plants in Britain and Europe.

The drink industry too whose finished product is both bulky and weighty, i.e. 'weight gain' is to be found in market areas since water is readily available.

In short then, the location of some industries in terms of perishability, others in terms of bulk and hence increased transport costs, are particularly sensitive to 'distance decline' (or 'distance decay') and the market place is the only feasible location from an economic viewpoint.

4. Transport

Fig 17.7 Road Network

Transport is an essential part of production. Its importance can be gauged from the fact that much of the tug-of-war that exists in industrial location between resources and markets is directly related to transport costs. Transport plays the following role in industry:

   (i)   Brings the raw material to the factory.
   (ii) Delivers the finished product to the consumer.
   (iii) Contributes to the overall costs of the finished product.

It is the cost of transport, for example, which is responsible for industries such as pulp and paper-making, being resource-orientated, as there is considerable weight loss involved in the finished product. Car assembly, on the other hand is market-orientated as the finished product is far more bulky than its components.

The cost of transport is related to the following factors:

(a) *terminal costs*, i.e. handling charges which are unrelated to distance.

(b) *haulage costs* or long haul costs which are directly related to distance.

The advantages and disadvantages of different forms of transport are discussed in Chapter 19.

5. Water

Fig 17.8 ›
Brewing: A Thirsty Industry

As water is readily available in developed economies, its importance as a factor influencing the location of modern industry is often overlooked. Apart from the fact that some industries depend on water for their power supplies and in industries such as brewing it is used as an actual raw material, water serves industry in a number of ways. Industry uses water to cool machinery, to wash raw materials and to dispose of effluents, etc. The quantity and quality of water are important to industry in general.

i) Quantity

To thirsty industries such as metal smelting, oil refining and pulp and paper making, the actual quantity of water available in a given area is vital. That one tonne of steel needs 60,000 gallons of water in its production helps to explain the waterside location (rivers, lakes, sea) of many of these plants. An electricity plant too which generates 300 megawatts can demand 12 million gallons an hour for cooling.

ii) Quality

The quality of water is often more important than its quantity to some industries. The soft water from the Pennines was of particular advantage to the Yorkshire woollen industry as soft water (being free of lime) facilitates lathering and helps the removal of grease from the wool. The quality of water is also significant to industries such as distilling, brewing and food processing.

The water requirements of modern industry then, both in terms of quantity and quality, are significant. For developing economies, many of which lie in semi-arid climatic regions, the expansion of industry faces obvious difficulties. In developed economies too the demand on water supplies is growing and the allocation of existing supplies is already a major problem in areas like California.

6. Labour

As a factor governing the location of modern industry, labour is difficult to evaluate. Its mobility together with the increasing automated nature of modern industry, certainly has reduced its importance. Its contribution however to the overall cost of a particular product (up to 60 per cent) is significant, gauged from the importance of the wage negotiations to which we are now accustomed. The high cost of labour also accounts for the introduction of mechanisation to reduce human input and the exploitation of female labour and 'cheap' Third World workers.

Manufacturing Industry

As a factor influencing the location of modern industry an industrialist has to take the following factors into account regarding labour:

(a) Quantity

Generally speaking, where labour is in plentiful supply, wages tend to be low, as workers have little bargaining power. Large numbers of workers, however are not the only consideration.

(b) Quality

The health of a workforce is of vital importance. This is particularly relevant in the case of developing countries, where, while large numbers are available for work, the level of output is often well below that of developed economies.

In spite of increasing automation, the level of skills is also important. Industries such as pottery and watch-making, etc. still demand a skilled workforce. Tradition is also relevant and though strictly speaking each worker has to learn a particular skill, certain regions have developed a certain 'atmosphere' which is more conducive to the development of those skills, e.g. steel in Sheffield, crystal in Waterford.

(c) Wage Demands

As wages can often account for a significant portion of the overall costs of a finished product, the actual wage demands of any workforce is an important consideration. In developed economies with high living standards the wage demands of even unskilled labour are considerably higher than those in developing economies. Hence the shift and growth of light engineering away from traditional industrial regions to centres in the developing world like Hong Kong and Singapore for products such as radios computers, toys etc.

(d) Industrial Relations

Skill and wage demands aside, a region with a poor record of industrial disputes will be avoided, as with keen competition even the shortest strike can jeopardise the whole future of any business. Hence the interest of governments and employers in national wage agreements which allow employers to plan for future costs and creates a greater stability, minimising the likelihood of strikes.

(e) Output

Whether fact or myth, some workforces such as those of Germany and Japan have built up a reputation as a highly efficient labour force renowned for both the quantity and quality of their labour force. Increased output lowers unit costs and makes products more competitive A reputation for quality also gives some industrial products a greater share of the consumer market.

Developing economies face major difficulties in attracting industry from the viewpoint of labour. Quantity (i.e. numbers) apart, they lack the tradition, experience, training, skills and efficiency associated with major developed economies.

In the scientific and technological age in which we live, it is understandable that the importance of geographical factors has declined. Many industries can now be described as *footloose* (not tied to any particular factor, geographical or otherwise). The changing demands of twentieth-century society, the development of substitutes for traditional raw materials, increased efficiency in transport, the spread of electricity and the mobility and changing skills of labour demanded by modern industry have all contributed to their decline. Despite their decline, modern industry can ill-afford to ignore geographical factors. The vast quantities of water needed for cooling in nuclear power stations continues to tie that industry to areas of abundant water supplies. In spite of the spread of electricity, its cost will remain a considerable influence and the electro-metallurgical and electro-chemical industries will continue to respect areas where power is cheap.

7. Capital

Capital is essential to all manufacturing industries, irrespective of size. Its presence in England in the eighteenth and nineteenth centuries was a considerable advantage to the growth of the Industrial Revolution in that country. Its availability in such large urban centres such as London, Paris, and New York helps to explain the industrial growth of these cities. Lack of domestic capital in developing economies on the other hand poses major problems for industrial development in those regions.

Today, in developed economies, capital enjoys a greater mobility and in consequence it has declined as a locational influence on industrial development. Its costs and availability remain important and the inability of many industries to service borrowed capital in the recession of the 1980s in Ireland and elsewhere resulted in their eventual closure.

8. Government Policies

In view of the contribution made by industry to the economic development of a country, it is understandable that governments should and do take an active role in decisions involving both the type and location of modern industry. Unlike the nineteenth century, most governments today take a more active role in decisions in relation to industrial location. This change has been made possible on the one hand, by the decline in the importance of geographical factors in deciding the location of industry and on the other, by a growing awareness of the shortcomings of the 'free enterprise' (*laissez-faire*) system. While the extent to which government policies influence the type and location of modern industry varies from one country to another, all governments try to create an economic balance and so aim for a more equitable distribution and diversification of industry.

Dynamic Regions

Dynamic regions which experience continual industrial growth often suffer from problems such as pollution, traffic congestion, etc.

> Agglomeration is when several industries choose the same area as their location in order to minimise costs.

The establishment of new industries, or expansion of existing ones are often discouraged by the imposition of disincentives (additional taxes, etc.) (see Chapter 21).

Manufacturing Industry

Maladjusted Regions

Regions which were once the very nerve centre of economic growth but have now lost their industrial magnetism (often due to the exhaustion of natural resources, e.g. coal) present a more difficult problem. Regional policies often include the development of new industries which can utilise the existing infrastructure (i.e. road, water, power, etc.) and the labour force with the aid of retaining programmes (see Chapter 21).

Peripheral Regions

The problem facing peripheral regions is that they lack industrial development and in consequence industrial investment is encouraged by various incentives such as grants, low interest loans etc. (see Chapter 21)

In conclusion then, it can be said that the general idea behind government policies relating to industrial development is two-fold.

1. To decentralise industry, i.e. to create a greater regional balance and avoid the concentration of industry in any one area.
2. To diversify industrial production, i.e. to develop new products, which while serving to reduce imports helps to create a greater self-sufficiency.

Environmental Impact of Industry

Unlike the nineteenth century, there is a greater awareness in many countries of the environmental impact of any industry. As industrial complexes are designed to be functional, they rarely make an aesthetic contribution to an area. Consequently, many governments and local authorities have introduced the need to obtain planning permission and in some countries zones are designated for new industries while other areas are set aside as 'green belts' where industrial development is prohibited.

'Out of sight, out of mind' is not a concept that can be applied to industry. Apart altogether from the aesthetic effects, many industries are a source of pollution of the air, water or simple noise. Laws governing effluent disposal, which can have serious ecological repercussions already exist in some countries, though the actual enforcement poses problems of another kind. As the treatment of effluents can often be extremely costly, firms who are not permitted to dump their waste into nearby rivers may decide against an otherwise ideal location.

There is therefore a considerable variety of factors which influence the location of modern industry in developed countries. Few industries ever attain the optimum location, if indeed such a location exists. Developing economies on the other hand are faced with major disadvantages. With limited domestic markets they are separated from the major international markets in the developed world by thousands of miles. Add to this an unskilled labour force, shortage of capital, lack of water (seasonal droughts) and underdeveloped infrastructure and the problems facing industrial development appear almost insurmountable.

A Price for Progress

At the beginning of this chapter we said that industrialisation while improving the living standards of the countries concerned has commanded a high price. It is interesting at this point to examine some of the less attractive side effects which result from industrial progress.

Future historians who wish to summarise the latter half of the twentieth century are unlikely to omit the word 'pollution' from their work. Apart altogether from the fact that manufacturing industries are devouring *stock resources* (forests, coal, oil, minerals) at an unprecedented rate, they are seriously jeopardising *flow resources*, some of which are vital to the existence of life on this planet, e.g. air and water.

Air Pollution

The amount of carbon dioxide we pump into the atmosphere has been increasing at an alarming rate since the Industrial Revolution. At present, about 20,000 million tonnes of carbon dioxide are pumped into the atmosphere each year. In Europe alone, 44 million tonnes of sulphur dioxide and 15 million tonnes of nitrogen oxides are pumped into the atmospheric dustbin. While the longer term effects of these gases are related to the greenhouse effect and acid rain, the most immediate effect is smog which can have devastating, if localised, effects. While government restrictions on industrial emissions and the domestic burning of fossil fuels are to be welcomed, atmospheric pollution on a global scale is on the increase as a result of industrial emissions. Local restriction on the pollution of a flow resource are of little overall significance. Breathing polluted air, homemade or otherwise is a high price to pay for an industrial wage packet.

Water Pollution

Pollution of water by industry, though often less visible and indeed sometimes invisible, is equally alarming. A nuclear power plant producing 1,000 megawatts may need up to one million gallons of water per minute in its cooling operation. Used water from those plants may be totally free of contamination but it enters seas and rivers at a considerably higher temperature. Increasing water temperatures like this has a considerable effect on salmon and trout spawning and results in bringing young fish prematurely into an environment in which they are unable to survive. Two million tonnes of crude oil, 4,000 tonnes of lead and 100 tonnes of mercury are discharged each year into the Mediterranean.

Effluents like mercury can quickly build up to serious levels though the amounts may seem harmless. Fish consume this mercury and are unable to pass it or break it down and it is collected in the tissue and passed on to humans at the end of the food chain. In 1953 in Japan, mercury poisoning in fish killed 52 people and, attacking the nervous system, crippled 100 more.

The price to be paid for industrialisation doesn't however end with pollution. Modern industry rises and falls as if guided by some economic Darwinism. Industries which set up in a particular region invariably bring a higher standard of living to the workers (and their families) in those areas. Their collapse however often results in a trail of depression, both social and economic. It is small consolation to the thousands of unemployed in the Nord region of France or the Sambre-Meuse area of Belgium, to be told that they now live in a maladjusted region.

The World's Major Industries

Before examining some of the world's major industries it is helpful to remember that in the broadest sense all industrial production can be classified as 'heavy' or 'light'.

Heavy Industries

Certain industries are classified as heavy when their finished products are literally heavy in terms of actual weight. Many of these, though certainly not all, often involve the primary processing of raw materials. Typical examples would include the iron and steel industry and the chemical industry.

Light Industries

Industries on the other hand whose products are literally light in weight are considered as light industries. As a general rule these industries often use partially processed 'raw materials'. Examples of light industries would include food processing and textiles.

A useful exercise while examining the following industries is to try to establish what factors exercise the greatest 'pull' on their location, i.e. are they 'resource-orientated', 'market-orientated' or more footloose in distribution – and also – why?

HEAVY INDUSTRY

Iron and Steel

◁ Fig 17.9
Centres of the Steel Industry

That the iron and steel industry is a symbol of industrial progress can be gauged from the fact that, with one or two exceptions, its growth has closely mirrored that of per capita wealth. Although the early iron and later steel industry was tied to raw materials, modern integrated iron and steelworks have adopted new locations as the sources of ore and energy have changed over time. Initially, iron-making was sited where there were outcrops of ore on the earth's surface and abundant wood for use as charcoal. This explains the presence of early iron works near forested areas such as the Forest of Dean (Britain). In fact, Sweden once produced one-third of the world's iron output. In 1709, Abraham Darby discovered that coke could be used to smelt iron ore. At that time it took eight tonnes of coal and four tonnes of ore to produce one tonne of iron so new furnaces were located on coalfields. The first areas to develop were in South Wales and the West Midlands (Britain).

As the advanced economies began to mature at the end of the nineteenth century, scrap iron and steel became available (from old plants and machinery). New methods of production which reduced the amount of coal needed, together with the exhaustion of ore mines has seen new integrated steelworks located on coastal sites, e.g. Port Talbot, Newport, etc. (Britain). The expected flight of the iron and steel industry from the world's coalfields, however, failed to materialise, at least on the scale predicted. Many of the old steel producing regions still retain an important place among the world's steel manufacturers. This traditional location pattern can be explained by:

i) Geographic Inertia

Iron and steel works involve an immense capital investment and modernising existing plants is often far more economic than a switch to a greenfield location (i.e. one on a completely new site).

ii) Market Orientated

Iron and steel is essentially a primary manufacturing industry. In consequence it attracted steel consuming industries in view of the material's 'loss of weight'.

The combination of geographic inertia and geographic momentum guarantees the industry a certain locational stability.

The Chemical Industry

Fig 17.10

Manufacturing Industry

Though the chemical industry has witnessed a phenomenal expansion in the twentieth century, its growth is directly tied to that of the Industrial Revolution. Its world distribution pattern is closely linked with developed economies and its growth has surpassed that of all others. Today the USA leads the way, closely followed by Russia, Japan, Germany and the UK.

The chemical industry has its origins in the bleaching of cloth in the early stages of the Industrial Revolution. The Leblanc method of using common salt, limestone and sulphuric acid produced an alkali which not only hastened the process of bleaching but also helped in the manufacture of soap, glass and paper for which there was a growing demand. As salt was the heaviest of the raw materials the industry became resource-orientated.

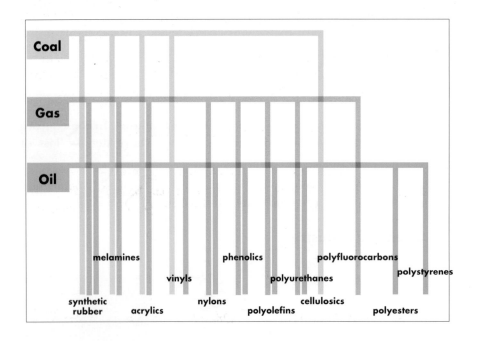

⟨ Fig 17.11
Products of the Chemical Industry

From this narrow base the industry expanded due to a greater variety of raw materials, e.g. coal tar, gas, oil and continued scientific research. The chart shows the variety of products which can now be extracted from coal, gas and oil by changing their molecular structure.

Location Pattern

Generally speaking the products from the chemical industry are expensive and as such, require a large, rich market. The high input of scientific knowledge and expertise further increases the pull effect of developed economies and hence its concentration in the 'industrialised world'.

Certain geographical factors however limit its location in these regions particularly the need for water as a raw material, as a coolant and for the disposal of effluents. The nature of the effluents also demands its isolation from built-up residential areas and large sites with portal access are also vital. Hence its concentration in the Cork Harbour region where 60 per cent of all chemical plants in Ireland are now located.

LIGHT INDUSTRIES

Motor Vehicle Production

Fig 17.12

There are now over 700 million motor vehicles in the world and the number is rising by more than 40 million each year. In the US alone, over 14 million people are employed in the motor industry or its many ancillary services.

| World's biggest motor manufacturers ||||||
Global ranking	Company	Revenues	Profits	Assets
		(in millions US dollars 1994)		
5	**General Motors** (Chevrolet, Pontiac, Oldsmobile, Buick, Cadillac, Opel, Vauxhall, Saab, Holden)	154,951	4,901	198,599
7	**Ford Motor** (Ford, Jaguar)	128,439	5,308	219,354
11	**Toyota**	88,159	1,185	98,037
20	Daimler-Benz (Mercedes)	64,159	650	60,365
23	**Nissan**	58,732	(1,672)	82,280
29	**Chrysler** (Chrysler, Dodge)	52,224	3,713	49,539
34	**Volkswagen** (VW, Audi, Seat, Skoda)	49,350	92	52,329
41	**Fiat** (Fiat, Lancia, Alfa Romeo)	40,851	627	59,128
45	**Honda**	39,927	619	34,708
55	**Mitsubishi Motors**	34,370	127	32,544
64	**Renault**	32,188	656	42,357
72	**Peugeot** (Peugeot, Citroën)	30,112	559	26,267
94	**BMW** (BMW, Rover)	25,973	427	24,971
	Source: Fortune, August 1995, plus other sources			

Fig 17.13

Leaving Certificate Geography

Manufacturing Industry

The US, Japan and Germany lead the world in automobile manufacturing. In the late 1930s, the United States accounted for about 80 per cent of world automobile production. Today that proportion is down to a quarter, but the top two US car makers, General Motors and Ford still sell more cars than anyone else. The biggest selling car in history however is the Volkswagen Beetle, the 'People's Car' whose cumulative sales have exceeded 21 million.

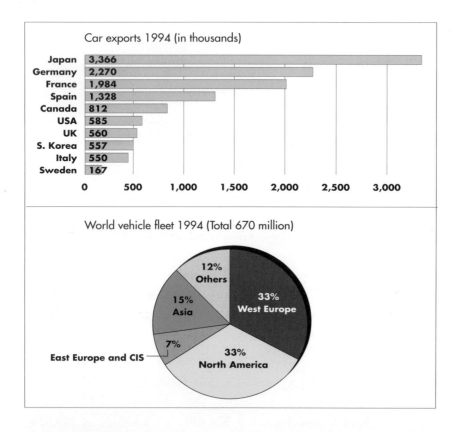

◁ Fig 17.14
Car Exports and Distribution

> **Factors influencing the location of the motor industry in Japan:**
> - The steel works are located around deep sheltered natural harbours.
> - The five major conurbations provide the workforce and large local markets for such steel-based products as cars.
> - The workforce is highly skilled and output per worker is estimated to be up to four times that of their European counterpart.
> - Well developed transport networks to allow the 'component' parts to be transported to the assembly plants when such a division between the two stages is necessary, e.g. Japan and Europe.

Because the cost of setting up tooling systems, basic car designs tend to persist for three to six years. It usually takes well over a billion dollars and at least five years to proceed from concept to production model.

The key to understanding the actual distribution of car manufacturing lies in the recognition of two actual stages of production, i.e. (a) the initial production of the component parts; (b) the actual assembly of these components.

The significance of this division lies in the fact that, as an assembled car occupies up to four times the space of an unassembled car, transport costs from the production area to the market place can be considerable. While the primary production stage may be resource-orientated, assembly is often market-orientated.

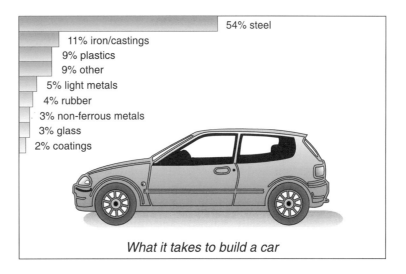

Fig 17.15 >
What it Takes to Build a Car

What it takes to build a car

USA:	Detroit (Ford, General Motors).
Europe:	Germany, Wolfsburg (Volkswagen), Cologne (Ford)
Japan:	Toyota; Nissan; Honda.
England:	Dagenham, Liverpool and the Midland Complex.
France:	Paris (Citreon; Renault, Peugeot).
Italy:	Turin, Fiat. UNO

Fig 17.16 >
Main Automobile Producers

The Textile Industry

Fig 17.17 >

Manufacturing Industry

The textile industry is widely distributed throughout the world. This distribution reflects the wide availability of raw materials, the slight weight loss if any, involved in production, the ease with which textiles can be transported and the undemanding nature of the skills necessary in their manufacture. Textile manufacturing provides an excellent example of a footloose industry.

History of Textile Products

The modern textile industry can be said to have its beginnings in the Industrial Revolution which produced the machinery necessary for mass production. The woollen industry of Yorkshire was the first to benefit, as it developed from a domestic industry based on local wool and the presence of soft water. In the nineteenth century, the cotton industry in Lancashire was revolutionised and Britain in that period monopolised the world's textile production. The twentieth century, however has witnessed a considerable diversity both in location and products. Many of the developing nations, developed their own industry after independence from their colonists. The combination of local raw materials (e.g. cotton) a cheap and plentiful labour force (the basic skills are easily learned) and minimum capital outlay all helped in its development. The manufacture of man-made fibres too, such as nylon, rayon etc., while primarily exclusive to developed economies, has eliminated whatever locational pull raw materials might have exerted in the past. Despite the use of modern machinery however, the industry still demands a considerable amount of human labour. A cheap hard working labour force is vital as labour costs can often exceed a third of the overall costs of the finished article. This helps to explain the competitive nature of products from the developing economies in the more advanced industrial countries.

Location Patterns

Though essentially footloose, markets tend to exert the greatest pull on the textile industry. New York for example has half the USA's clothing factories and a third of the industry's labour force.

The great textile centres in Europe are London, Paris, Milan and Rome. Markets alone haven't made these cities the leading textile producers. They are also the centre of fashion houses and, as trends change, it is vital that manufacturers stay in close contact with designers etc.

Locational Factors

1. Markets: Largely unaffected by geographical factors, textile industries set up in market centres, not merely for convenience but also to be in touch with fashion trends.

2. Cheap Labour Force: Much of the textile industry remains manual. A cheap hardworking labour force is essential. Large cities, themselves market places, often provide this cheap labour force, as immigrants often crowd into these areas. Jewish immigrants from Russia and Poland almost created the 'rag trade' for which New York and London are now famous.

3. Political Influence: Governments can have a considerable influence on the textile industry, in production limits if not in actual location. The restriction placed on cheap imports from the developing regions of South-East Asia by the European Union is a typical case in point.

Questions

ORDINARY LEVEL

1. Amongst the factors that influence the location of manufacturing industry are Markets, Labour, Government Policy, Transport, Raw Materials.
    (i) Select any **three** of the factors listed above and in the case of **each**, explain its influence on industrial location. In your answer refer to specific examples you have studied. *(60 marks)*
    (ii) Manufacturing industry has both positive and negative effects on the environment. Examine this statement. *(20 marks)*

    Leaving Cert. 1995

2. (i) Any farm **or** any manufacturing industry may be viewed as a system with inputs, processes and outputs. Examine this statement with reference to an example you have studied.
    (60 marks)
    (ii) Describe briefly the possible negative effects of **either** agriculture or manufacturing industry on the environment. *(20 marks)*

    Leaving Cert. 1994

3. (i) Any manufacturing industry may be viewed as a system of inputs, processes and outputs. Explain this statement with reference to any industry which you have studied. *(60 marks)*
    (ii) Examine some of the ways in which modern manufacturing industries are responding to the growing global concern for the environment. *(20 marks)*

    Leaving Cert. 1992

HIGHER LEVEL

1. Manufacturing Industry has given way to the Service sector as a major employer of workers and has become increasingly 'footloose' in locational terms.
    (i) Explain the term 'footloose industry' and discuss the factors which might influence the location of **one** such manufacturing industry you have studied. *(40 marks)*

    Leaving Cert. 1995

2. Farms and manufacturing industries may be viewed as system with inputs, processes and outputs, all of which are inter-related.
    Select **either** Farming or Manufacturing Industry to examine the accuracy of this statement. *(100 marks)*

    Leaving Cert. 1993

3. (i) The distribution of manufacturing industry has changed over time. Older industries are still to be found concentrated in particular regions, while newer industries are more dispersed. Examine this statement referring to appropriate examples.
    (60 marks)
    (ii) Describe and explain how the development of manufacturing industry in developing economies may differ from the experience in developed countries. *(40 marks)*

    Leaving Cert. 1992

CHAPTER 18 — *Tourism*

< Fig 18.1
Doolin, Co. Clare

Tourism has experienced a steady growth since the nineteenth century and in the latter half of the twentieth century its growth has been little short of phenomenal. Spain for example has witnessed an annual increase in the amount of visitors from 100,000 in 1946 to 53 million in 1992. The number of tourists expected to visit the Mediterranean in the year 2025 will be 260 million. The variety of relief and climate together with a wealth of historical cultural and artistic heritage makes Europe the most popular tourist destination. It accounted for more than half of all tourist arrivals in the world, 64 per cent in 1990. Between 1980 and 1990 tourist arrivals in Europe increased by an average of 3.5 per cent per year: during the same period income from tourism increased by 8.3 per cent per year. In 1996 tourism accounted for eight per cent of Ireland's GNP courtesy of 4.6 million tourists.

Most of Europe's, and indeed the world's, tourists are city dwellers and the beneficial effects of getting away from life in the 'concrete jungle' are obvious. The effects on the regions to which they go are, however, less obvious. While the economic effects of this *invisible export* to a country's balance of payments are clearly positive, the side effects of mass tourism, especially on less economically developed regions can often be disastrous and irreversible.

The Growth of Tourism

Mass tourism is a twentieth-century phenomenon whose growth can be linked directly to the emergence of developed economies. This is borne out by the fact that at present 80 per cent of the world's tourist expenditure is accounted for by the USA, Canada, Germany and nine other European countries – all developed economies. Though the rich travelled to spas and other health centres as early as the seventeenth century, developed economies only made their appearance with the advent of the Industrial Revolution. It was the various improvements which followed in transport, wages and working hours that allowed tourism as we understand the term today to develop.

> A tourist is defined as a person who spends at least one night away from home on vacation.

REASONS FOR GROWTH OF TOURISM

1. Improved Transport

(a) Railways

The invention of the steam engine in the nineteenth century was mainly responsible for the growth of many of the seaside resorts with which we are familiar today. Railways not only enabled large numbers of people to travel long distances but they considerably reduced the cost of travel. In 1841, for example, it cost £1.50 to travel on the Mail Coach from Sligo to Dublin. In 1866 the same journey cost 37p by train.

(b) Cars

The invention of the motor car, and, in particular its mass production on assembly lines which considerably reduced its cost, greatly facilitated the movement of people to less accessible places.

(c) Air

The development of air transport has had a considerable effect on the growth of twentieth-century tourism. Combining speed with comfort, it has brought distant and often inaccessible regions within easy reach.

2. Increased Income

The wages of workers in industrialised economies have shown a steady increase in real terms since the nineteenth century. Increased spending power enables many people to spend more on recreational activities.

3. Improved Working Conditions

The reduction in working hours, together with the introduction of annual 'paid' holidays, has increased the leisure time for many people.

4. Lifestyle

While wages and working conditions have improved, the living and working conditions in the concrete jungle (see Chapter 12) place enormous pressures on the average person. This in turn has created a social climate which makes the need to 'get away from it all' a necessity as much as a luxury.

5. Education and the Media

Through education and the media (television, newspapers etc.) people today have a greater interest in cultural and historical attractions. The growth in educational school tours today provides a typical example.

6. Advertising and Promotion

Many foreign tourist resorts are household names as a result of constant advertising and travel programmes on television such as 'Wish You Were Here'. The growth in tour operators and charter flights has made foreign holidays cheaper and easier to organise.

7. Governments

The importance of the economic contribution from tourism to a country's economy can be gauged from the fact that most developed economies now have appointed a Minister for Tourism.

Government sponsored bodies too have been set up and Bord Fáilte was established in this country in 1955. In some countries government investment has led to a massive injection of capital and the Languedoc-Roussillon project in the South of France has completely altered the economy of that region. The European Union (EU) too has encouraged tourism as part of its commitment to create a 'People's Europe'. Apart from financial aid through the European Investment Bank and the European Regional Development Fund, the European Union has also reduced or removed many difficulties that hinder travel within its borders, e.g. passports, etc.

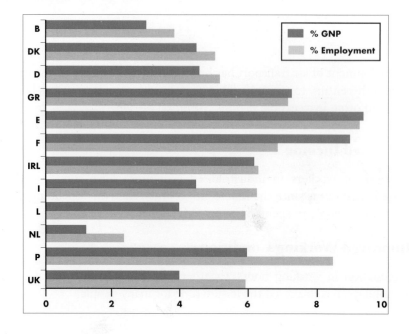

Fig 18.2
Economic Reasons for Government Support of Tourism

- Tourism is a major source of direct employment. The present target in Ireland is to raise the present number to 120,000 by the year 2000.
- Tourism has a major multiplier or spin-off effect. It benefits all economic activities from farming to transport and as a result creates additional employment.
- As an 'invisible export' it brings in large amounts of foreign currency which helps the balance of payments and provides vital capital to aid development in less developed regions, e.g. the Mediterranean.
- Tourists are often attracted to peripheral regions, e.g. the west of Ireland, the south of Italy. These regions are remote with poor socio-economic conditions. This isolation, coupled with a poor infrastructure and lack of raw materials, does little to attract secondary industrial development. The growth in tourism therefore helps to diversify economies which are often over-dependent on the primary sector.

Tourist Attractions

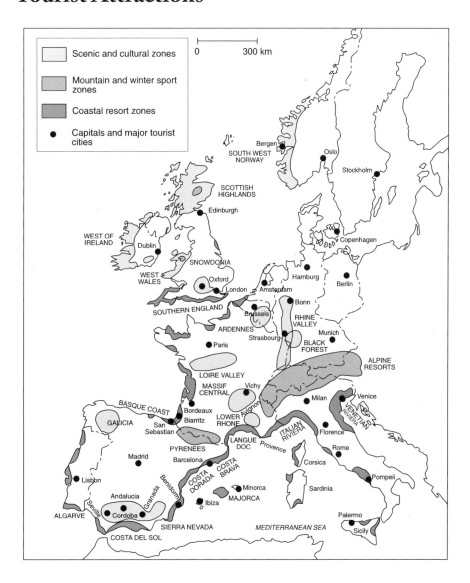

Fig 18.3
European Tourist Areas

The Physical Environment

1. Relief and Drainage 2. Climate 3. Vegetation

1. Relief and Drainage

Relief

< Fig 18.4
Grand Canyon, Arizona

Relief is a major tourist attraction. The presence of long, sandy beaches on the east coast of Spain helps to account for the litoral (on the shore) nature of much of Spain's tourism. Inland too, the rugged relief attracts tourists who enjoy holidays of a different kind. The spectacularly glaciated uplands of Switzerland, the fiorded coast of Norway, the Niagara and Victoria Falls, the volcanic mountains of Vesuvius, Etna and Fuji, the Rhine Gorge and Grand Canyon all attract thousands of tourists each year.

Drainage

It was the presence of water which led to the early development in modern tourism.

< Fig 18.5
Kylemore Abbey, Co. Galway

The beneficial health effects of mineralised water (spas) were realised by the medical profession as early as the seventeenth century, and hence the growth of spa towns such as Bath and Tunbridge Wells. In the west of Germany today 400 of the 1,300 resorts are officially registered as health resorts. The growth in coastal tourism too is directly related to the presence of the sea, which apart from climate is probably the single most important physical tourist attraction. Apart altogether from its beneficial health effects, there has been a massive growth in aquatic sports such as swimming, cruising, canoeing, angling, water skiing and sub-aqua diving. Inland waterways are also important; many find complete relaxation in cruising inland waterways or angling for an entire holiday, e.g. on the Shannon. Water has enhanced the scenic beauty of countries such as Norway, Ireland (Killarney Lakes), the Lake District of England and the Alpine slopes of Northern Italy.

Climate

Fig 18.6 > Mediterranean Beach

The attraction to hot climatic conditions is a twentieth-century phenomenon. In the nineteenth century, milk-white skin was considered an attractive physical attribute. Today the bronzed look is in vogue and has given rise to a 'cult of the sun'. This is evident too in the growth of sun-beds, quick tanning oils and even pills which produce a bronzed skin colouring.

Climate exerts the single greatest influence on tourism. This is evident by the fact that the regions which have experienced the greatest growth are those where the summers are hot and dry. Mediterranean regions which lie under the influence of high pressure in summer guarantee these sun-drenched conditions and are often in marked contrast to the more changeable weather conditions experienced in the higher latitudes of Western Europe (where most of the tourists come from). The growth of winter sports tourism in countries like Switzerland, Austria and Andorra is directly related to the climatic conditions which blanket these regions in snow, and centres such as St Moritz and Zermott are now major winter sports resorts.

Vegetation

As a component of the physical environment, vegetation exerts a small but significant influence on tourism. Besides the recreational facilities however such as walking, camping and picnicking which forests offer, flora is home to most of the world's fauna. The interest which wild animals generate can be gauged from the numerous wildlife parks already in existence in many countries including Ireland (Fota in Co. Cork), and the growth in popularity of Safari holidays.

Tourism

Fig 18.7
Safari Holidays are on the Increase and can provide Valuable Tourist Revenue for Developing Countries

Human Inputs

Historical and Cultural

Fig 18.8
The Parthenon in Athens

Historical and cultural influences are major tourist attractions. Archaeological remains of castles, cathedrals and monuments are to be found in abundance throughout Europe. In this respect Mediterranean countries such as Italy and Greece are littered with historical, cultural and artistic attractions making Europe the greatest 'open air museum' in the world. All the great historical periods are represented by buildings which now stand as tombstones to dead civilizations.

Great cathedrals like Notre Dame (Paris) testify to a period in which the Catholic Church played a more dominant role and famous museums like the Louvre (Paris) and Prado (Madrid) house the great works of the artistic giants of the Renaissance period.

Historical variety is equalled by the cultural diversity evident in the national dress, music, dance and language of different countries.

Facilities

Fig 18.9 > Hotel Complex under Construction

Many of the regions which are now experiencing a tourist explosion remained inaccessible and undeveloped until recently. The capital investment by many governments and individuals in improving the infrastructure, building apartment blocks, hotels, swimming pools, etc., has played a major role in their growth. In spite of the fact that many Mediterranean countries are relatively undeveloped, their tourist resorts are like 'oases of development' offering tourist facilities which are on a par with those from the 'source area' (where the tourists come from).

Cost of Living

Unlike previous centuries when only the rich could afford to travel long distances, international tourism is now available to most classes. Besides the attractions offered by the physical environment many Mediterranean countries are less developed and, in consequence, their cheaper cost of living is an added bonus to foreign tourists.

It should be remembered however that many of the physical and human attractions offered by the Mediterranean are not exclusive to that region. The Mediterranean climate is found on the American and Australian Continents which also have their own unique cultural attractions. The phenomenal growth in Mediterranean tourism, expected to reach 260 million by the year 2025, is explained by its proximity to the large affluent market of the nearby

European Union where people treat holidays more as a necessity than a luxury. It is not uncommon for people to take two, or even three, holidays a year. This, together with the ease of access to the Mediterranean by air, rail and road, makes concerns that the region is fast becoming a 'Paradise under Pressure' more understandable.

Effects of Mass Tourism

It is difficult to calculate the exact effects of mass tourism on any given region or country. What can be said however is that mass tourism does have major economic physical, social and cultural consequences in every tourist region for good or bad.

These effects may be:

- Economic
- Physical
- Social
- Cultural

1. ECONOMIC EFFECTS

Because tourism is labour intensive, its capacity for direct employment in accommodation (hotels, guest houses) catering (cafés, restaurants), entertainment (fiestas, bars, discos) and transport is enormous. As an economic activity too its 'spin-off' or multiplier effect is unequalled. Agriculture and fishing benefit due to the increase in demand for food from the influx of people, while craft industries enjoy an increased demand for souvenirs and clothes. The construction industry which employ numerous tradesmen (plumbers, carpenters etc.) benefit from the growth in the construction industry. The tertiary or service industry receives its own spin-off with the demand for photographic and sports equipment, while cars, mopeds and bikes are in constant demand. The economic importance of tourism can be gauged from the fact that it forms the single biggest industry in Florida, accounts for over 40 per cent of the total value of Spanish exports and directly employs over five per cent of the Swiss labour force. It also forms the single biggest growth sector in the Irish economy and presently contributes eight per cent of GNP.

Its economic effects however are not altogether beneficial. As an economic activity, tourism is essentially seasonal. In the EC countries in 1994, for example, over 60 per cent of all holidays were taken in the months of July and August and for Mediterranean countries that figure rose to over 70 per cent. As a result, most employment is part-time and because the local labour force often exceeds demand, wages are low and hours are long. The influx of tourists too inevitably results in inflated prices for food, drink and even land which local inhabitants often find difficult to afford.

Many of the tourist complexes are foreign owned and as a result, profits are often taken from, rather than invested in, the region. The most damaging economic effects however are less dramatic but more serious over a longer period of time, especially in less developed regions. The traditional economy of these regions is often based on agriculture. As tourism grows a dual economy of agriculture and tourism develops. As tourism increases, the local economy becomes as dependant on tourism as it once was on agriculture. But tourism is notoriously unreliable and the absence of snow in mountainous areas or the outbreak of sickness in a Mediterranean resort can spell economic ruin.

2. Physical Effects

The predatory nature of mass tourism is perhaps best highlighted by the physical scars it leaves on its victim areas. Tourism has the uncanny ability to select the most vulnerable areas, usually the unspoilt and least developed. Dramatic changes take place in the infrastructure with improvements in roads, airports, water supplies, electricity and sanitation. Hotels and apartment blocks spring up, crowding the area as they compete for land space with bars, restaurants and discos. Loud music and flashing neon lights complete the transformation from a quiet sleepy village to a miniature Las Vegas. Little thought is given to the aesthetic effects which such monstrosities create. Sanitation facilities are rarely able to cope with the tourist invasion and consequently most sewage enters the nearby seas untreated, only to return to the beach via the tide and destroy the very attraction on which tourism depends. Pollution is not confined to the sea. Littered beaches, trampled vegetation, overcrowding, noise and air pollution (cars, mopeds etc.) invariably accompany mass tourism.

The physical effects of mass tourism are not altogether detrimental. The improvements in transport (road, rail, air), social facilities such as swimming pools and communication services (telephone), benefit both the local inhabitants as well as the tourist.

3. Social and Cultural Effects

Fig 18.10
Flamenco Dancing

The social effects of mass tourism in a region are harder to measure than either the economic or physical effects. One reason for this is that the social effects are slower to surface and are often less dramatic in their immediate effect. On the positive side tourism provides many jobs both directly and indirectly. This is particularly significant in less developed regions whose economy is often based on subsistence primary activities. The seasonal nature of many tourist resorts means that employment is seasonal, which in itself creates many problems. Local people employed in menial jobs are often blatantly exploited, while the co-existence of local poverty and tourist affluence does little to create social stability. In consequence many of these regions rank high among those associated with violence and crime.

On the positive side however local customs may benefit from an injection of outside interest and create a type of cultural renaissance in a region. The reverse however is also possible and the demands created by mass tourism, often reduces local culture to 'stage level'.

Tourism – The Future

In view of the fact that the 'tourist' is essentially associated with developed economies, and that over 50 per cent of Europeans still do not take an annual vacation, its growth potential is enormous. This viewpoint is further enhanced by the fact that as developed economies enter the technological age, more free time will inevitably follow. In order that both the tourist and the tourist region gain the maximum benefit, recreational planning is essential. It is vital that infrastructural development precedes the mass influx of tourism. Greater consideration must be given, from an aesthetic viewpoint, to the indiscriminate erection of apartment blocks and hotels. No less important in the longer term are considerations towards social justice, so that the economic benefits and facilities at present enjoyed by a small section of the population are made available to all.

Recreation planning too must take into account that every region has its tourist quota, i.e. the optimum amount of people which a region can adequately care for, irrespective of its stage of development. When this is exceeded it invariably leads to physical, social and cultural decay which far outweighs the short-term economic gains.

Case Studies

1. The Mediterranean
2. Switzerland – Tourism in mountainous areas
3. Cities
4. Ireland – Tourism in the countryside

1. THE MEDITERRANEAN

Until the 1950s and 1960s most of the people who holidayed in the Mediterranean were relatively wealthy. These were the only ones who could afford the cost of travel and hotels. Tourists were few in number so their presence did little to affect the character of the environment in the regions in which they stayed. All this changed from the 1960s however with the arrival of package holidays and cheaper air travel. More and more people in Europe had jobs that provided an increased amount of paid leisure time and preferred to spend their holidays in the 'frying pan' conditions of Mediterranean sunshine than in local resorts where the weather was unreliable. New motorways made travel to Southern Europe easier and quicker. Visitors from around the world came to enjoy the physical and cultural attractions of the world's largest 'open air museum'. The result was a dramatic increase in the number of holiday makers travelling to countries like France, Spain, Italy and Greece. Spain alone had to cope with over 55 million tourists in 1996 while the Mediterranean is expected to attract 260 million by the year 2025. When we realise that between 70 per cent to 90 per cent of these numbers are attracted to coastal locations it is understandable that the Mediterranean is fast becoming a 'Paradise under Pressure'.

Reasons For Growth in Mediterranean Tourism

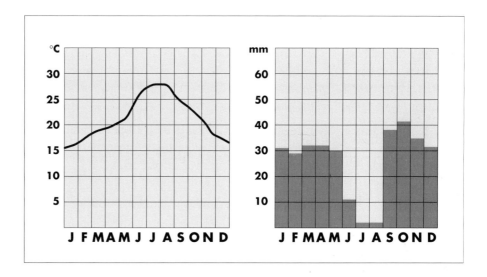

Fig. 18.11 >
Mediterranean Temperature and Rainfall Chart

Climate

The Mediterranean climate with its hot dry summers and mild wet winters is the single greatest attraction for tourists. Worshipped for thousands of years by ancient civilizations, the sun is again in vogue as its cult followers 'head for the Med' where the high pressure and cloudless skies that dominate the region in summer guarantee hot, sunny conditions. The sun tan is the outward sign of conversion for these worshippers of the new 'bronze age' as they spend hours in adoration, sizzling in frying pan conditions, oblivious to the dangers of skin cancer.

Relief

In spite of its relatively small area, the topographical variety which the Mediterranean has to offer is staggering. Miles of sandy beaches usually sheltered from the westerly winds lie along the coast. The attraction of the beaches is further enhanced by the tideless Mediterranean with its almost lukewarm, blue waters.

Inland, the mountainous regions, studded with glacial lakes offer an unrivalled topographical utopia to scenic imbibers, while the snow-covered slopes of the Alps and Pyrenees attract more active holiday-makers.

Cultural

< **Fig. 18.12**
The Library of Ephesus in Turkey

The Mediterranean is often regarded as the cradle of European civilization. Ruins from the Greco-Roman period together with the great medieval and Renaissance cathedrals draw tourists from all over the world.

A number of other factors also combine to account for this phenomenal growth. Most governments in Mediterranean countries, realising the economic potential of tourism have taken an active part in the development of tourist resorts. Loans and grants are provided to encourage the construction of facilities such as hotels, apartments, etc. Infrastructural developments in roads, water, electricity and sewage all help to woo the visitor who is accustomed to facilities associated more with developed regions. The advent of package holidays too has dramatically reduced the cost of travel and accommodation while the price of food and drink, though increasing, still lags behind that of Northern Europe.

Pollution – The Price of Success

In all, 18 countries surround the Mediterranean Sea with a total population of over 350 million people. Of those 100 million live on the coast. This increases during the summer as another 100 million or 35 per cent of the world's tourists visit on their holidays. Because the entrance through the Strait of Gibraltar is only 13 kilometres (eight miles) wide the Mediterranean is essentially a lake (the word Mediterranean actually means 'the middle of land') and this together with the lack of tides makes it easy to pollute.

Sewage

While sewage adds nutrients to the sea, its purification removes oxygen which results in disease. Typhoid, cholera, dysentry and hepatitis are now common in many areas.

> *Annual Pollution in the Mediterranean*
> - 85 per cent of all sewage is untreated
> - Two million tonnes of crude oil
> - 60,000 tonnes of detergents
> - 4,000 tonnes of lead; 100 tonnes of mercury

Fig. 18.13 The amounts which enter the Mediterranean each year

Oil

One quarter of the world's oil pollution occurs in the Mediterranean. The major source of this is tankers washing out their tanks at sea and the 60 oil refineries located on the coast. Apart from affecting watersports and beaches, oil has been found in the tissue of fish and shellfish.

Industrial Waste

Financial gains from tourism have resulted in industrial development – much of its waste being dumped untreated into rivers which flow into the Mediterranean. The large port of Fos to the west of Marseilles has a big industrial complex. Much of its waste goes into the sea.

Agricultural Waste

Increased tourist numbers increases the demands on agriculture and hence the use of fertilisers increases. Today phosphates and nitrogen which run off from the land and into the sea create problems on beaches. They encourage the growth of algae. When the algae die they are washed onto the beaches and rot causing a foul smell.

The simple fact now is that the 'world's biggest open air swimming pool' is fast becoming the 'world's biggest sewer'. At present more than 28 per cent of the major resorts are unsafe for bathing. Many species of fish like mussels and oysters are almost totally unsuitable for human consumption while the danger of dysentery is rising at an alarming rate.

A Solution – Mediterranean Action Plan

In 1982 the United Nations devised the Mediterranean Action Plan. This was a series of strict anti-pollution laws designed to reduce the waste entering the sea. Traditional national rivalries were set aside in favour of a concerted action to clean up the Mediterranean. Eighteen countries signed the plan.

The Mediterranean Action Plan
- Naples, Marseilles, Athens plan new sewage works.
- Fos, the industrial port in the south-east of France has reduced its discharge into the sea.
- Mercury and DDT are completely banned.

- Long-term monitoring of pollution is carried out at over 80 sites.
- Other harmful substances such as chromium and arsenic require special licences and are strictly controlled.

While it is in the interest of each of the 18 countries to guarantee the future of the Mediterranean, few can doubt that the action plan (as was the Blue Plan of 1979) is well intentioned, the stark fact remains that tourism, as with other sectors of the economy, is primarily developed to create profit – and a short term one at that. To solve the present sewage problem alone almost £3,000 million pounds is needed. The obvious question is 'who will pay?' – especially when the poverty of the countries involved is taken into consideration. More important, can the developing North African countries, whose tourism is still in its infancy, be expected to limit potential tourism growth to compensate for the excesses which have resulted from the more developed tourism of countries such as Spain. The Mediterranean Sea provides one of the classic dilemmas facing twenty-first century development – will it be profit at all cost? Only time will tell.

2. Switzerland: Tourism in Mountainous Areas

Tourism was already a growing economic activity in Switzerland when Thomas Cook organised the first package tour there in 1863.

Its original attraction was as a health resort and even today many still go to breathe in the pure mountain air. Towards the end of the nineteenth century however, its reputation as a winter playground began to establish itself as skiing became popular. Today tourism is a vital cog in the Swiss economy as it:

⟨ **Fig. 18.14**

- Employs up to 160,000 people – over five per cent of the total workforce.
- Attracts more than 10 million people.
- Accounts for almost 10 per cent of foreign currency earnings.

Advantages For Tourism

1. Accessibility

The central geographical position that Switzerland occupies on the European mainland is the most important factor which accounts for the growth of the Swiss tourist industry. As a land-locked country it is surrounded by densely populated affluent countries, e.g. Germany, Italy, France etc. Although it is a mountainous country it is easily accessible by road, rail and air and is particularly attractive in winter as the Mediterranean enters its rainy season.

2. Physical

Fig 18.15
Ski Slopes in Switzerland

In scenic beauty Switzerland has few rivals on the European mainland. The Alps, apart from their scenic value boast Europe's highest peaks many of which exceed 4,000 metres and offer obvious challenges to mountain climbers. Elsewhere the more gentle slopes provide some of Europe's finest ski resorts.

Distance from the sea together with the effects of altitude ensure that temperatures in winter remain below freezing with an adequate snowfall, at least in the mountainous regions. As more than one third of all visitors now arrive in winter, snowfall is essential and its absence in lower mountain resorts can have disastrous results for the economy of these regions. St Moritz, Zermatt and Davos are the principle winter resorts. Elsewhere the scenic beauty, tranquillity and refreshing purity of the air make lakes like Lucerne, Geneva and Thun major attractions.

3. Human Input

The Swiss Government is concious of the catastrophic effects tourism can have on the environment and now enforces strict planning regulations. The general tendency is to concentrate on quality tourism rather than quantity (mass) tourism, while the buildings must be integrated into the local landscape. As a result, Switzerland can now boast of some of the world's finest (and most expensive) hotels. Access to the ski slopes is by cable cars and ski lifts (aided by the development of HEP) while the ski resorts themselves provide equipment and expert tuition. Apart from skiing, the resorts offer other activities such as golf and horse-riding, while night time activities are centred on the casino or numerous cosy bars, night clubs and discotheques.

Tourism is a serious business for the Swiss and at present there are more than 1,600 hotels as well as many thousands of guesthouses and chalets. The country has been divided into ten regions by the tourist authorities and the tourist year is divided seasonally into summer and winter.

Tourism

The Effects of Tourism in Mountainous Regions

Tourism in the mountainous areas of Europe affects both the environment and the lives of the local people. The main problems result from the increasing numbers of tourists who create a demand for more ski lifts, hotels, restaurants, self-catering chalets and ski-runs. In order to build these and other facilities, areas of forest often have to be cut down. Trees however are vitally important on many mountain slopes. Their roots bind the soil together and their leaves and branches cushion the impact of rain. In Italy and Switzerland, hundreds of square kilometres of forest have been destroyed. Once the trees have gone the soil on the slopes becomes unstable and mud slides are common during heavy rains or when the snow melts.

Large mud slides can destroy buildings, cut railway lines, block roads, bring down power lines and wash away bridges. This happened in the 1980s in Italian valleys, affecting towns such as Tartano, Sondrio and Bergamo near the Swiss border. Now trees are being planted as a matter of urgency to make slopes safer.

The increase in hotels and apartment blocks often results in the local people becoming outnumbered by the visitors. This can dramatically change the character of traditional villages where local traditions of dress, food, dance and music are reduced to the level of tourist 'shows'. Competition results in rising land prices. Local people then can often no longer afford to buy land as they are priced out by large hotel groups.

On the positive side however, the improvements in infrastructure, e.g. new roads, railways, sewage and electricity are major benefits to local people in remote places. It also means more jobs for local people as ski instructors, guides and staff in hotels and restaurants. Craft industries, e.g. woodcarving, clocks, watches and embroidery, also benefit from an increased market. This economic injection has the added advantage of reducing or stopping the out-migration from these areas, especially of young people (see Chapter 21).

3. Tourism in Cities – Paris

Europe's cities such as London, Paris and Rome and Athens are among the most historic and beautiful centres in the world. As a result they attract millions of visitors each year.

Paris

The 'city of light' and the 'jewel of France', Paris originally developed on a group of islands, Ile de la Cité in the Seine.

Between the twelfth and fourteenth centuries it was surrounded by high walls which were replaced by the Grand Boulevards in the reign of the 'Sun King', Louis XIV. Today, it

‹ 18.16
Cathedral of Notre Dame, Paris

has a population of almost nine million and is visited by over two million tourists each year.

Its main attractions include:

- Beautiful buildings and squares including the Arc de Triomphe, the Eiffel Tower, the Champs-Élysées and the Place de la Concorde.
- Historical Buildings – Such as the Cathedral of Notre Dame.
- Museums and Art Galleries – including the Louvre, probably Europe's most famous art gallery.
- Nearby attractions – such as the Palace of Louis XIV at Versailles and the beautiful woodland areas of the Bois de Boulogne.

The Effects of Tourism on Cities

Environmental Problems

As the number of visitors increases, so does the problem of actually getting around. Rome is typical of a large tourist city where terrible traffic jams have become common. With so many cars, lorries and buses crammed on to Rome's narrow streets, air pollution has become a serious problem. Many vehicles still use leaded petrol and the concentration of lead in the air can reach dangerous levels. Some research suggests that the high levels of lead in the air are linked to brain damage in children. This pattern of air pollution is true of most of Europe's main cities.

Tourist traffic not only adds to air pollution, it also contributes to noise pollution for people living near roads and railways. Because certain parts of cities offer more attractions, this causes overcrowding on public transport particularly the London and Paris underground railways which are already packed beyond their capacity at peak times.

The sheer number of visitors can wear away stone steps as has happened at Athens or even too many hands touching a statue can cause erosion.

As tourist numbers increase there is pressure to build more hotels. This takes up much needed land and one solution to offset this is by converting old buildings such as warehouses into restaurants and hotels.

Social Problems

Large numbers of visitors can change the character of small towns. Visitors tend to impose their own cultural values on local people who may come to resent tourists. Too many tourists can provoke over-charging by taxi companies, hotels, restaurants or souvenir shops which take advantage of the demand. One of the more negative social effects however is the growing incidence of violence and crime as tourists are seen as easy targets for ready cash.

Cultural Problems

Local customs may benefit from an injection of outside interest and create a type of cultural renaissance. The reverse however is also possible and the demands created by mass tourism often devalues the cultural heritage reducing it to the level of mere 'theatre'.

City planners are well aware of the problems tourists bring to their areas. In Paris, they have been working to try to ensure that the visitors enjoy their stay but do not clog up the rest of the city. The plan includes:

- The preservation of the historical cultural centres of Paris by limiting the height of the new office blocks and by making narrow boulevards into one-way streets.
- Building a new high speed metro called RER.
- Building three new ring motorways and urban expressways.
- Experiments with schemes which restrict the number of tourists to the main attractions during the height of the season. This helps to reduce long queues and encourages visitors to spread themselves over a wider part of the city.

Many cities now agree on the need for Environmental Impact Evaluation Surveys. These are carried out prior to most new tourist developments. Such surveys highlight both the costs as well as the benefits that can be expected from development proposals. Ultimately, cities have to make a choice between the needs and preferences of the local people and those of the tourists.

4. Tourism in the Countryside – Ireland

Lying on the periphery of Europe, the importance of tourism to the Irish economy was realised as early as the 1950s when Bord Fáilte was established in 1955. Since then tourist numbers grew by up to 50 per cent helped by a growing affluence and improvements in transport. This growth was interrupted between 1969-72 as a result of the 'northern troubles' and again in the early 1980s, the result of a global economic recession. Growth in the 1990s has been spectacular with the number of visitors topping 4.6 million in 1996. This is 10 per cent up on the 1995 figures and over two million more than the figures for 10 years ago.

The Importance of Tourism For Ireland – 1996	
Revenue from Tourism	£1.445 billion – up 12% on 1995
% GNP	8% – with a target of 120,000 for the year 2000
Tourist Numbers	4.6 million – up 10% on 1995

Ireland's Tourist Attractions

Unlike the Mediterranean region, with its 'frying pan' conditions, or the snow covered slopes of Alpine regions, climate is not a particular attraction for Irish tourism. What Ireland does possess however, is a wealth of natural and cultural attractions unsurpassed by any other country in Western Europe.

Scenic Landscape

In terms of scenic beauty, few countries have as much to offer as Ireland. The rugged scenery of the west of Ireland, a central lowland dotted with lakes, canals and rivers and an indented coastline with miles of sandy beaches.

Fig 18.17 >

Climate

While the maritime climate cannot offer blue skies and scorching sunshine, it does offer moderate temperatures which are often a welcome change from the oppressive heat of many European and American cities.

The perennial rainfall too results in a green vegetation cover and Ireland is known throughout the world as the 'Emerald Isle'.

The attraction of this scenic environment which is spared climatic extremes is enhanced by the low density of population, and low levels of pollution where the tourist can 'get away from it all'.

Outdoor Activities

The variety of the physical environment together with human inputs has resulted in a huge variety of outdoor activities.

Fishing in coastal waters, or in the numerous lakes and inland waterways attracts thousands of anglers from other countries. There is growing interest too in golfing, boating, horse-riding and walking holidays. Unlike sunbathing, these are relatively unaffected by climatic conditions and so provide obvious opportunities for extending the tourist season beyond the traditional summer period.

Historical Attractions

In terms of historical attractions, Ireland could be said to be an open air museum. The ancient burial sites at Newgrange actually predate the pyramids of Egypt. While the countryside is awash with early Christian settlements, e.g. Gallarus Oratory (Co. Kerry) and Glendalough (Co. Wicklow). Most Irish cities date back to the Viking and Norman invasions with their associated castles. Elsewhere, stately homes and gardens offer more serene attractions.

Tourism

Fig 18.18 Newgrange

Cultural Attractions

Ireland is known throughout the world for its 'Céad Mile Fáilte' – its hospitality unrivalled anywhere in Europe with the possible exception of Greece. It is this warm welcome to a country with a rich heritage of traditional music and dance which provides the abiding memory for most visitors. Unfortunately cultural attractions like these are fragile and once broken are rarely, if ever, retrievable.

Tourism – The Future

The large numbers of people who emigrated from Ireland since the Famine in the mid 1840s to the present day inevitably means the return of large numbers of tourists who wish to visit families and friends. A recent survey estimated that up to 35 per cent of overseas visitors to Ireland arrive for that reason. This, however, should create no room for complacency. Our tourist facilities still remain inadequate by comparison to our European competitors. The rising cost of living and motoring together the growing temptation to make a 'fast buck' does little to encourage the return of a tourist who is offered a bewildering choice of holidays in a competitive market.

Add to this the increasing pollution levels and the fragility of the Northern political situation and the recent growth in tourist numbers could witness an equally dramatic downturn.

Questions

ORDINARY LEVEL

1. (i) 'Tourists move out of high income industrial regions and are attracted to coastal, mountainous or rural areas.'
   Explain the statement referring in your answer to examples from Western Europe. *(30 marks)*

   (ii) Tourism can have a negative impact on an area. Explain **three** reasons why this is so. *(30 marks)*

   (iii) Examine the importance of tourism as a means of creating employment. *(20 marks)*

   Leaving Cert. 1996

2. Ireland's Tourist Regions

A	Ivernia
B	South East
C	Shannonside
D	Ireland West
E	Lakeland
F	East and Dublin
G	North West
H	Northern Ireland.

   With reference to any **one** of the tourist regions of Ireland with which you are familiar:
   (i) Name that region and describe some of the advantages **and** disadvantages as regards the attraction of tourists. *(40 marks)*

   (ii) Suggest how that region could be further developed to take advantage of the growth in tourism. *(20 marks)*

   (iii) 'Large scale development of the tourist industry can result in socio-economic problems for a region.'
   Explain this statement. *(20 marks)*

   Leaving Cert. 1994

3. The Government Plan (1989) to develop tourism includes the following aims:
   - To double tourism earnings to £1,020 m per year
   - To create 25,000 new jobs
   - To provide better quality and value of service
   - To promote Ireland more competitively in existing and new markets
   - To spread the benefits of tourism widely and to extend the tourist season

Tourism

(i) Suggest ways in which this Plan might be achieved during the 1990's. *(60 marks)*

(ii) Explain why the high number of Irish people who choose to holiday abroad each year is a cause of concern for tourist authorities and for the government. *(20 marks)*

Leaving Cert. 1993

HIGHER LEVEL

1. Tourism is now an important international industry and is growing year upon year.
   (i) Examine how tourism can benefit a region or country **and** how it can cause problems. *(60 marks)*
   (ii) Briefly describe the development of another service industry which you have studied. *(40 marks)*

   Leaving Cert. 1994

2. Tourism is now the single greatest industry in the world.
   Promotion of tourism can be attractive for developing countries. It can also, however cause damage to their societies, their economies and their environments. Examine this statement in detail. *(100 marks)*

   Leaving Cert. 1992

3. The tourist industry worldwide has grown dramatically in recent years. Indeed, by the year 2000, it may well become the largest single industry in the world.
   (i) With reference to **two** contrasting examples, describe and explain the variety of attractions of regions for tourists. *(50 marks)*
   (ii) Successful long-term development of tourism in any region requires careful planning.
   Examine this statement, with reference to particular examples. *(50 marks)*

   Leaving Cert. 1991

CHAPTER 19 — *Transport and Trade*

Fig 19.1 ›
The Westlink Motorway at Ballymount, Dublin

Transport is an essential part of production. There could be no large manufacturing centres without transport facilities, because large industrial towns must draw their raw materials from a wide area. Transport is also an essential element in trade. Without transport, each area of a country would have to be self-sufficient. It is the surplus of commodities in one area and their shortage in another that forms the essential basis for trade. The movement of those goods however depends on a transport network.

The Role of Transport

- Essential in linking people, resources and activities and to help in the exchange of goods.
- In spite of a relative decrease in transport costs, transport remains a major factor in industrial location (see Chapter 17) and in determining agricultural land use (see Chapter 13).
- It is sometimes used as an indicator of wealth and economic development as measured by the number of cars or tractors per 1,000 people. Developed countries for example have only 25 per cent of the world's population but they have 88 per cent of its rail traffic and 72 per cent of its cars and lorries.

Factors Influencing the Development of Transport

1. PHYSICAL

While there are many factors which influence the selection of a land route between two places, the physical landscape is always a major consideration and routes which keep construction costs to a minimum are generally favoured. Two aspects of the physical landscape are of particular importance, i.e. relief and drainage.

Relief

< Fig 19.2
Winding Alpine Road in Grossglockner, Austria

When examining the influence of relief on the construction of transport links it is convenient to think of both altitude and gradient. This is best seen in high mountainous regions such as the Alps. Because temperatures decrease with altitude, many Alpine routes often became impassable in winter. This led to the construction of tunnels such as the Mont Blanc tunnel in the 1960s. Steep gradients too can often be overcome with the construction of zig-zag bends which spread the gradient over a longer distance. Lowlying relief with gentle gradients however must be regarded as positive as it considerably helps in the construction of transport links and reduces construction costs.

Drainage

As with relief, drainage may be both a positive and negative influence on transport links. Well drained lowlands which have an absence of lakes allow road and railways to take the shortest, most economic route. River valleys also often contribute to the construction of roads in upland regions.

Swampy areas and lakes, however, force transport to take indirect routes while rivers often necessitate costly bridging and force routes to the edge of the floodplains.

2. ECONOMIC

Because transport is an essential part of production and an infrastructural necessity for trade, it follows that transport networks are best developed in the most densely populated parts of the developed world as in the lowlands of Western Europe and the USA.

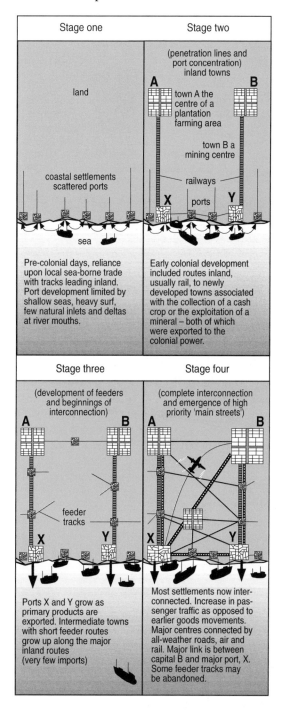

Fig 19.3
The Growth of a Transport Network in a Developing Country

In 1963, Taaffe, Morrill and Gould suggested a model to show how a transport network might evolve in a Third World country. Their model, which was originally based on studies in Nigeria and Ghana, has since been applied to Third World countries, especially in South America, and Malaysia. In many cases only the second stage has been reached, principally because following independence these countries had insufficient money to improve and modernise their network beyond the main urban areas. To reach stage three and four, a country needs the development of manufacturing industry and the emergence of an affluent sector in the population. Stage one shows a country with a number of small ports scattered along the coast. Over time a number of inland routes were built to penetrate the interior probably to tap inland resources. These routes were often built by European powers in their overseas colonies. As the ports grew, intermediate towns with short feeder routes grew up along the major inland routes. In stage four most of the settlements are interconnected.

3. POLITICAL INFLUENCE

Governments have a major say in the development and pattern of transport networks. Because transport is an essential requirement for industry and trade and because of the high cost of developing an efficient transport network, it is generally recognised that active government involvement is required. But apart from economic reasons, governments may establish major routes for

Leaving Certificate Geography

political and military reasons or to assist national unity. The present upgrading and extension of the Irish road network for example is a direct result of government planning aided by EU funding.

4. TRANSPORT COSTS

Transport costs are influenced by a number of factors including the following:

Type of Commodity

Generally speaking, bulky goods which are neither fragile nor perishable, e.g. coal, timber, iron ore are less costly to transport than goods which require special packaging or refrigeration.

Type of Transport

In spite of recent developments in size and speed of craft, air transport remains the most expensive form of transport. Because goods have to be brought to airports, this transhipment increases costs further. Generally speaking, only goods of high value are moved by air.

Distance

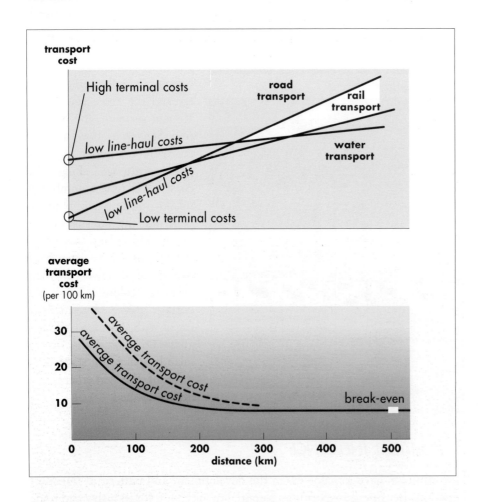

< Fig 19.4
Variations in Transport Costs

While transport costs obviously increase with distance, a doubling of the distance does not result in a doubling of the cost. This results from the fact that terminal costs at both ends remain the same regardless of distance. This is referred to as the 'tapering of freight rates'.

Size of Shipment

While haulage costs increase with distance, they decrease with the amount of cargo handled. Transport companies try to achieve high load factors (the percentage of the vehicle which is occupied). Where there is mass movement of goods or people in one direction many vehicles often make the return journey empty. This helps to explain the cheap flights to Britain from Ireland at Christmas because most movement is in the opposite direction.

Government Policy

Governments may influence transport costs by subsidies. Even though the load factor varies considerably between the West and East of Ireland, CIE tries to maintain its charges at the same level so that remote areas are not subjected to further disadvantages.

Transport costs then are made up of three components.

- Running costs, e.g. fuel, wages, etc.
- Terminal costs, e.g. loading, unloading, documentation, etc.
- Overhead costs, e.g. cost of vehicle, vehicle repair, terminal upkeep, etc.

Types of Transport

ROADS – PERSONAL FRIEND OR PUBLIC ENEMY?

Fig 19.5
Traffic Congestion in Dublin City

Transport and Trade

While the Romans were probably the first great road builders and many of the routeways they selected are still used today, the great improvement in road transport came with the development of motorised transport. Road transport dominates the modern world for a number of reasons.

Advantages of Road Transport

Flexibility

Roads are the only form of transport which provide a door to door service. Because roads are easier to construct and can climb steeper slopes than railways, they can readily adapt to changing conditions of urban expansion, etc.

No Timetable

Road transport of either goods or people are not subjected to timetable schedules, which can significantly reduce delivery time.

Terminal Costs

Because lorries can deliver door to door there is less need for terminal facilities (warehouses, etc.). The absence of this *break of bulk* (the need to unload and load the goods again) reduces costs as well as time.

Track

While modern road building can be extremely expensive – more than £2 million per kilometre for primary routes, the cost is usually borne by governments. This considerably, if artificially, reduces the cost of road transport for individual haulage companies.

The advantages of road transport are evident in the following statistics:

- There are now (1997) over 700 million motor vehicles in the world and the number is rising by over 40 million each year.
- In Britain 84 per cent of passenger kilometres are now travelled by road.
- Ninety per cent of inland freight goes by road (in Britain).

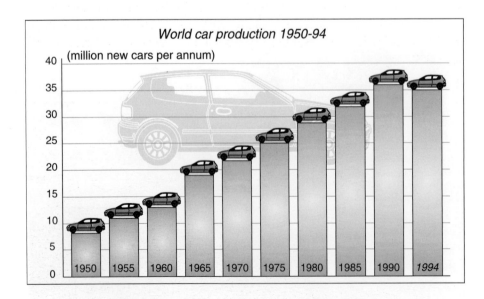

Fig 19.6

Motorised transport is taken for granted today and the car has become a personal friend. Unfortunately the personal friend seems to have become a public menace.

Disadvantages of Road Transport

Line-haul Costs

The carrying capacity of even the largest lorries (juggernauts) is small when compared to other forms of transport (trains, air etc.). This fact, together with the cost of fuel, drivers' expenses and wear and tear of the vehicle means that running costs (line-haul costs) are very high. Road transport is therefore more suited to small loads over short distances.

Road Construction

Because roads are extremely costly to construct and maintain, funding is usually by way of Government departments. While toll roads and bridges can offset costs to some extent, the initial capital outlay is enormous and must be borne by governments who raise the money from general taxation.

Traffic Congestion

In spite of road improvements traffic congestion is a major problem in all urban complexes. Parking facilities, one way systems, ring roads and by-passes have all failed to improve, let alone solve the problem which is essentially due to the rising number of vehicles.

In New York, horse-drawn traffic used to move at 18.5 kilometres per hour in 1907. Today, car traffic now moves at a snail's pace of eight kilometres per hour.

Accidents

Motorised transport is responsible for over 300,000 deaths a year worldwide, plus many millions of injuries – four million a year in the US alone.

Pollution

If present trends in car ownership (not counting lorries) continue there could be up to three billion cars in the world by 2050. In the US, for example, there are 730 vehicles for every 1,000 people while in Germany the figure is 519 per 1,000.

Fig 19.7

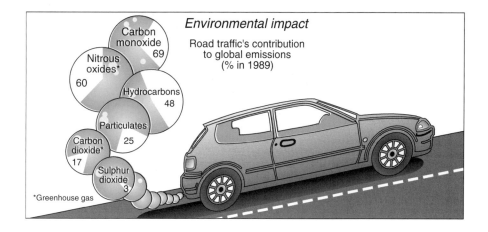

Carbon monoxide, nitrous oxides, lead and other toxic emissions make vehicle exhaust fumes among the most pressing of modern environmental problems.

The average car also emits four times its own weight in carbon dioxide every year. Motor transport worldwide accounts for a fifth of man-made emissions which contribute to global warming.

In some urban areas there is evidence that traffic pollution can cause significant amounts of respiratory disorders among people living nearby, with children and the elderly especially at risk.

Add to these problems the loss of farmland and wildlife habitats, the scarring of beautiful landscapes, noise pollution and the growing social phenomenon of 'road rage' and it is probably understandable why many people see the car as public enemy number one.

RAILWAYS

‹ Fig 19.8

Railways were a product of the Industrial Revolution and solved the problem of carrying bulky commodities over long distances at high speed (thereby replacing the canals). Their impact on nineteenth-century economy and society was revolutionary and the transformation of the American 'wild west' was completed in two generations, as the 'iron-horse' carried European structures and values to a less than enthusiastic native American civilization. But just as the canals suffered from competition from the railways, the railways in turn were ousted by the motorised transport, air and pipeline transport of the twentieth century. By around 1930, the rate of rail construction was equalled by the rate of closure for the first time in over one hundred years.

Advantages of Rail Transport

Bulk Carriage

The major advantage of rail transport over road is that it can move large amounts of bulk over long distances far more cheaply than road freight.

Time and Safety

As trains have their own tracks, there is less congestion which considerably reduces time – especially over relatively long distances, when trains travel well in excess of 150 kph.

Because they are tied to the track, trains are also less affected by weather conditions (ice, fog, etc.) and with less congestion (than roads) their safety record is much better than road transport.

Line-haul Costs

Rail transport can carry greater bulk (passengers and goods), and needs fewer employees and less fuel so its running costs are considerably lower than that of road transport.

Pollution

Because one train can carry the equivalent of many juggernauts (depending on the amount of carriages), there is far less atmospheric pollution.

Disadvantages of Rail Transport

Fixed Track

Trains are tied to a fixed track and are therefore unable to offer a door to door service. Consequently, transhipment is necessary which is often both time-consuming and costly.

Terminal Facilities

Warehouses, loading equipment and overhead costs (engines, etc.) are very high.

Construction Costs

Railways are expensive to construct and because locomotives cannot haul heavy loads up gradients of more than 1:100 they are less suited to highland areas where bridges, tunnels and embankments have to be constructed.

Timetable

Because transhipment is necessary in train transport, considerable amounts of time is necessary for loading and unloading. This inevitably leads to timetable schedules which can result in considerable delays even over short distances.

In spite of these disadvantages, rail transport remains an economic proposition for moving bulky goods over medium to long distances. The piggyback facilities now offered by some rail services combine the advantages of both rail and road transport.

In this system, truck trailers are carried on flat wagons between train stations. They are then off-loaded and hitched to trucks which bring them to their destination. Increasing traffic congestion on roads, however, will ensure an increasing role for a fast commuter service both on and below (underground) the surface.

Fig 19.9
The Dart at Killiney, Co. Dublin

Pipelines

Pipelines have been used to transport water since ancient times. The spectacular growth in their use in the twentieth century can be attributed to the growing dependence on oil and natural gas in the developed economies. Because, as a method of transport, their advantages far outweigh their disadvantages, the growth is likely to continue into the twenty-first century.

Advantages of Pipelines

Bulk over Long Distance

Pipelines can transport large quantities of oil and gas over long distances at a consistent rate without interruption. At present 'The Friendship Line' links the Soviet oil fields in the Urals to Poland, Czechoslovakia and Hungary. Oil unloaded at Rotterdam can be pumped throughout Europe through a system of pipelines. Ireland is now being connected to the European natural gas grid as the life-span of the Kinsale gas field is relatively short.

Low Maintenance Costs

While pipelines are costly to construct, they are relatively cheap and easy to operate. Maintenance costs of the track (pipeline) and wages are considerably lower than those of either road or rail.

Environment

Underground pipelines result in little if any loss of farmland and cause minimum disruption to wildlife habitats. Even when overground they occupy considerably less space than either roads or railways. Pipelines too are least affected of all methods of transport by adverse weather conditions, have no problems with traffic congestion and are not subject to timetable schedules like trains, boats or aircraft.

Disadvantages of Pipelines

Cost

While running costs may be low, construction costs are often very high. Expensive equipment and a skilled labour force are needed and the actual transport of the pipelines into remote and difficult terrain can considerably add to the cost.

Slow Movement

Pipeline transport is less likely to cause ecological disasters (as in tanker disasters) but the actual movement is slow.

Fixed Track and Capacity

Pipelines are the least flexible of all transport systems. Their capacity is fixed so that an increase in demand requires a new pipeline.

In spite of these limitations, pipelines enjoy a growing popularity as a form of transport. Today they account for over 15 per cent of freight in France while in the USA the figure is over 20 per cent.

In Germany crushed coal immersed in water is moved by pipeline while the saeters (summer mountain pastures) in Switzerland use pipelaits to transport milk to the farms below.

AIR TRANSPORT

Fig 19.10
Dublin Airport

Dramatic changes have taken place in air transport since the Wright brothers' first powered flight in 1903. Aircrafts have increased in size and jet airliners travelling faster than the speed of sound have reduced planet earth to a global village.

Yet in spite of these advances, air travel still remains costly and is still reserved for people and valuable cargo.

Advantages of Air Transport

Speed

The greatest advantage of air transport is its speed over long distances. Today jets carrying over 300 passengers can travel at over 500 mph while Concorde can fly across the Atlantic in less than four hours.

No Track Maintenance

Apart from terminal costs, air transport has no track to construct or maintain (unlike road or rail). This gives it other advantages. It is independent of adverse weather conditions in flight as there is no track and there is maximum flexibility so that it can travel by the most direct route.

Topography

Apart from suitable airport sites, air transport is independent of the local topography. This allows it to operate in regions where it would be extremely difficult to construct road or railway lines. Consequently, air transport is particularly useful for transporting people and supplies to remote settlements. In Norway for example STOLs (Short take-off and landings) are extensively used to link settlements separated by a difficult physical environment.

Disadvantages

Cost

In spite of spectacular advances in aircraft size and speed, actual costs remain high. Construction costs from the planning to the production stage are very expensive. Add to this the cost of maintenance, fuel and terminal costs and it is difficult to see how prices can be reduced relative to other forms of transport.

Airport Sites

Extensive flat solid ground with a preferably flat hinterland is needed for runway construction, take off and landing. Areas subject to fog have to be avoided as this will affect the running of the airport.

Transhipment

The physical requirements for airport sites, together with safety factors often result in airports being located outside the cities which they are meant to serve. This necessitates transhipment and delays (due to unloading, traffic congestion, etc.) thereby reducing air transport's greatest advantage – speed.

Safety

While air transport has a relatively good track record for safety, the number of fatalities continues to increase. The number increased by 10 per cent between 1995 and 1996. While this increase is primarily due to the continual increase in the number of passengers, there is growing concern about the increasing use of airspace over large urban areas. The amount of passengers using London airports grew from 31.1 million in 1976 to 62 million in 1988. The estimated figure for the year 2000 is 123 million.

WATER TRANSPORT

Water transport was formerly the most important means of transport and in spite of the competition from road and rail, it still remains the cheapest form of movement for bulk commodities.

Inland Waterways

Fig 19.11
Barge on the Grand Canal, Dubin

The Industrial Revolution produced a need to transport bulky commodities over long distances. As road surfaces were unreliable and limited in carriage capacity, canals appeared as a means of commercial transport in the late eighteenth and early nineteenth century. The monopoly enjoyed by the canals however was short lived with the competition from railways. However, since the 1970s there has been a renewed interest in inland waterways. This began with the increase in road and rail charges by the rise in oil prices following the 1974 Middle East War and the realisation that, in spite of its slowness, water transport remains the cheapest for bulky goods.

The EU have now introduced huge 4,400 tonne barges propelled by push-barges which can carry the equivalent of 110 rail trucks or 220 x 20 tonne lorries. To accommodate these super-barges the EU governments have financed the widening and deepening of canals and the linking of the Mediterranean, North and Black Sea.

OCEAN TRANSPORT

The oceans of the world are the 'global highways' which link the continents of the world. They offer routeways whose inflexibility is governed only by the presence of ports.

Requirements of Major Ports

Physical
- Deep sheltered harbours with low tidal range.
- Low-lying area to allow for the construction of terminal facilities.
- Preferably ice-free to allow movement throughout the year.

Economic

Productive hinterland to ensure large freight movement both out of and into the port. Rotterdam in the Netherlands provides an excellent example. Other major ports include Marseilles, Hong Kong, London, New York, etc.

Communication

All ports are 'break of bulk points' where goods are loaded and unloaded. An efficient transport network then of road, rail and pipelines is essential to guarantee an efficient delivery of goods. A port that stores goods which are later collected to be delivered to other areas is known as an *entrepôt*.

Recent changes in ocean shipping have witnessed the growth of some ports and the decline in others. The increase in ship size, especially purpose-built bulk carriers for commodities such as oil and iron ore has meant that wider and deeper estuaries are needed. Some ports have responded to this challenge by building outports to receive the larger vessels and provide more room for terminal facilities, e.g. Europort – outport for Rotterdam, Cuxhaven for Hamburg. Because water transport is slow, two innovations have managed to shorten the *turnaround* time (the time taken to unload and load cargo). The first was the development of roll-on-roll-off methods where lorries loaded with freight are driven on board so reducing the need for cranes. The second was the introduction of *containerisation* in which goods are packed into containers at the factory, taken by train or lorry to the container port and loaded by specialist equipment. Some of the larger ships can now hold up to 4,000 containers. While this has considerably increased efficiency it has decimated the dock labour force in many ports.

Advantages of Water Transport

- Greatest capacity for carrying bulk.
- Running costs are very low. The track is free – needing no maintenance. Improvements in engines have reduced fuel needs and more automation has reduced the need for crew.

Disadvantages of Water Transport

- Slow Movement. The size of the vessels, the bulky nature of the commodities combined with a track (water) which resists speed, all combine to make water movement the slowest form of modern transport.
- Terminal Costs. Deep water berths (often needing constant dredging) docking facilities, expensive machinery (cranes) for loading and unloading, storage facilities for commodities such as oil and grain all combine to make terminal costs high.
- Physical Factors. Silting is a major problem often necessitating costly dredging. High tidal ranges too can restrict the accessibility of some ports while frozen seas is a problem for others, e.g. St Laurence Seaway and Gulf of Bothnia (Baltic).
- Pollution. The high quantities of oil being transported by super-tankers has inevitably led to accidents which have had catastrophic results for both marine life and beaches (see Chapter 18).

		Water	Rail	Road	Air	Pipeline
PHYSICAL	Weather	Storms, fog, frozen water	Snow blocking lines, cold frozen points heavy rain – land slides	Fog, ice, snow drifts	Fog, ice, snow at airport	Not greatly affected
	Relief	Deep sheltered harbours, canals need gentle sloping lowland	Gentle slopes essential	Avoids valley flooding, forced around lakes and estuaries	Flatland for terminal	Important in construction only
ECONOMIC	Speed/Time	Slowest of all	Fast over medium distances	Fast over short distance	Fastest over long distance	Fast as flow is continuous
	Running or Haulage Costs	Relatively cheap	Relatively cheap medium distances	Cheap over short distance. Rapid increase with distance	Most expensive of all	Cheapest of all
	Terminal Costs	Expensive portal facilities	Expensive building and maintenance of station & track	Very expensive for vehicle and track	Very expensive to build planes and airports	Very expensive to build
	Number of Routes	Declining as ships increase in size	Decreasing - lack of flexibility	Many routes great flexibility	Limited by suitable airports	Very limited - not flexible
	Goods/ Passengers carried/ congestion	Ideal for bulky goods - little congestion	Good for bulk, little congestion except inner city rail	Limited bulk, major congestion in urban areas	High value freight, limited congestion	Liquid freight, no congestion
ENVIRONMENTAL	Environmental problems	Danger of major accidents	Noise and air pollution	Noise and air pollution	High noise pollution	Very little. Visual eyesore when on surface

Fig 19.12
Physical, Economic and Environmental Factors affecting Transport

Trade

Trading results from the uneven distribution of raw materials over the earth's surface. Because no country is able to produce all its requirements, it exports those commodities of which it has a surplus and imports commodities which it does not produce and for which there is a demand. In one sense then, trade is based on area specialisation. It allows areas to concentrate on economic activities in which its advantages are greatest compared with other areas. This is referred to as the *law of comparative advantage*. This law operated in nineteenth-century Britain. Britain could have produced more of its own food but having a greater comparative advantage over other countries in manufacturing, it opted to concentrate on industry where greater profits could be made.

Leaving Certificate Geography

Large countries like the United States, with varied resources are vast enough to be fairly self-sufficient. While they have a large volume of internal trade, foreign trade is small in relation to their population. Small countries tend to have large foreign trade. New York for example is supplied with oranges from Florida (internal trade) but Dublin has to import these from Spain (foreign trade).

The level of economic development of a country is another factor which influences the volume of trade. Developed countries have manufacturing industries which require imported raw materials and large urban populations which can afford a varied diet, all this results in a large volume of foreign trade. Less developed countries however often have to live on what they can produce.

Their few exports are often confined to 'cash crops' (see Chapter 20) so they cannot afford to import a large volume of goods.

> Characteristics of International Trade
> - Almost 50 per cent of the world's trade is between developed economies.
> - The developed countries have relatively little trade (by volume) with the Third World but export more to them than they import.
> - There is little trade between Third World countries.

GOVERNMENTS AND INTERNATIONAL TRADE

The relationship between the value of goods exported by a country and the value of its imports in any given year is known as its balance of trade. An excess of exports is referred to as a favourable or positive trade balance, while the opposite is an unfavourable or negative balance.

BALANCE OF PAYMENTS

This refers to a country's total payments and receipts in all economic transactions with the rest of the world.

Governments may sometimes intervene to ensure an equitable balance of trade. One method of doing this is by the use of tariffs which are taxes placed on goods imported into a country. Tariffs can have a number of advantages:

- Provide government revenue.
- Help to reduce the import of those commodities on which the tariff is placed.
- Help to protect home production against foreign competition and so protect employment.

MAJOR TRADE ALLIANCES

1. The European Community (EC) or European Union (EU)

The European Union consists of 15 countries at present:

Germany, Italy, France, Belgium, Netherlands, Luxembourg, Denmark, Ireland, Britain, Greece, Spain, Portugal, Sweden, Finland, Austria.

2. The European Free Trade Association (EFTA)

EFTA consists of Norway, Sweden, Denmark, Britain, Austria, Portugal and Switzerland.

3. North American Free Trade Association (NAFTA)

The NAFTA countries are Canada, the United States, Mexico.

4. Latin American Free Trade Association (LAFTA)

Brazil, Argentina, Paraguay, Uruguay, Colombia, Ecuador, Peru and Mexico comprise the Latin American Free Trade Association.

Questions

ORDINARY LEVEL

1. (i) Select **one** advantage **and one** disadvantage of **each** of the following modes of transport: pipeline, canal, containerisation, air, rail. *(40 marks)*

   Leaving Cert. 1994

2. Canal, rail, air, pipeline
   (i) Identify the main advantages of **each** of the above modes of transport. *(20 marks)*
   (ii) Select **two** of these modes of transport and examine the role played by **each** in the economic development of **any** country which you have studied. *(60 marks)*

   Leaving Cert. 1991

3.

Development Indicators			
	West Germany	Brazil	Bangladesh
Gross Domestic Product per person (US$ per year)	13520	2214	144
Food Intake per person (calories per day)	3500	2800	2000
Percentage Employed in agriculture	8	50	90
Percentage adult literacy	99	60	23
Life expectancy (years)	73	60	48

   (i) Explain, with reference to **two** examples which you have studied why transport costs are an important factor influencing the distribution of manufacturing industry. *(40 marks)*
   (ii) Explain **two** ways in which advances in technology have changed the transportation of people and goods during the twentieth century. *(40 marks)*

   Leaving Cert. 1989

Transport and Trade

HIGHER LEVEL

1.

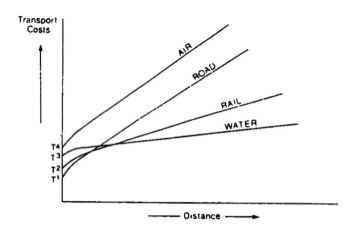

(i) Explain fully what this graph shows about the effect of distance on the cost of transport. *(40 marks)*

(ii) With reference to countries, regions or cities which you have studied, examine **three** ways in which the development of transport links has encouraged economic growth.
(60 marks)
Leaving Cert. 1994

2. (i) 'Transport binds the world's economies and societies together'. Account for the importance of transport both economically and socially in the twentieth century. *(60 marks)*

(ii) Traffic congestion is a problem in many towns and cities.
In **any** town or city than you have studied outline the nature of the traffic congestion problem and discuss attempts to overcome it. *(40 marks)*

Leaving Cert. 1993

3. Examine, with reference to appropriate examples, the importance of each of the following in the development of transport networks:
– The physical environment.
– Level of economic development.
– Government policies. *(100 marks)*

Leaving Cert. 1992

CHAPTER 20
Worlds Apart – The First and Third Worlds

Fig 20.1
'There is enough for people's need, but not for people's greed'
Mahatma Gandhi

Many of the preceding chapters have, in one sense been a study of the developed world, the 'First World' in which modern farming techniques enable many of these countries to produce vast surpluses of food and where forestry, fishing and mining are highly mechanised.

Industrialisation too has enabled many of the people there to enjoy high living standards and most can expect to live to over 70 years. There is another world however, the developing 'Third World', inhabited by 75 per cent of the world's population, a world which possesses a mere 20 per cent of the world's income and where a quarter of all children die before reaching five years of age, a world in which 30 per cent of its people suffer from malnutrition and over 10,000 die every day from starvation and related diseases'.

Leaving Certificate Geography

Worlds Apart – The First and Third Worlds

What has created this frightening division of the world's wealth? Does it simply reflect the uneven distribution of the earth's resources or are the causes more complex? Is there any connection between the affluence of the developed 'North' with the poverty of the developing South? And what of the future? Does this unequal distribution of wealth pose a serious threat to world peace?

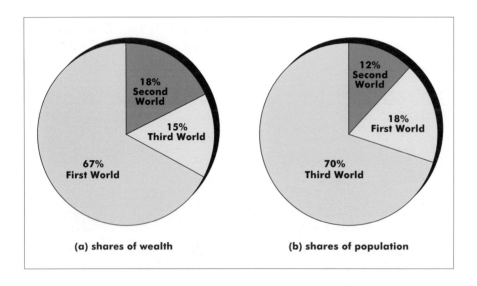

‹ Fig 20.2
Wealth and Population Distribution

Developing Economies – The Search for a Definition

One of the difficulties in defining terms such as 'developed economies' and 'developing economies' is the absence of an acceptable criterion on which a definition can be based. One of the more accepted definitions is based on the yardstick of per capita real income, i.e. the earnings of a country's population in relation to the cost of living. Countries in which the per capita real incomes are below $1,000 are generally regarded as poor or developing countries. Countries on the other hand which have per capita incomes in excess of $3,000 can be regarded as rich or developed countries. Definitions based on average incomes however (obtained by dividing the gross national product (GNP) by the number of people of a particular country) often hide unequal distribution of that wealth. Besides, many countries have per capita incomes which lie between the two extremes, and in consequence, some would argue against a two-fold division of the world's economies.

For all its shortcomings however, GNP does provide a useful criterion and while terms like highly developed countries (HDC), moderately developed countries (MDC) and less developed countries (LPC) might be more accurate and more sensitive, sharp divisions between per capita incomes do exist, as the following charts show:

Fig 20.3 > Per Capita GNP (US$)

GNP per person	1950	1960	1980	1988
Industrialised countries	4,130	5,580	10,660	17,080
Moderately developed	640	820	1,580	1,930
Less developed	170	180	250	320

Fig 20.4 > Inequality. The Developed and Developing Worlds Compared

	Developed World	Developing World
World's population	25%	75%
World's energy consumption	85%	15%
World's G.N.P.	83%	17%
World's food grain	70%	30%
World's export earnings	82%	18%
World's education spending	89%	11%
World's health expenditure	94%	6%
World's science & technology	95%	5%
World's industry	92%	8%

The term 'Third World' is generally applied to the developing world today. The traditional political divisions between the West and East has created the division between the First and Second Worlds. The Third World is an economic rather than a political division.

The Third World – The Origins of Poverty

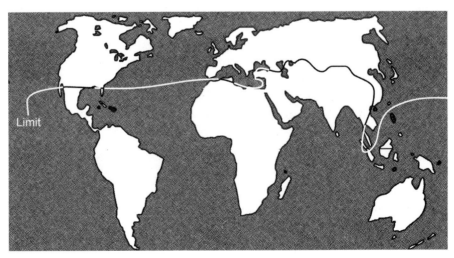

Fig 20.5 > The North/South Divide

Generalisations invariably hide exceptions. Certain countries, like Brazil, Mexico and Egypt might well object to their inclusion in a category which also includes countries such as Sudan, Ethiopia and Bangladesh. Such objections are often based on the rapid industrialisation which is taking place in these and other countries. Industrialisation aside, low per capita income is a common feature of these regions, often averaging £1.50 a week. What is relevant at this stage is the origins of that poverty.

One of the most notable features of Third World countries is their geographical distribution. Almost all have tropical locations and it is tempting to point to climate as being the major physical obstacle to economic development. Climate, however, for all its importance has never been the sole arbiter of the world's wealth. Many developing countries seem to have limited natural resources – but so also have developed countries like Denmark and Japan. It is obvious then that the causes of the Third World are more complex and demand a closer analysis. One method of analysis is to examine the physical environment and then proceed to human inputs, both past and present.

PHYSICAL ENVIRONMENT

Location

The relative location (as opposed to the absolute location) of Third World countries is generally one of isolation from the main market centres of the developed world. This isolation results in considerable transport costs which directly affect all imports and exports. This problem is compounded by the lack of a developed transport network within the countries themselves. Consequently, the distribution of imports and the collection of exports is impeded.

Minerals and Energy

Many of the Third World countries are rich in minerals and energy resources. In fact, about 70 per cent of the developed world's imports of fuel and minerals come from the Third World. The presence of minerals and energy resources however does not in itself guarantee wealth. The exploitation of minerals needs high levels of technology and skill and much of their contribution to employment and profits comes from the processing stage. In most cases however these resources are exported in their 'raw' state and in consequence developing economies benefit very little from their presence.

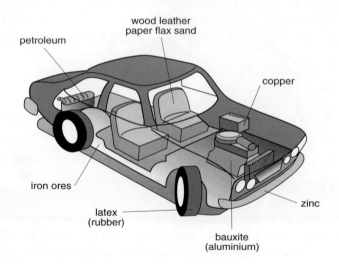

‹ Fig 20.6

Types of Raw Materials from the Third World used in the Car Manufacturing Industry

Relief and Drainage

Most developing countries have a lowland relief and while some possess mountainous areas (e.g. Peru) they also have considerable tracts of lowland. They compare well with developing countries then and in no way can relief be held responsible for the lack of economic development. Rivers on the other hand which generally reflect climatic conditions are very different to those of the developed countries in the more temperate latitudes. The seasonal nature of the rainfall in the tropics (summer maximum) results in considerable variations in regimes. The importance of water to economic activities such as manufacturing industries has already been discussed (see Chapter 17) and in consequence would appear to represent a major handicap to development in this sector. This disadvantage however can be partly off-set by a concentration on industries which are 'less thirsty'.

Climate

While 'climatic determinism' must be avoided, the influence of climate on human activities cannot be over-stressed. The problems caused to agriculture by seasonal drought are obvious as are its effects on industry, whose consumption of water can be enormous. The indirect effects of these tropical climates are no less considerable. They often provide ideal breeding grounds for pests and insects (locusts, tsetse-fly, mosquito etc.) whose effects on crops, cattle and people often prove disastrous. Even where man comes to the aid of nature's deficiencies (seasonal drought) and builds dams for irrigation purposes, the stagnant water behind the dam often provides ideal breeding grounds for disease (as discussed in Chapter 13). The perennial high temperatures too do little to encourage maximum output from the labour force.

Soils

Soil, more than any other natural resource, is particularly sensitive to climate. Tropical climates with their seasonal droughts are a particular handicap both to the development and fertility of soil. In fact, many of the recent famines in Africa and South-East Asia can be attributed in part, to the lack of rainfall and the poverty of the soils. The problem is then often aggravated by desertification (see Chapter 7). In consequence many of the river valleys in developing countries are densely populated as their flood-plains often provide the only regions of fertile soils. It would be incorrect however to form the opinion that tropical soils are unproductive. About 60 per cent of all exports of agricultural goods come from the Third World. The EU and USA for example get all their coffee, cocoa, tea, bananas, etc., from developing countries.

Vegetation

The forests of the tropical regions provide the world with vast quantities of hardwoods and rubber as examined in Chapter 15.

HUMAN INPUTS

Generally speaking then, the physical environment of developing countries cannot be considered negative in promoting economic development. About 60 per cent of all world exports of both agricultural and mineral commodities (other than oil) came from these regions. The origins of the problems confronting the Third World must, therefore, lie elsewhere, i.e. human inputs. To quote the historian Keith Griffin:

> 'Underdevelopment as is encountered today in Spanish America and elsewhere is a product of history. It is not the primeval condition of man, nor is it merely a way of describing the economic status of a 'traditional' society.
>
> Underdevelopment is part of the same process which produced development ... it is only from an examination of the forces of history, i.e. of the historical uses of power, both political and economic that one may obtain an insight into the origin of underdevelopment.'

Let's examine then the historical uses of power, both political and economic and see if they do provide that insight into the origins of underdevelopment.

Political

Colonialism – a Past with a Presence

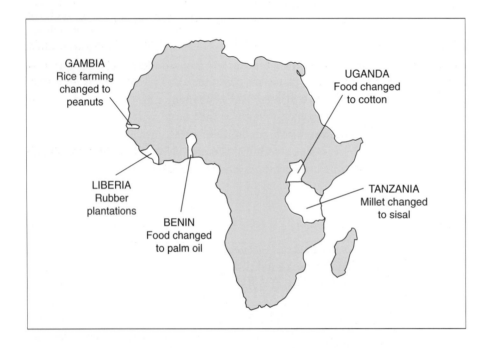

< Fig 20.7
Switches to Cash Crops

In Chapter 17 we noted how the Industrial Revolution of the eighteenth century gave rise to the developed economies of Western Europe. Since then that revolution has influenced all countries to a greater or lesser extent. The Industrial Revolution soon developed an insatiable appetite for raw materials and markets for finished products. To feed it, the industrial countries looked beyond their national boundaries, and so began an age of colonisation which was to have disastrous effects on many countries, particularly those in the tropics.

Before Colonisation

Prior to the colonial era, countries which now belong to the Third World had subsistence agricultural economies. While land holdings differed from one region to another, most people had access to sufficient farm land. They grew their basic food requirements and cottage or domestic industries provided them with clothes and agricultural implements. The colonial era, however, changed all that.

During Colonisation

The colonisers took over much of the best farmland, set up large estates on which they grew cash crops such as coffee, tea, sugar (cane) cotton, rubber etc. The native workforce was either forced onto marginal land, or employed on these estates and in many cases European grain was imported to feed them.

The colonial history of Ireland is not dissimilar. During the sixteenth and seventeenth centuries, the ownership of much of the land of Ireland was transferred to English settlers in what were known as the plantations of Laois and Offaly, Munster and Ulster.

Thus the self-reliant agricultural economies of these countries was distorted and cash crops, rarely of any value to the native population, replaced the traditional food crops of the staple diet.

In return these colonies were rewarded with manufactured goods from the developed world and being cheaper, they gradually eliminated the production of the traditional cottage industries. Figure 20.7 shows the switch from traditional foodstuffs to cash crops in some of the African countries.

Elsewhere, Asian countries like Malaya changed to rubber and India switched to tea and cotton. Some countries proved to be extremely rich in minerals such as gold, copper and tin, all in great demand by the industrial predators of the developed world.

After Colonisation

Though many of the countries involved achieved independence in the nineteenth and twentieth centuries, little has changed.

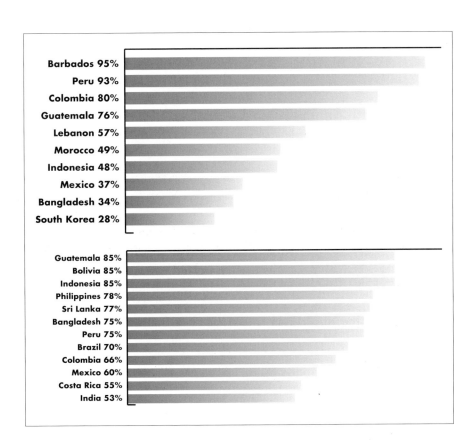

Fig 20.8
Top: Percentage of Land Belonging to Top 10 per cent of Land Owners

Bottom: Percentage of Rural Households with No Land

Land ownership in many cases was transferred to a powerful elite. A United Nations Survey of 83 developing countries for example showed that three per cent of all land owners owned 80 per cent of the land and that many of them were descendants of the old colonial class, as in South America.

Independence made little change in agricultural production and many of the new states continued to produce and export cash crops to pay for the imports of manufactured products. In the Philippines today over half of all farming land is used to produce export crops such as sugar, bananas, rubber, coffee, etc., while the average individual eats less than the average calorie requirements.

Brazil is the fourth largest food exporter in the world, yet over five million Brazilians are starving.

Calories – Necessary Requirements

It is impossible to generalise on the exact food requirements for each person (it varies with age, body size, climate, etc.) but it is generally agreed that about 2,400 calories per day are required for normal body functioning.

Undernourishment, or starvation is the lack of sufficient quantity of food. Malnutrition is the lack of sufficient quality of food.

Agricultural techniques remain backward in the developing world and many farming systems are monocultural and easy prey to its inherent dangers. The cash cropping imposed by the colonial systems proved to have longer lasting effects. Many cash crops which replaced traditional cereals, gradually exhausted the soils of vital minerals through gross overcropping.

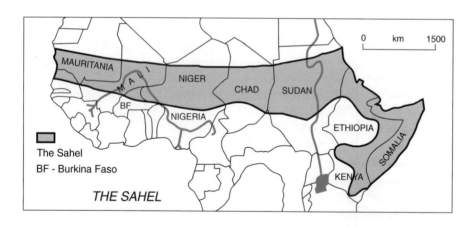

‹ Fig 20.9

The Sahel

Unable even to support grass growth, many of these regions fall victim to soil erosion. The Sahel region of Africa provides a typical example. In the 1950s, part of this region – 63 times the size of Ireland – was turned into a beef exporting zone. As beef numbers soared by careful, monitored veterinary schemes, overgrazing and over-cropping followed. With the ecological balance broken, the drought in 1968 had disastrous consequences. The desertification which followed was blamed on the lack of rain, but the real cause lay in the interference with the delicate ecological balance – the disaster was manmade. The agricultural ills continue to spread like a cancer to other economic activities. Almost 80 per cent of all Central American forests have been destroyed since 1975 so that the land can be used to produce beef for the US. At present the forests are being cleared at the staggering rate of one acre per second, both for agricultural needs and timber products.

Colonisation proved no less disastrous in other economic sectors. Indigenous craft industries could only offer token resistance to the mass production from Europe's industrial giants and soon collapsed. India for example, the largest exporter of textiles in the world in 1700, became by 1850 a net importer of Lancashire cotton goods. Ireland's cottage industries suffered a similar fate in the nineteenth century.

The legacy of colonisation then, is fundamental to the understanding of the origins of the Third World.

Economic

Pattern of Trade – A Raw Deal?

> 'The north-south trade is a one-way street. Unless the South exports to the North, it cannot in turn pay for the North's exports to the South ... only if the North provides better access to its markets can it expect to export.'
>
> (The Brandt Report)

In spite of the dependence of the developed economies (the North) on the raw materials of the developing world (the South) – trade in terms of value is totally imbalanced in favour of the developed world. The vast bulk of Third World exports remain in their 'raw' and consequently cheap state. Their imports on the other hand are mainly manufactured and in consequence, costly products. Even when Third World countries export finished products such as textiles, etc., to the developed world, carefully designed tariffs and quotas ensure that the share of the market will be limited. The simple fact is that the developed countries control world trade. If a Third World country exports a 'raw' product like fruit to the EU, a tax of five per cent is charged on entry. But if the same country wants to export tinned fruit, the tax increases to 39 per cent.

This economic trade imbalance is further aggravated by the fluctuation in prices which these countries receive for raw materials – more particularly when the monocultural nature of this commodity dependence is taken into consideration.

Cotton Prices per kg. on World Market	
1990	182 cents
1992	130 cents
	(Source: World Bank)

Fig 20.10

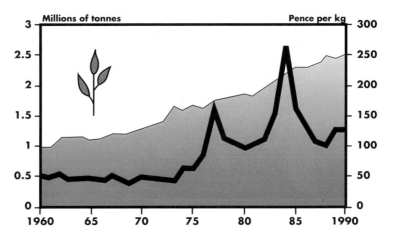

Fig 20.11
World Tea Harvest and London Trading Prices

In the 1960s, Zambia was receiving $4,565 per tonne for its copper. In the 1980s that price had dropped to $1,820 per tonne. Some Third World countries responded to this fall off in prices by producing more.

This however only helps to drive prices down further and the problem of inflation, to which developing countries are particularly sensitive makes the situation becomes even more drastic. In 1960 for example, the cost of one tractor was equivalent to the price of four tonnes of rubber. By 1982 it took 20 tonnes of rubber to buy the same tractor.

Finance – A Debt leading to Death?

Any attempts made by developing economies to improve their infrastructure and indeed their economy as a whole invariably requires large sums of money. Lack of finance means that this has to be borrowed from foreign governments, international agencies and private banks. Between 1974 and 1979, the debts of developing countries to various financial institutions more than doubled. When an economic recession affected the North in the early 1980s, fewer cash crops were bought from the South. Many developing countries found that they were unable to maintain their repayments and had to borrow more money simply to pay the interest on the loans.

In 1989 alone, the Third World paid out $52 billion more in debt services than it received in new investment and loans. This crippling debt is precisely why developing countries are unable to help themselves.

Far from actively helping the South, the repayments demanded by the North is actually preventing developments. The debt is leading to death.

Military Spending

‹ **Fig 20.12**

While the servicing of international (financial) debts may be outside the immediate control of developing countries, the same cannot be said for the level of military spending. Over 55 per cent of all arms purchased are bought by the countries of the South and arms imports now account for 75 per cent of their total trade. Many Third World countries are politically unstable and more than 50 of them are controlled by military governments. Understandably then, arms rank high among imports. In many countries death from a repressive government or civil war is often the greatest threat to life – as witnessed by the recent events in Rwanda.

Yet these resources spent on arms could transform the social and economic conditions in the south.

- The cost of one fighter jet would cover the cost of vaccinating more than three million children.
- British operating costs for two days during the Gulf War were equal to the UK government's total allocation to African famine relief in the same year (1990).
- The cost of one Tornado bomber would feed four million African people for one month.

In many developing countries then, the price of military spending is an increase in human misery, poverty and war. Since most arms production comes from developed economies, it is up to the North to give the lead.

However, in a world where military strength is confused with security, and economic might is confused with right, it is difficult to imagine when, if ever, this will occur.

Causes of The Third World – Exploding The Myths

OVER-POPULATION AS A CAUSE OF THE THIRD WORLD

One of the most common arguments for the lack of development in the Third World has been over-population. Reduced birth rates, it is claimed, will lower population and so reduce poverty.

The fact is that parents in developing countries have large families because they need them to work and as a type of insurance policy in sickness and old age. A reduction in family size will only come about when the social and economic conditions allow people to do so (as discussed in Chapter 12). High birth rates are the result of poverty, not the cause of it.

INCREASED FOOD PRODUCTION – ELIMINATE HUNGER

Unfortunately increased food production will not necessarily eliminate hunger. The Green Revolution – 'making two blades of grass grow where one grew before' was the result of the introduction of hybrid seeds producing high yields of millet, maize and rice. The result has been a dramatic increase in food production often surpassing that of population increase.

< Fig 20.13

But food is distributed not on the grounds of need but on one's ability to pay for it. So while Brazil remains the fourth largest food exporter in the world, over five million Brazilians are starving.

CHARACTERISTICS OF DEVELOPING ECONOMIES

1. Primary Activities

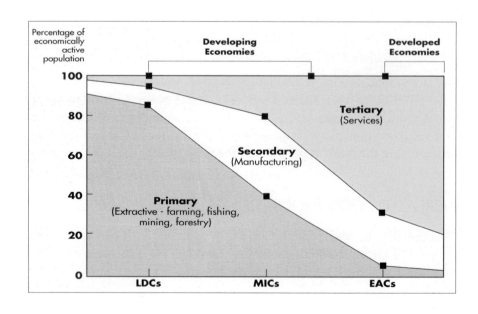

< Fig 20.14

Occupations of Lesser Developed Countries (LDCs), Middle Income Countries (MICs) and Economic Advanced Countries (EACs)

Leaving Certificate Geography

Primary activities dominate the economy, especially agriculture. Despite employing a high percentage of workers however, output is rarely sufficient in either quantity or quality. The major characteristics of agriculture include:

(i) It employs a high percentage – often in excess of 70 per cent of the workforce – most of whom are female.

(ii) Land holdings are unequally distributed. Many rent tiny holdings from landlords who own vast estates.

(iii) The production of cash crops for export dominates the best land.

(iv) Agricultural techniques are backward and, with little mechanisation, yields are low.

(v) Food is often insufficient in quantity and quality and is lacking in vital vitamins. Consequently, diseases such as marasmus, kwashiorker, beri-beri and anaemia are widespread.

(vi) Farming is often monocultural.

2. Secondary Activities

Manufacturing industries on a large scale do not exist. Hampered by the lack of various geographic and non-geographic factors industry:

(i) Employs a small percentage of the workforce.

(ii) Lacks a developed infrastructure and skilled workforce.

(iii) Lacks capital for investment.

(iv) Lacks major market outlets.

3. Tertiary Activities (Services)

In all developing economies, the tertiary sector is totally inadequate. In consequence, then, it lacks vital resources such as:

Water

Fig. 20.15 **Irrigation**

Lack of water is a major problem in areas of seasonal drought. Today, many African and South-East Asian countries are classed as 'water-poor economies'. Poor water supplies and sanitation, together with poverty and ignorance have proved to be a lethal cocktail. Today, cholera (an extreme form of diarrhoea which can kill within hours) is widespread throughout Africa and South-East Asia.

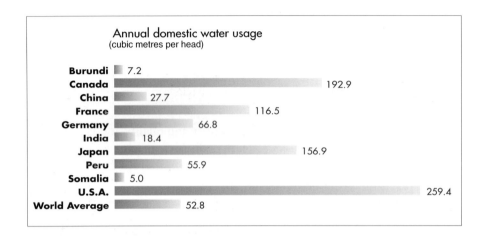

< Fig 20.16

Sanitation

Sanitation facilities are totally inadequate in most developing countries. In Bangkok (Thailand), six million people have virtually no sewerage system while in Calcutta (ten million) only a third of the city dwellers have access to sanitary facilities.

Hospitals

Medical facilities, though improving, are totally inadequate and the rates of patients to doctors, can be as high as 50,000 to one.

Education

While the percentage of adult illiterates in developing countries has been decreasing and now stands at about 50 per cent, the absolute number has been rising. In developed economies the illiteracy rate is about one per cent, in Asia it is 47 per cent but in Africa it is 72 per cent.

Births and Deaths

Most developing countries are in either Stage 2 or 3 of the demographic cycle – so while death rates are beginning to fall, birth rates remain high resulting in a rapid increase in population (see Chapter 11).

Cities of The Third World

One of the most serious problems confronting developing economies is the growth of its cities. Pushed from the land by inadequate land holdings, increased rents and unemployment, people drift to the cities in the hope of employment and better services.

Cities of The Third World

JJ Lean, author of *Rich World Poor World*, wrote:

> 'Trucks loaded with poor peasant immigrants come down night after night from the north-east of Brazil, two weeks journey and many men, women and babies travelled in open trucks sitting on planks lashed across the truck, with the family possessions tied up in a blanket at their feet. The trucks dumped the homeless families under the viaduct, I can see across from my house and then set out north for another load. A day or so later there would be a few more shacks crowded into our already overcrowded favella.'

Fig 20.17

Fig 20.18

Percentage who live in Shanty Towns	
Rio de Janeiro	30%
Sao Paulo	25%
Bombay	40%
Calcutta	60%

CALCUTTA

Over 100,000 people in Calcutta live and sleep on the streets. For 60 per cent of its inhabitants life is slightly more comfortable – a bustee with one room, no bigger than the average bathroom. These dwellings are made of wattle with tiled roofs and mud floors – materials which are unlikely to combat heavy monsoon rains.

In Brazil these dwellings are known as favellas (wild flowers) so called as they have replaced these flowers on the hillsides. Given such conditions, why then are cities in the Third World growing at twice the rate that European cities grew in the nineteenth century?

Nineteenth-century urbanisation was 'migration-led', i.e. the vast majority of new urban dwellers were from rural ares. This is less true of today's developing countries. Although the popular image of the rural poor streaming into shanty towns on the edge of Third World cities are correct, the relative share of migration in the growth of urban populations is smaller in proportion to natural births. Contrary to what we may imagine, in much of the developing world today, the modern city is more healthy than were the

Leaving Certificate Geography

cities of Europe and North America in the nineteenth century. The birth rate remains higher and the death rates lower.

PROBLEMS IN CITIES OF THE DEVELOPING WORLD

Housing

The main problem facing cities in the developing world is how to house the population. Inward migration and natural increase has far outpaced the ability of governments to provide adequate housing. In Addis Ababa, the capital of Ethiopia, it is estimated that as many as 90 per cent of the city's population live in slums. In Sao Paulo (Brazil) the population grew by a staggering half million each year between 1970 and 1985. Urban dwellers then are faced with three choices: to sleep 'rough' on the pavement, to rent a single room (if they have the resources) or to build a shelter for themselves (favella or bustee).

Many opt for the latter and build without permission, on land they do not own. In time some squatter settlements may develop into residential areas with basic service like a water supply, sewage systems and refuse disposal.

Services

Few cities in the developing world have adequate water supplies or sewerage mains. Rubbish is often dumped on the streets and rarely collected. Heavy rains often block drains leading to obvious health hazards.

Drinking water is often contaminated with sewage which inevitably gives rise to outbreaks of cholera, typhoid and dysentery. The uncollected rubbish is an ideal breeding ground for disease.

The atmosphere falls victim to uncontrolled emissions. Lahore, in Pakistan suffers smoke pollution ten times worse than New York. In Calcutta, it is estimated that three out of five people suffer from respiratory diseases caused by air pollution.

Crime

Crime is a fact of life in all the major cities of the Third World (and in many rural areas, e.g. Rwanda). As in the developed world, crime is closely associated with drugs. It is estimated that about 200 of the 500 favellas of Rio de Janeiro are dominated by drug gangs.

Children are always the most vulnerable and child labour, often more akin to slave labour, is more common than child education. In Brazil alone there are millions of abandoned homeless children most of whom live on the margins of the big cities.

Developing Economies – A Solution

If the problems facing the Third World are complex and interrelated, so too are the solutions. As many of the problems which are now facing the Third World countries were experienced in Europe in the eighteenth century, it

Developing Economies – A Solution

would be tempting to suggest that developing economies should model development on the European experience.

However, development of this kind has brought its own problems of pollution and maladjusted areas. Besides, the economic development achieved in the North through industrialisation was achieved at the expense of the Third World (through exploitation of its raw materials, labour force and markets). In consequence the economic environment today is very different to 200 years ago. Third world countries therefore will have to find their own ways to develop in a manner which best suits their own particular resources.

While the ultimate economic goal may be obvious, the methods used to achieve it are less clear. Some countries like Nigeria have opted for rapid industrial expansion in the belief that the wealth generated will 'trickle down' and ultimately be beneficial to all – a theory known as 'Trickle-Down Economics'. Tanzania on the other hand has opted for a more self-reliant approach, and while growth may appear less spectacular, it avoids the excesses which have occurred in Nigeria – affluence on the one hand and grinding poverty on the other.

Whatever the correct solution, it is obvious that in view of the present scale of world poverty, the most important need is to feed the world's people. While the argument as to the ability of the world to support an indefinite growth in population will undoubtedly remain a matter of intense academic debate for many years to come, it is interesting to examine the following facts.

To feed everyone adequately, the world needs to produce the equivalent of 500 lbs of grain per person per year. On average the amount of grain marketed in the world each year is about 1,300 tonnes per person – enough to provide the necessary 500 lbs of grain for over six billion people.

Recent research in the Netherlands gives little cause for pessimism in the near future. This showed that the earth could sustain the production of over 32 billion tonnes of grain a year – 25 times more than at present.

(*Source* – UN Fund for Population Activities)

So the inability of the world then to feed its population is a myth. What is a fact is that at present food supplies are unequally divided. Paul Harrison in his book *Inside the Third World* wrote:

> *'Westerners who are concerned about the pressure of population on the world food resources should remember that their typical family of four are consuming more grain than would a poor Indian couple with eighteen children. Much of the protein wasted on livestock eaten by the west comes from the poor countries – oil seeds and peanuts from West Africa, fishmeal from Peru, soya beans from Brazil and so on. I remember the hunger of little boys, in Kano, Northern Nigeria – their sad faces covered by the flaky skin of kwashiorkor. Yet, on the outskirts of the town huge pyramids were being piled up of sacks of protein-rich ground nuts for export to the north.'*

When one considers that it takes five kilograms of grain to produce one kilogram of meat, and that the developed world on average eats 25 per cent more calories than they need, it becomes obvious that a change of diet would not merely benefit the rich north, where heart disease, diabetes, arthritis and gallstones are often related to over-eating or eating the wrong kinds of food, but would also release vast quantities of food to the poorer South.

Worlds Apart – The First and Third Worlds

It is unlikely that the eating habits of the developed world will show any drastic change in the short term, however, and as an immediate solution the developing world will themselves have to produce their own food requirements. This can be done in a number of ways – as discussed in Chapters 13 and 14.

Examine the following table carefully and rearrange the items in order on the basis of which you think would help to solve the developing world's food supply problems on a permanent basis.

1. Training and instruction on better farming techniques.
2. Land reform – a more equable distribution of land.
3. A change form cash cropping to more basic food requirements.
4. More education in the right foods and what constitutes a balanced diet.
5. Use more efficient agricultural implements.
6. Better seeds and animal strains.
7. Less control of their agricultural products (cash crops etc.) by the developed economies.
8. An immediate food aid programme by the developed world such as that organised by Live Aid Concerts in 1985.

Sufficient food however only creates the basic human conditions on which development can be built. The need to industrialise is central to development itself. Industry helps to:

1. Provide employment for an increasing population and alternative employment for those released from primary activities (especially agriculture) as mechanisation becomes more widespread.
2. Reduce imports of costly manufactured products.
3. Diversify economic output and thereby release it from its dependence on monocultural activities.

Industrialisation on a scale comparable with that of the developed world faces many difficulties both geographical and non-geographical – among which are the lack of a skilled mobile labour force, a developed transport network, power resources, capital and markets. Capital and markets are particularly important, as without them industrialisation on any major scale is impossible.

CAPITAL

To enable the expansion of productive economic activity capital resources must be released for investment. Consumption however already claims most of the resources, so that any switch towards investment is likely to reduce consumption which will further aggravate the present problems. While development itself must ultimately be the responsibility of the governments of the developing economies themselves, aid and co-operation from the developed world is vital.

of the developing economies themselves, aid and co-operation from the developed world is vital.

IRELAND'S AID TO THE THIRD WORLD

'To those who say we should look after our own first, that charity begins at home, I say this. Our concern for underprivileged people in the Third World is no different to our concern for underprivileged people here in Ireland, it is simply an extension of that concern.'

(Garret Fitzgerald, Taoiseach, 1983)

Despite a UN resolution that developed countries should provide one per cent of GNP (0.1 per cent for official government to government aid) Ireland's contribution was a mere 0.23 per cent of GNP in 1983. In 1992 it had dwindled to 0.16 per cent before increasing to .31 per cent for 1997.

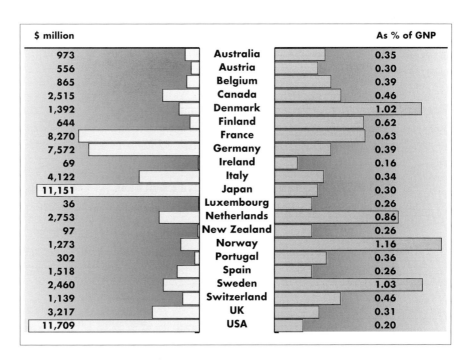

Fig 20.19 > Aid to the Developing World

Throughout the 1980s, Ireland's contribution to the ODA (Official Development Assistance) budget rarely exceeded £30 million annually, yet our income from semi-state contracts in the developing world, (e.g. ESB in Lesotho) amounted to £38 million in 1982 alone. The harsh reality is that we earn far more from the Third World than we give in aid.

Bilateral aid is aid given from one country to another. At present Ireland's bilateral aid is concentrated on Lesotho, Tanzania, Zambia and the Sudan. Multilateral aid is given by a number of countries through international institutions such as the United Nations.

Worlds Apart – The First and Third Worlds

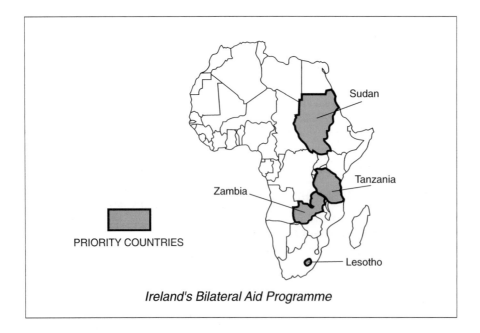

Fig 20.20

Ireland's Bilateral Aid Programme

TIED AID

Financial assistance is often given on the condition that the receiver buys goods from the lending country – goods which are often of little real value to the developing nations. US aid to Peru in the 1980s was given on three conditions:

1. Peru had to buy American jets instead of French ones.
2. Peru had to allow the US fish her traditional waters.
3. Peru could not interfere with a US oil company drilling for oil on her territory.

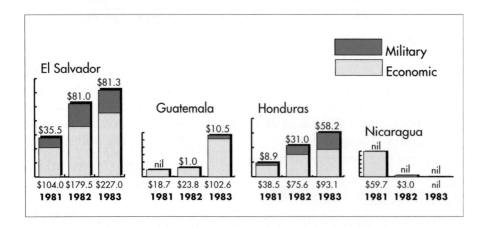

Fig 20.21
US Aid to Selected Countries

Today many of the donor countries insist that the receiving country buys its imports from them. Aid is often given in the form of loans which must be paid back with interest, often at very high rates. Of the $272.9 billion borrowed recently by Latin America in one year $170.5 billion or 62.5 per cent went on simply paying the interest on loans already borrowed. Only 8.4 per cent was used for investment in local projects like building skyscrapers for rich businessmen, but of no use to the poor subsistence farmers who make up the majority of the population.

NGOs – No Strings Attached

The Non-Governmental Organisations (NGOs) who work in the Third World give unconditional aid, i.e. 'no strings attached'. Among these are Trocaire, Gorta, Concern, GOAL and Oxfam.

Each organisation has its own particular line of action. Concern and Gorta send volunteers to work on relief and development projects. Trocaire uses money (not volunteers) to support projects organised by local groups. What unites these groups is their commitment to the elimination of hunger. Of these organisations, a World Bank Report stated:

> *'The effectiveness of NGOs is the result of many factors, commitment to poverty relief, freedom from bureaucratic procedures and attitudes, scarce funds which force concentration on priorities and their small size, which makes it easier to understand and respond to the needs of local communities.'*

At present, NGOs are also involved in the setting up of a programme of debt relief.

They want:

- A limit put on foreign debts which Third World countries owe.
- Some of the debt to be written off.
- Debts to be rescheduled, which means that these countries are given a longer time to pay off the debt and often at a lower interest rate.

Markets

Even if the Third World countries use some of this aid to develop industry on a large scale and process its own raw materials, they still needs market outlets for their finished products. At present the developed countries exercise a stranglehold on these markets by carefully designed tariffs and quotas.

Multinational Companies – A Solution or a Problem?

One 'ready-made' solution to the industrial problems confronting the Third World would appear to be offered by multinational corporations. Their industrial expertise, ready capital and access to the markets of the industrial North would appear to offer an immediate solution.

But while multinational corporations may provide immediate employment, their long term effects may not be so beneficial. In his book *Inside the Third World*, Paul Harrison describes the effects of a new plastic shoe industry in a developing country:

> *'The PVC material for the shoes was imported as were two plastic moulding machines. Forty workers were employed to produce one and a half million pairs of shoes a year. They lasted longer than locally made leather shoes but were not dearer, so they sold well making a large profit for the manufacturer and decent wages for the workers.*
>
> *Meanwhile the local producers – some 5,000 of them who earned their living making leather shoes – went out of business, as did the local suppliers of leather equipment. In all some 800 lost their livelihood while 40 jobs were gained, the import bill increased and the profits went abroad.'*

Worlds Apart – The First and Third Worlds

The real reason why multinational corporations invest in developing economies then, is that labour costs are cheap and vast profits, to all intents and purposes, tax-free, are made. It is virtually impossible for the developing country to keep track of profits. In Columbia for example, an investigation into multinational corporation practices showed that pharmaceutical firms were earning 79 per cent profits instead of the officially declared six per cent.

The developed world must pay a just price for the raw materials it depends on from the developing world. Failure to do so helped to bring about the oil crisis in the 1970s with the formation of OPEC which in turn helped to create an economic recession in the affluent North. Developed economies must allow developing countries a share in their markets just as they enjoy a considerable share of markets in the developing world. Unless need can be distinguished from greed, and might from right, we may fail to appreciate the most important fact of all – that the human race is locked together in a spaceship – planet earth – and its survival can only be guaranteed by co-operation not competition.

Questions

ORDINARY LEVEL

1.

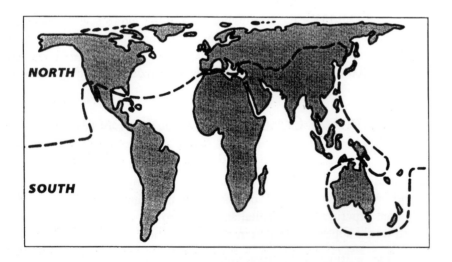

North:	South:
25 % World population	75% World population
85% Energy consumption	15% Energy consumption
89% Education spending	11% Education spending
70% Food production	30% Food production
92% Industry	8% Industry
94% Health spending	6% Health spending
82% Export earnings	18% Export earnings

Examine the map and data which accompany this question.

    (i)   Explain what you understand by the terms 'North' and 'South' in today's world. *(12 marks)*

    (ii)   Name **two** countries in the 'North' and **two** in the 'South'. *(12 marks)*

    (iii)   With reference to the data above, describe **three** differences between the 'North' and 'South'. *(24 marks)*

    (iv)   Explain any **two** causes of these differences. *(32 marks)*

Leaving Cert. 1996

2. Among the factors which hinder world development are
   - Famine
   - Rapid population growth
   - Military spending
   - Unfair trade
   - Conditional (tied) aid
   - Foreign debt

    (i)   Select any **three** of these factors and explain how they have hindered development in the poorer countries of the world. Refer in your answer to examples you have studied. *(60 marks)*

    (ii)   'The greater Dublin area is very developed compared to other Irish regions.' Examine this statement. *(20 marks)*

Leaving Cert. 1995

3. (i)   'Aid given by countries of the developed world has been of enormous benefit to countries of the developing world. However some of this aid is unsuitable, comes with conditions attached and is not properly used.'
   Examine this statement, referring to an example **or** examples you have studied. *(60 marks)*

   (ii)   Explain how Ireland earns more money from its links with developing countries than it gives them in aid. *(20 marks)*

Leaving Cert. 1994

4. The Food and Agricultural Organisation estimates that there are over 500 million seriously undernourished people in the world today. Each year, about 20 million people die from hunger and of these about 17 million are children.
   Look at the statements below, which have all been put forward as explanations of the World Food Problem.

   (i)   People are hungry because of a scarcity of food in the world.

   (ii)   Hunger results from overpopulation; there are just too many people for food-producing resources to sustain.

   (iii)   If the countries where so many are hungry did not produce agricultural exports for foreign consumers, then they could produce more food for local people. Export agriculture therefore, is the chief enemy.

   (iv)   To help the hungry, developed countries should increase their foreign aid programmes.

   By referring to examples which you have studied, examine which of these statements you disagree with and which you agree with. Explain your answer carefully. *(80 marks)*

Leaving Cert. 1993

Worlds Apart – The First and Third Worlds

5.

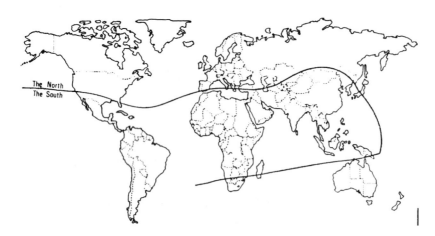

Examine this map.

(i) Which continent contains most of the countries of the Developed World? Name **four** of those countries. *(15 marks)*

(ii) The map illustrates a problem regarding the use of the terms 'North' and 'South'. Explain. *(10 marks)*

(iii) Describe and explain some of the major difficulties experienced by people living in developing countries. *(30 marks)*

(iv) 'International aid is a waste of time and money.'
Briefly examine arguments both for and against this statement.
(25 marks)

Leaving Cert. 1992

HIGHER LEVEL

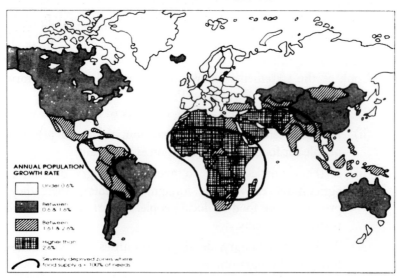

1. Study the map and answer the following:-
   (i) Account for the success of some developing countries in reducing their population growth rates. *(30 marks)*

(ii) Explain the underlying reasons for some of the socio-economic problems of the 'severely deprived zones' shown and suggest a strategy by which these problems could be overcome. *(70 marks)*

Leaving Cert. 1995

2. 'We live in a world with great inequalities among human societies in terms of social and economic development. Any lasting solution is likely to involve difficult decisions for developed countries'.

   (i) With reference to examples which you have studied, examine **three** of the major causes of the economic inequalities referred to in this statement. *(75 marks)*

   (ii) Comment briefly on what is meant by the reference to 'difficult decisions for developed countries'. *(25 marks)*

   Leaving Cert. 1994

3. (i) Physical factors are just one set of causes of famine in Africa. With reference to a specific country or specific countries, describe some other causes of famine in Africa. *(60 marks)*

   (ii) Critically examine food aid from western countries as a means of coping with famine in an African country you have studied. *(40 marks)*

   Leaving Cert. 1993

4. 'Periodic reports in the international media, highlighting acute, regional instances of famine, undoubtedly help to concentrate short-term relief efforts on such regions and raise the level of public awareness of the fact that there is a world food supply problem. If this problem is to be effectively tackled however it must be seen in the context of long-term political and economic inequality on a global scale, rather than in terms of localised food production difficulties'.
   Examine the above statement, with reference to appropriate examples which you have studied. *(100 marks)*

   Leaving Cert. 1990

CHAPTER 21
Striving For a Balance – Problem Regions

< Fig 21.1

Having focused on the problems facing the Third World in the last chapter, it seems almost paradoxical that one should refer to problems in developed economies. That such problems exist highlights to some extent the inadequacy of using per capita incomes as a yardstick for development. However just as developing economies such as Brazil have pockets of affluence like the Copacabana strip, developed economies have their pockets of poverty such as the slums of Harlem (New York). Inequalities of welfare exist in every part of the globe and the slum tenements of New York, the high unemployment rates in the 'has been' industrial regions such as the Sambre-Meuse (see Chapter 16) and the grinding poverty of the Mezzogiorno (Southern Italy) represent major problem regions albeit on a national level.

PROBLEM REGIONS: A SEARCH FOR A DEFINITION

In spite of the high level of per capita income enjoyed by developed countries, as an average income it often masks regions which are seriously deprived either socially or economically (or both) and where the standards of living are well below the national average. The difficulty of actually labelling a particular area, a 'problem region' however stems from the various interpretations of the word 'problem'. Since every region is confronted with some kind of drawback, be it physical, social or economic it could be argued that problem regions exist everywhere.

> In the socio-economic sense then, the term problem region is reserved for those areas 'which are confronted by major problems that have a serious effect on the standards of living for the people who inhabit them'.

The question of actual definition however still remains. Problem regions which would emerge based on such yardsticks as housing shortages or high crime rates might be very different to those which would emerge based on high levels of pollution. Even when an accepted yardstick has been agreed, the question of percentages and a time-scale must also be taken into consideration.

Once particular areas have been identified, the implementation of actual boundary limits needs careful and sensitive consideration.

Since problem regions are generally the recipients of government aid, areas with similar though less acute problems will inevitably protest at their exclusion. Subtleties of definition and demarcation (drawing lines) apart, the rapid economic and social changes that have taken place in developed economies have created major problem regions for certain countries. These areas are now a matter of major concern to most governments.

Types of Problem Regions

At present it is possible to identify three types of problem regions in Western Europe.

1. Underdeveloped regions or less developed regions (peripheral regions).
2. Congested, dynamic or core regions.
3. Maladjusted or depressed regions – regions of industrial decline.

In view of the negative economic effects and social stress caused by political unrest in some areas, one might justifiably consider such areas as problem regions.

LESS DEVELOPED REGIONS OR UNDERDEVELOPED REGIONS

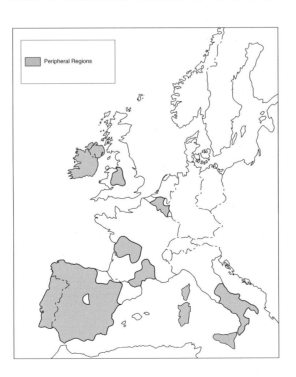

Fig 21.2

Examples

Mezzogiorno (Southern Italy); Western seaboard of Ireland

Central Massif (France); Norrland (Sweden)

Andalucia (Spain)

Origin of the Problems

Many of the problems which affect less developed regions are directly related to obstacles posed by their physical environment.

Physical Environment

1. Geographic location.
2. Minerals and energy.
3. Relief and drainage.
4. Climate.
5. Soils.

1. Geographic Location

Underdeveloped regions tend to have a peripheral location. The disadvantages posed by this isolation – absolute in terms of distance and relative in terms of poor transport are obvious.

Economic development faces major difficulties as it is removed from the central market axis. From an agricultural viewpoint, perishable products like those of market gardening require faster, and therefore, more costly modes of transport along with more careful handling and packaging. The perishable nature of fish poses similar difficulties. Transport costs are equally relevant in the case of industrial development. Because these regions rarely offer market centres of significant buying power, their isolation from major market centres results in extra transport costs being a major prohibitive factor for industrial development. Hence these regions are said to be victims of 'distance decline' or 'distance decay'. They are regions where the law of *comparative advantage* rarely operates.

2. Minerals and Energy

Most of these peripheral regions lack significant deposits of minerals and energy resources. Even when they do exist, e.g. Norrland (Sweden) they are often processed or utilised outside the region and consequently, rarely provide the region of origin with the maximum economic return.

3. Relief and Drainage

Most peripheral regions tend to have upland or mountainous terrain and offer few attractions for commercial farming on any significant scale. Difficulties are also posed by such topography for developing a comprehensive transport network which is an essential ingredient for economic development.

4. Climate

As problem regions exist in the north, south and west of Europe, generalisations on climate are difficult. It can be said however, that climatic conditions often border on the extremes.

In the Nordic countries, latitude combined with altitude reduces mean temperatures which seriously restricts agriculture. In the Mediterranean regions of Southern Europe, high temperatures and a seasonal (summer) drought creates major agricultural difficulties. In other areas exposed to westerly winds, excessive rainfall and reduced levels of insolation (due to cloud cover) pose similar problems.

5. Soils

The combination of unfavourable relief and climatic extremes has serious effects on the formation and quality of soil. Many areas, particularly in the Mediterranean region, are subject to soil erosion, while the glaciers of the

Quaternary Ice Age comprehensively stripped large areas of Northern Europe (e.g. Norway) of its soil cover.

Generally speaking, peripheral regions possess a limited supply of natural resources which in turn has seriously retarded economic growth. In the words of the geologist Elie de Beaumont they have become a *pole repulsif*, i.e. a reservoir of labour and economic resources to be despatched for the benefit of other parts of the country. Consequently, primary activities dominate economic life and with limited infrastructural development, industry on a large scale is usually unknown.

Human Factors

From the viewpoint of government aid, peripheral regions were largely ignored until the 1950s and so they failed to benefit from the increased social services and infrastructural development which more advanced regions enjoyed. It was mainly from these regions that the rural exodus took place which saw the agricultural workforce of the EC being halved between 1950 and 1970. Many of those who left were young and energetic leaving behind the old and more conservative.

A Search for a Solution

In one sense it could be argued that these problem regions, by continual out-migration, solve their own problems. With less pressure on existing resources, those who remain will be better off. There are even those who advocate a policy of 'strategic withdrawal' from these regions based on the belief that their eclipse is predetermined by a type of economic Darwinism. They argue that it is futile in the long term to inject massive state capital into these regions and besides, such capital injections are carried out at the expense of the more advanced regions. On the other hand, there are three courses of action open to governments who believe in greater regional equality.

1. Investment in the Public Sector

Investment in the public sector essentially involves the improvement of the basic infrastructure of the region. Transport and power are the main targets for improvements. The Cassa per il Mazzogiorno was set up in 1950 to improve conditions in Southern Italy and undertook a grand scheme of road building (autostradas), dam-building and the construction of aqueducts.

Governments can also give a major economic and social injection to these regions by moving various government services to these regions as the Irish government did through its policy of decentralisation, for example, moving the Department of Education to Athlone. The injection of a new, and often young, labour force stimulates the social and economic life of regions which are more accustomed to being drained of human resources.

2. Inducement to Private Industry

Private industry can be induced into problem regions by both 'carrot and stick methods'. Capital grants and/or tax concessions can be offered to industry already operating or willing to invest in less developed regions (carrots). It can also be done by penalising the further expansion of industry in the existing core or dynamic regions (sticks).

Striving For a Balance – Problem Regions

3. Inducements to the Agricultural Sector

Various grants have been introduced by some governments and by CAP to encourage farmers to invest in disadvantaged areas and to invest in forestry and marginal land (see Chapter 15).

The European Community has three structural funds which play a vital role in closing the gaps between the more advanced regions and those which are less favoured.

(1) The European Regional Development Fund (ERDF)

The ERDF was set up in 1975 and is the main instrument for promoting the development and structural adjustment of underdeveloped regions.

(2) The European Social Fund (ESF)

The ESF was established in 1958 and while it does contribute to regional programmes, its main goal is the elimination of long-term unemployment and the integration of young people into working life.

(3) The European Agricultural Guidance and Guarantee Fund (EAGGF)

The determination of the EU to create a greater economic convergence can be gauged from the fact that the financial resources of these three structural funds have been increased substantially from ECU 7 billion or approximately 19 per cent of the Community Budget in 1987 to ECU £14 billion or 25 per cent of the budget in 1993. It was from such funding that Ireland recently received over £6 billion for regional aid.

For those who advocate a 'laissez-faire' approach and condemn any government interference in peripheral regions, a study of dynamic or core regions is worthwhile.

DYNAMIC CORE, OR CONGESTED REGIONS

Examples

Paris Basin (France)

Plain of Lombardy (Italy)

Randstad (Netherlands)

Ruhr Conurbation (Germany)

Dublin Region (Ireland)

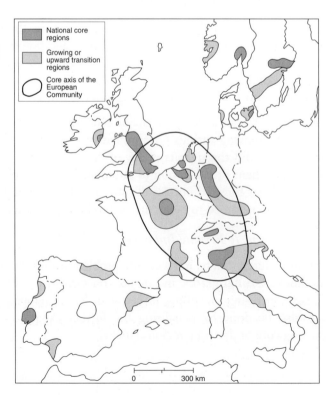

Fig 21.3

A well-defined core region has emerged within the European Community along the London-Milan axis and includes the London, Paris and Plain of Lombardy regions. Other core regions such as Hamburg and Copenhagen exist in separate centres. Core regions themselves can vary in size from single cities to large areas like the Plain of Lombardy in Italy.

These areas generally tend to have the following in common.

- High density of population.
- A concentration of growth industries.
- Intensive tertiary activities.

Reasons for Growth of Core Regions

The essential reason for the growth of core regions is due to a favourable physical environment.

1. Geographical Location

While peripheral locations suffer from isolation, core regions lie at the very nerve centre of economic activity. So the domestic market is excellent both in quantity and quality. High population density improves the attractiveness of the market which is further enhanced by a well developed transport network. Apart from an excellent domestic market, these regions enjoy access to other market centres. The attraction of these regions with accessible markets and a developed infrastructure to industry is obvious. Industrial momentum follows, and with it growth.

2. Minerals and Energy

Many core regions developed due to the presence of significant deposits of minerals and energy resources, e.g. Ruhr. Others, either by virtue of their coastal location or a well developed transport system had easy access to these essential ingredients.

3. Relief and Drainage

Most core regions tend to have a lowlying, undulating, well drained topography. This offered obvious advantages for the development of commercial farming. The lowlying relief also facilitated the development of a transport network essential for economic development.

4. Climate

While core or dynamic regions exist in every Western European country and as such experience varying climatic conditions, they generally avoid climatic extremes. The transitional-continental climates, for example, of the Paris Basin and the Plain of Lombardy have been particularly suited to arable farming.

5. Soils

The attractiveness of these regions for agriculture is further enhanced by fertile soils. The Paris Basin, for example, is blessed with humus-rich 'limon'-type soil. Here the fine dust or loess which was blown westward from the German uplands at the end of the Ice Age was checked by the mild, moist climatic conditions and deposited cover a large area. The Plain of Lombardy, once a sea-arm of the Adriatic was infilled by fluvial, glacial and fluvio-glacial deposits and now forms one of the finest and most fertile flood-plains in Western Europe.

Urbanisation

The attraction of these regions then for human settlement are obvious. Throughout the nineteenth century, the number and size of farm families increased as population moved from stage 2 to stage 3 of the demographic cycle (see Chapter 11). In the beginning the extra hands were welcome but soon the additional labour exceeded the farm's ability to support them. The law of diminishing returns had set in; farms became over-populated. This problem was now aggravated by a second technology. The Industrial Revolution brought mechanisation, the tractor and the combine harvester etc. While farm production increased the demand for labour fell. The flight from the land had begun and with it a growth in cities. Urban growth then was essentially 'migration-led'. As cities continued to grow so did their attraction for the location of industry. The savings to be made from agglomeration (agglomeration economies) are obvious. Industries which use the products of neighbouring businesses as raw materials for their own products considerably reduce transport costs.

As cities grew however, so did their problems. The average population density of the USA at present is about 15 people per square kilometre. In New York City some 55,000 live in each square kilometre. In Canada, where the average is a mere 1.5 persons per square kilometre, Montreal has a density of 52,000 per square kilometre. The demands that such high densities have placed on a limited environment are enormous and have inevitably led to many problems.

Urban Problems

As cities tend to expand from a particular focal point, city centres are usually the oldest part and have long suffered from a lack of investment. Unemployment increased and as traditional industries declined, new and more footloose industries spread to the suburbs. In Britain, for example, unemployment rose from 33 per cent in 1951 to 51 per cent in 1981.

Social Problems

Most inner city housing tends to be of poor quality and high density. Even where slums have been cleared, they have often been replaced with poorly-built high-rise flats. The combination of poor housing, unemployment, and poverty, together with the fact that many city areas have concentrations of ethnic minorities can cause tension, resentment, and even riots. Crime rates tend to be higher so that Dublin, for example, which has one-third of the population accounts for over half of all crime.

Environmental Problems

Cities inevitably suffer from noise and air pollution caused by heavy traffic and the few factories which remain near city centres. Visual pollution too in the form of derelict factories and houses together with wasteland does little to foster community spirit. Traffic congestion is also a major problem, as the number of cars and lorries continues to increase.

Solutions to Dynamic (Core) Regions

If the essential reason for the existence of dynamic regions is the growth of industry, then a simple solution would appear to be to discourage industrial growth through the imposition of higher taxes, etc.

Decentralisation can also be encouraged by offering grants and tax concessions to industries willing to set up in other regions.

Such 'carrot and stick methods' however may prevent an increase in urban problems, but they do little to solve existing ones. Besides, some would argue that intervention at this level penalises regions which have obvious industrial attractions in favour of other areas. When such measures were introduced by the French government to control the growth of Paris it resulted in new industries being set up just outside the city. It took no account however of the growth in the tertiary sector, a significant growth sector in any city. Consequently, the population of Paris grew by over three-quarters of a million between 1962 and 1968 despite the absence of growth in the manufacturing sector.

The reality is that cities, and not merely industry, exert a magnetic effect on people. Consequently, the long term solution would appear to be the setting up of alternative 'poles of development' similar to the *metropoles d'equilibre* decided on by the French Government in 1965.

Meanwhile the present growth in the major cities can be tackled by the creation of overspills. Measures of this kind anticipate growth, allow for planning in advance, recognise the attractions which cities offer and will help to create a more balanced spatial distribution of population. Alternative poles of development too will create a greater spatial industrial balance and governments themselves can lead the way by moving various departments to these regions.

MALADJUSTED OR DEPRESSED REGIONS – REGIONS WITH DECLINING INDUSTRIES

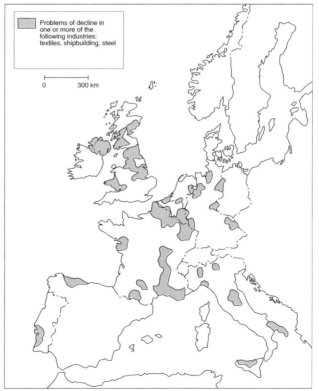

Fig 21.4

Examples

1. Sambre-Meuse (Belgium)
2. Nord (France)
3. West Midlands (Black Country, England)
4. South Wales
5. North-East England
6. Cork Region, Ireland

Maladjusted, depressed, or regions of 'resource decline', can be simply defined as 'has been' industrial areas. Once thriving industrial centres, they are more recognised today as areas with high unemployment, a middle-aged labour force whose skills have outlived their usefulness, disused factories, derelict housing and an outward migration of youth.

Origin of the Problems

Unlike industry in the twentieth century, nineteenth century development was much more closely linked to geographical factors. Coalfields exercised a considerable pull effect, as did some raw materials such as iron (see Chapter 17). Industry was therefore, far more 'resource' or 'power' orientated than it is today, and regions such as the Sambre-Meuse, Nord, etc., which possessed the vital ingredients, developed as major centres of heavy industry.

Geographical inertia further boosted their importance, and a skilled labour force, enjoyed higher incomes if not the most salutary working conditions. The twentieth century has witnessed a decline in the importance of coal as a fuel. The increasing use of oil and gas, the development of power grids and improvements in transport all helped to release the iron grip that these areas exerted on industrial location.

It would be incorrect however to suggest that the eclipse of these regions was due to geographical factors alone. Growing technology, the changing demands of markets, the narrow industrial base of many of the regions concerned, changes in government policy and the emergence of more attractive locations (e.g. Third World) also contributed to their demise. In the 1980s for example the Cork region lost Dunlop, Ford and the Verolme Dockyard. Sunbeam, a major textile employer on the northside of the city was also lost, while in the 1990s the rationalisation of Irish Steel has considerably reduced the workforce.

A Search for Solutions

For all the problems which confront maladjusted regions, solutions do exist. In the first place they possess an existing infrastructure – an essential brick for industrial construction. A traditional industrial labour force can be retrained. Closed factories can be reopened and outmigration can be stemmed by increasing job opportunities. As potentials for 'poles of development' maladjusted or depressed regions are considerably more attractive than peripheral regions.

PROBLEM REGIONS: AN OVERVIEW

In unplanned or mixed economies, the existence of problem regions is inevitable. If they are ignored, they are destined sooner or later to damage the economic and social fabric of a country as a whole. Problem regions however, can never be treated in isolation. Care must be taken that the solutions for one problem region will not result in the creation of problems for other areas. The ultimate solution must be found in overall national planning and not temporary regional remedies.

Case Studies – Problem Regions in France

1. An Underdeveloped Region: The Central Massif.
2. A Depressed Region: The North East or Nord-Pas de Calais.
3. A Dynamic (Core) Region: Paris Basin.

LESS DEVELOPED OR UNDERDEVELOPED REGIONS – CENTRAL MASSIF (FRANCE)

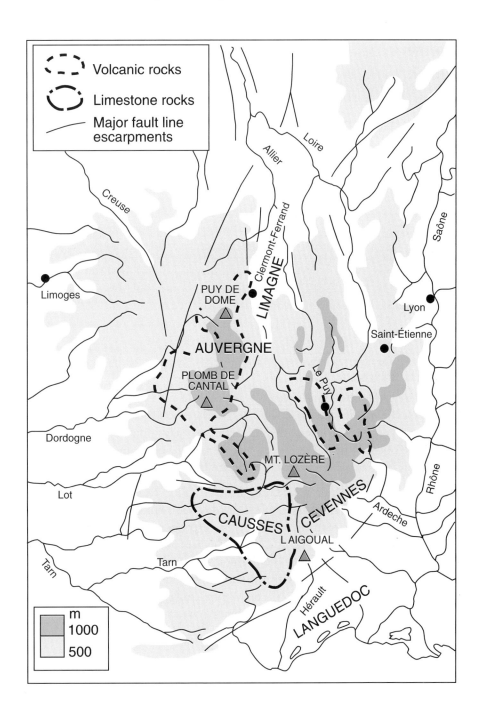

Fig 21.5

Striving For a Balance – Problem Regions

If maladjusted and dynamic regions experience problems associated with past or present development, the Central Massif owes its problems to the lack of development. A study of this region shows that many of its problems are related to its physical environment.

Relief

The region is an elevated plateau with an average height of about 1,000 metres. It acts as a watershed for a number of important rivers many of which flow westwards reflecting the general decrease in altitude in that direction. The topography is further complicated by faulting and fluvial erosion which has created spectacular gorges especially in the limestone area.

This topography poses obvious difficulties for the development of commercial farming and an efficient transport network.

Climate

Elevation and distance from the sea have combined to account for the climatic extremes which are often associated with the region. Snow can lie on the ground for up to six months in some areas and the number of frost-free days is often less than 100.

Precipitation too is heavy at 1,200 millimetres and usually falls as snow in winter. These climatic extremes compound the problems for agriculture.

Markets

The unattractive physical environment results in a poor domestic market where demand is low due to out-migration and low income levels.

Though France has a relatively compact shape the region suffers from the 'friction of distance', not in the absolute sense but by virtue of an underdeveloped transport network.

The Economy

The development of successful commercial farming is hampered by an unattractive physical environment. In spite of rural depopulation, there are still too many farmers and over 50 per cent of all farms are less than 20 hectares. The loss of young people through migration means that the farming population is old and conservative. Generally speaking, farming is mixed and underdeveloped. Wheat, barley and rye are grown in the more fertile valleys while the Charolais and Limousin cattle are fattened for beef.

Remoteness from major marketing centres, poor transport facilities and limited raw materials militated against the development of industry on a large scale. The coalfields here then assumed a disproportionate importance and supported small though significant industries such as iron and steel, chemicals and textiles. The decline in the importance of coal has significantly affected these industries and with the exception of centres such as Clermont-Ferrand, home of the giant Michelin tyre company, many of the more mountainous centres continue to experience economic decline.

In 1962, the Somival, a planning corporation was set up at Clermont-Ferrand to tackle the problems of the region. As a result, significant changes occurred in the structure and output of agriculture. Farm size increased as the number of holdings fell from over 200,000 in the 1960s to about 130,000 in the 1980s. Irrigation schemes and the introduction of cash crops (such as tobacco) helped to modernise crop and animal production.

Areas where land is more marginal have been earmarked for afforestation and over 20 per cent of the entire area is now under trees, 80 per cent of which is privately owned. Tourism has been encouraged with the development of regional and national parks and its potential for growth is enhanced by its relative proximity to the huge, affluent, urban market of Paris.

A Maladjusted or Depressed Region – Nord-Pas De Calais

The fragility of industrial development based on a finite resource, (coal) is particularly highlighted by the presence of maladjusted or depressed regions. The Nord region, possessing the vital ingredients of coal, a coastal location, the early development of canals and proximity to the iron ore of Lorraine was the centre of industrial development in the country.

By 1950, it supplied over 50 per cent of all French coal and over 40 per cent of its total energy requirements. Based on the coalfield, the region became the centre of the iron and steel, engineering and textile industries. Its dramatic rise as an industrial centre was equally matched by its spectacular decline and by the 1960s it had become a classic example of a maladjusted region similar to the Sambre-Meuse region of Belgium. Fracturing and faulting which occurred during the Alpine mountain-building movement together with the exhaustion of the more accessible seams dramatically increased the cost of mining. Cheaper coal imports together with competition from oil and gas further aggravated the problem and resulted in the wholesale closure of pits around Valenciennes, Bethune and Brusy. By 1960, up to 2,000 jobs per year were being lost and when the government decided to run down coalmining in 1960, that number jumped to 7,000 per year. The decrease in coalmining resulted in a contraction of associated industries and today the iron and steel industry is firmly concentrated in Dunkirk where the Unisor plant, which has a capacity of eight million tonnes developed using imported ores. The traditional textile industry has experienced similar problems. The 1970s witnessed a major decline due to a fall in domestic demand and an increase in imports from countries where labour costs were low. During the 1980s, major restructuring saw the introduction of high performance machinery which seriously reduced employment. By the mid-1980s over 150,000 jobs were lost though Roubaix, Lille and Tourcoing still remain important textile centres.

Urban renewal, improved communications, the development of a broader industrial base and its geographic location in the centre of a large affluent market will ensure the survival and eventual re-emergence of this region as one of the more dynamic areas of Western Europe.

A Dynamic (Core) Region – Paris Basin

Markets

The Paris Basin is centred on the large, affluent, urban market of Paris. It also enjoys easy access to the rest of the country via a series of sills or gaps such as the Gap of Poitou. France's central location in the EU means that the region has land ties with all the major EU countries. Access to affluent markets has stimulated economic growth and contributed to economic development more than any other factor.

Striving For a Balance – Problem Regions

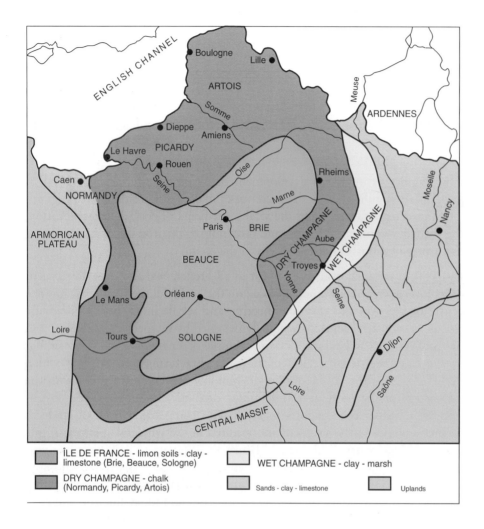

⟨ **Fig 21.6**

Relief

Relief is a major physical input. Its basin-shaped structure, resembling a series of concentric saucers owes its origin to the uplift which accompanied the Alpine mountain building movement. The huge land area is drained by the Seine with its extensive dendritic pattern. The Seine's navigable nature which makes Paris a major port has considerably helped in the economic development of that city.

Climate

The climate of the Paris Basin changes from maritime in the west to transitional in the east. The area therefore avoids climatic extremes and while perennial precipitation is particularly important to industry, high summer temperatures coupled with summer rainfall provide vigorous growing conditions which are particularly significant for arable farming.

Economy

Relief, climate, soil and accessibility to markets have combined to make this region one of the major agricultural regions on the continent. Market gardening and dairy-farming dominate the land use close to Paris reflecting the perishable nature of the produce. In the west, Beauce, and in the north, Brie, the emphasis on arable farming of wheat, barley, maize and sugar beet is reflected in the larger farm sizes and the fertile limon soil cover. Further east, sheep and vines utilize the poorer limestone soils of the Champagne region.

Proximity to the Nord coalfield, and to the port of Le Havre together with its central location and extensive rail network made Paris a major industrial centre as early as the nineteenth century. Today it is the single greatest industrial complex in Western Europe and with a population of almost nine million it has become a mecca for market orientated industries. It produces over 80 per cent of all French cars (Renault and Citroën), over 70 per cent of precision and electronic products and over half of the aircraft industry's products. The affluent market has also made the city a magnet for the jewellery and perfume industries. Paris claims to be the fashion centre of the world. Paris dominates in the tertiary sector also and has a virtual monopoly on banking and insurance. With a total employment of over four million, 60 per cent are in tertiary employment, which is not surprising as the city dominates the communication, administration, commercial and cultural life of the country, and enjoys an all-year round, thriving tourist trade.

Developed regions like the Paris Basin might more aptly be referred to as overdeveloped areas. Economic success has resulted in major physical and social problems. In the inner parts of the city which are the older areas, many houses lack basic facilities and are in decay. Overcrowding here is being relieved by out-migration as people move to the suburbs. Traffic congestion is also a major problem as over three million commute each day, one third of whom use private cars.

The long term solutions to the city's problem were suggested by the 'Schema Directeur' plan in 1965. Inner city redevelopment, ring motorways, improved rail transport and a growth corridor westwards along the Seine were all suggested to cope with future population growth estimated to reach 14 million.

Questions

Ordinary Level

1. Within the European Union, there is a major industrialised region – the so-called 'Industrial Corridor'. Its main industrialised areas include
   - The English Midlands
   - Randstad, Holland
   - The Rhine-Ruhr of Germany
   - The North Italian Plain

   (i) In the case of any **one** of these ares, describe and explain its industrial development. *(60 marks)*

   (ii) Describe some of the methods by which governments attract industry to their less advantaged regions. *(20 marks)*

   Leaving Cert. 1994

Striving For a Balance – Problem Regions

2.

Examine the map above, which shows those areas which are eligible for assistance from the European Regional Development Fund.
These areas can be grouped into **two** types:
• Rural underdeveloped regions
• Regions which have declined from former industrial prosperity.

    (i)   Name **two** specific regions belonging to **each** type.   *(20 marks)*

    (ii) Select **two** regions – one from **each** type.
         Examine in detail the cause of their present problems and indicate briefly how these problems might be reduced. *(50 marks)*

Leaving Cert. 1991

HIGHER LEVEL

1.   'Some regions of Western Europe are clearly identifiable as core regions, while other regions are more marginal.'
   Assess the validity of this statement with reference to any **two** countries of your choice.   *(100 marks)*

Leaving Cert. 1996

2.   Many European countries have developed a number of types of problem regions. Among these are core growth regions, old industrial regions in decline and underdeveloped regions.
   With reference to examples from a country of your choice, describe the major characteristics of **each** of these types of regions.
   Explain how the problems experienced have been or could be overcome.   *(100 marks)*

Leaving Cert. 1995

3.   'Peripheral regions within Europe suffer many disadvantages when compared to core regions'.
    (i)   Explain clearly what is meant by the terms 'peripheral regions' and 'core regions' within Western Europe.
         Give named examples.   *(30 marks)*

    (ii) With reference to **two** peripheral regions in Western Europe, describe the main problems faced by these regions and discuss attempts to limit the effects of the problems.   *(70 marks)*

Leaving Cert. 1993

4. Study this diagram, which illustrates the Core and Periphery Model of economic development and answer the following:

Core and periphery

PERIPHERY

Core
Little unemployment
High wages
Good communications
Many services
Intensive use of land
etc., etc.

High unemployment
Low wages
Poor communications
Few services
Extensive use of land

(i) Describe and explain how the model can be applied to Western Europe. *(70 marks)*

(ii) Examine some of the measures which can be adopted in order to reduce such sharp regional inequalities. *(30 marks)*

Leaving Cert. 1992

5. Within the European Community there are a number of highly developed 'core' regions, which have become very prosperous.

(i) With reference to agriculture **and** to manufacturing industry, describe and explain some of the reasons for the success of such core regions. *(60 marks)*

(ii) The success of these 'core' regions has itself led to problems, both within themselves and for other, less developed regions of the community. Examine briefly some of these problems. *(40 marks)*

Leaving Cert. 1991

CHAPTER 22

Ordnance Survey Maps and Aerial Photographs

‹ Fig 22.1

Map Reading

THE NATIONAL GRID

To find position or location on an Ordnance Survey Map we use the National Grid. This involves a network of 25 squares or sub-zones each of which has a letter of the alphabet.

The sides of each square are divided into 100 equal parts numbered 00 to 99. These lines are known as co-ordinates.

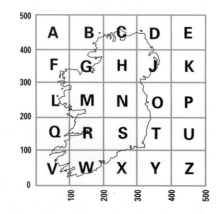

‹ Fig 22.2
The National Grid

348 *Leaving Certificate Geography*

Map Reading

> Vertical lines are called Eastings.
>
> Horizontal lines are called Northings.

When giving the location of a feature we must use the following order:

(a) Give the sub-zone letter.

(b) Give the easting coordinate.

(c) Give the northing coordinate.

Four-Figure Reference

If you want to give the location of a large area such as the square marked X (Figure 22.3) or the lake (shaded) we get the following:

Sub-zone letter   = M

Easting Co-Ordinate   = 62

Northing Co-Ordinate   = 48

Therefore, Lake X   = F6248

What would be the reference then for the lake in Square Y?

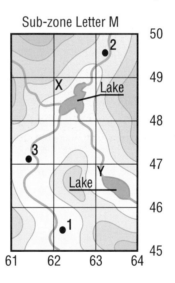

Fig 22.3

Six-Figure Reference

If we wish to get a more exact location we use a six-figure reference. Imagine that the distance between any two eastings (or northings) is divided into tenths.

Examine Figure 22.3 again, the point mark 1 has a reference code of M622455.

Exercise

What would be the six-figure reference for points 2 and 3?

Now examine the map extract on page 379.

Give a four-figure reference for Waterford Airport.

Give a six-figure reference for the Standing Stone north-east of Tramore.

Ordnance Survey Maps and Aerial Photographs

SCALES

> Scale is the relationship between a distance on a map and the corresponding distance on the ground.

Types of Scale

i) Statement ii) Linear iii) Representative Fraction (R.F.)

1. Statement

This method simply states (hence statement) the scale. The scale used in the Discovery Series in the Leaving Certificate exams is two centimetres to one kilometre.

2. Linear

The linear scale is a line, divided into kilometres which represents the actual distance.

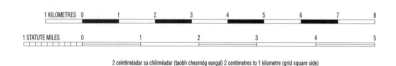

< **Fig 22.4**
Linear Scale

3. R.F. (Representative Fraction)

Because maps are used internationally it is necessary to show the scale in a way which can be understood by people who use a different system of measurement. This method states the distance on a map as a fraction of the corresponding distance on the ground. So 1:50,000 simply means that for every one unit on the map (e.g. 1 cm) there are 50,000 of those units on the ground (i.e. 50,000 cm).

So, if: 1 cm = 50,000 cm,

then   1 cm = 500 m or

         2 cm = 1 kilometre.

HOW TO MEASURE DISTANCE AND CALCULATE AREA

Distance

There are two types of distance:

Straight Line Distance ('as the crow flies') Between Two Points

To measure this distance, place the edge of a piece of paper on the two points and mark the paper.
Now place the paper's edge on the linear scale and read off the distance. You can also use a ruler and count the number of centimetres, remembering that every two centimetres equals one kilometre (2 cm = 1 km).

Map Reading

Fig 22.5
Measuring Straight Line Distance

Curved Distance

To measure a curved distance (e.g. road, river, railway) a number of methods can be employed, such as a piece of string or a dividers. One of the most accurate methods however is to use a pencil and the edge of a piece of paper. We break the curved distance into a series of straight lines, marking each straight line section on the paper's edge by turning the paper.

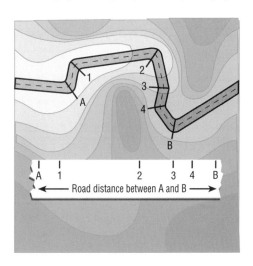

Fig 22.6
Measuring Curved Distance

The total distance involved will now be a straight line. Now place this on the linear scale and read off the actual distance.

Area

Regular Areas

To measure regular areas such as the entire area of a map extract.

1. Measure the length and convert into kilometres.
2. Measure the width and convert into kilometres.
3. Now multiply the length by the width.

N.B. Your answer will be in square kilometres.

Leaving Certificate Geography

Ordnance Survey Maps and Aerial Photographs

Irregular Areas

To measure an irregular area such as a lake:

1. Count off each full square between the grid lines – each square is one square kilometre.
2. Where a half square or more is included, count it as a full square.
3. Ignore all areas less than a half square.

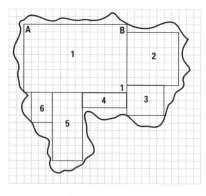

Fig 22.7
Measuring an Irregular Area

If more accuracy is required then the feature can be traced on to square paper and the boxes counted using the same method.

Exercise

Examine the map extract on page 379.

1. *Calculate the distance along the N25 from where it enters the map at S557097 to the roundabout at S598109. Use the linear scale on page 350.*
2. *Calculate the area in square kilometres of the entire map extract.*
3. *Calculate the distance 'as the crow flies' from the Post Office at S606112 to the Post Office at S579018.*

DRAWING SKETCH MAPS

N.B. A sketch map is a freehand drawing of a map and so tracing is not allowed.

- Give your sketch map a title.
- Draw the map to the same shape – though not necessarily to the same size.
- Draw in some eastings and northings (lightly) – they provide a useful guide for accurate location.
- Where relevant, sketch in the coastline.
- Draw in the required information.

If you were asked to indicate 'the road network of the area shown on the extract' you should show all primary and secondary roads.

There is no need to show third class or other roads. If time permits, it is worth shading (pencil) upland areas and drainage features (lakes) which have obvious influences on road direction.

There is no need, nor will any extra marks be allocated for the use of colours.

- Show direction – by using an arrow pointing north.
- Give the scale of the map – a statement is sufficient.

Map Reading

DIRECTION

There are two methods of stating direction.

1. Compass Points
2. Angular Bearings

Fig 22.8 (left) Compass Points

Fig 22.9 (right) Angular Bearings

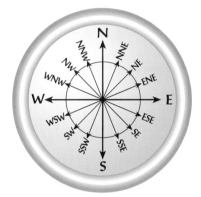
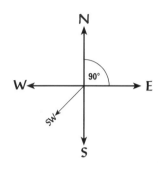

Bearings are measured in a clockwise direction from north which is 0°. A person then travelling eastwards is going at 90°.

Exercise

Examine Figure 22.9. What would the angular bearing be of a person travelling South-West?

Examine the map extract on page 379.

In what direction would a person be looking at the Lifeguard Station on Tramore Strand from the pier at S576004?

Give your answer in

(i) Compass Points
(ii) Angular bearings.

SHOWING ALTITUDES AND SLOPES

The relief or topography (shape of the land surface) is perhaps the most important physical feature on any OS map. Apart from its obvious influence on drainage (dictating the direction and flow of rivers) it also has a major influence on human geography, i.e. communication, settlement and land use.

There are a number of methods of showing height. These include:

Colour Layering (Layer Tinting)

This is a method by which colours are used to show height. Dark green shows land heights up to 100 metres. Lighter shades of green and then brown are used to indicate increases in altitude.

While it is useful to indicate upland or highland regions at a quick glance its main fault is that it gives the impression that all slopes are a series of steps.

Trigonometrical Stations

These are places marked by a triangle with a dot at the centre which indicates the exact height of that spot above sea-level. They are places where the surveyors set up their instruments, when mapping the surrounding area.

Ordnance Survey Maps and Aerial Photographs

Spot Heights

These are heights which were calculated by surveyors and are shown by black dots.

Contour Lines

These are lines on a map which join all places of the same height above sea level.

Contour lines are an excellent method of showing topographical features because they can show the height, shape and slope of an area.

The recognition of slopes then is easy. Once contour lines are drawn at regular intervals, 'the closer together contour lines are, the steeper the slope.'

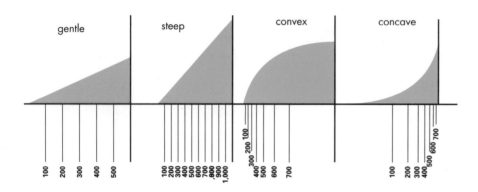

< Fig 22.10
How Contour Lines Show Slope

Using Contour Lines to Show Landforms

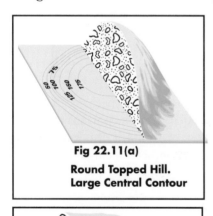

Fig 22.11(a)
Round Topped Hill.
Large Central Contour

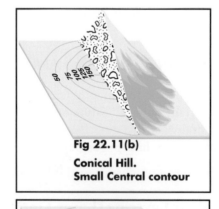

Fig 22.11(b)
Conical Hill.
Small Central contour

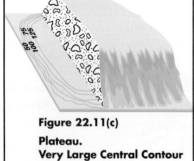

Figure 22.11(c)
Plateau.
Very Large Central Contour

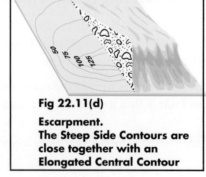

Fig 22.11(d)
Escarpment.
The Steep Side Contours are close together with an Elongated Central Contour

< Fig. 22.11

Fig 22.12
Round-Topped Ridge.
Long elongated Central Loop

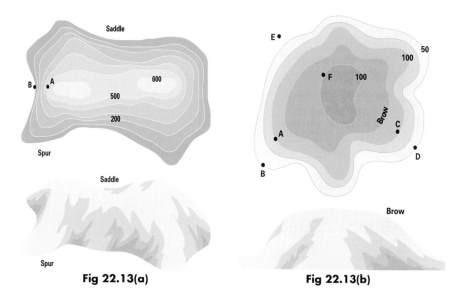

Fig 22.13
Contours to Show Landforms

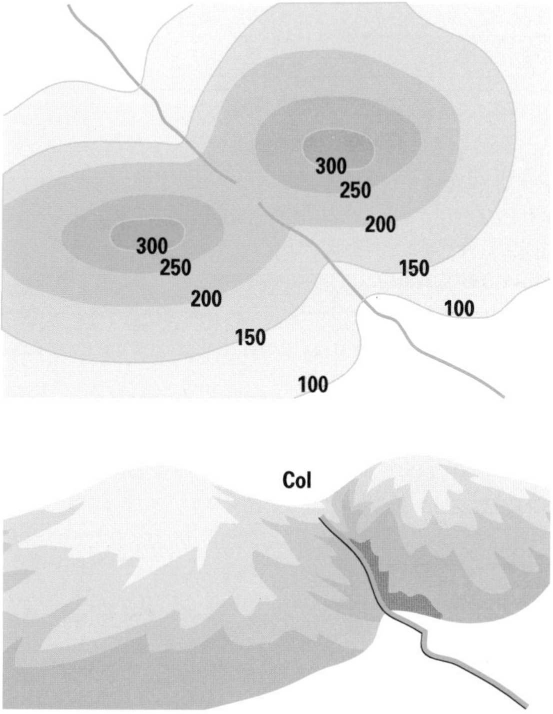

Fig 22.13(a)   Fig 22.13(b)

Fig 22.14
Contour Lines of a Col

Ordnance Survey Maps and Aerial Photographs

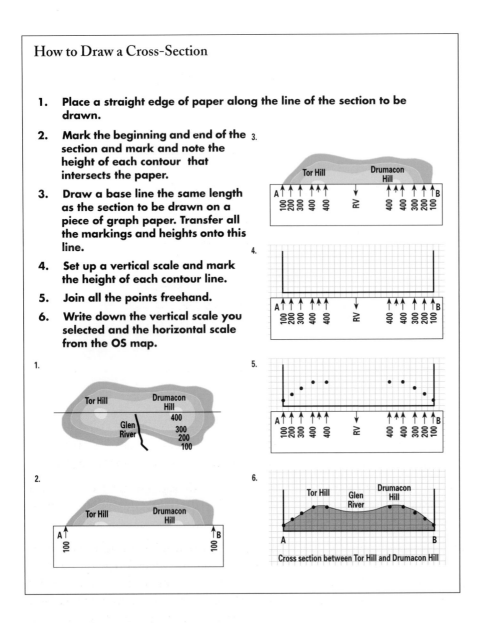

Fig 22.15
How to Draw a Cross-Section

ESTABLISHING INTERVISIBILITY

Put in its simplest form, establishing intervisibility means finding out whether two points on a map can be seen from each other. While the answer may be obvious in the case of Figure 22.16(a), a study of Figures 22.16(b) and 22.16(c) shows that the question may demand more careful examination.

While the three hills A, C and B are of comparable height in Figure 22.16(b), A and B are intervisible, but in Figure 22.16(c) they are not. If you examine, Figure 22.16(c) carefully you will notice that hill B is further from hill C, than it is in Figure 22.16(b). So in trying to establish intervisibility, both the heights and distances between the hills are important. The simplest method that should be used to establish intervisibility is by drawing a skeleton section.

This is similar to a cross-section except only the two end points are marked together with the highest point in between. These are then plotted on graph paper. Join the two end points with a straight line. If this line passes above the intervening point, then the points are intervisible.

Map Interpretation

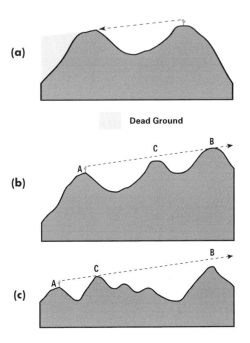

Fig 22.16
Establishing Intervisibility

Map Interpretation

When looking at an Ordnance Survey map, for examinations or for any other purpose, it is useful to use the factors above to analyse the map systematically.

All the information provided on an OS Map can be divided into two categories: physical geography which includes relief and drainage and human geography which includes communication, settlement and land use.

The extent to which we can move from map reading to successful map interpretation of landforms is dependent on our knowledge of physical geography. Relief has a major influence on drainage and is itself influenced by drainage. It is the slope of the land which determines the speed and direction of river flow. It also helps to explain patterns of drainage and the presence or

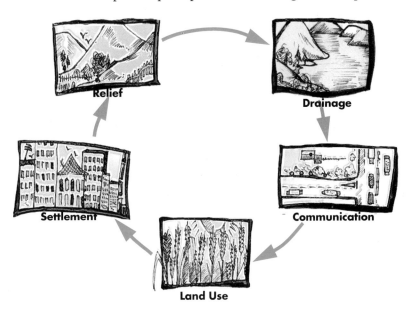

Fig 22.17

Leaving Certificate Geography   **357**

absence of lakes. The same rivers however also help to shape the relief over which they flow by the processes of erosion and deposition.

In short, there is an interplay between relief and drainage. Relief and drainage together have a major influence on human geography. People are drawn to low-lying, well-drained relief.

Once these areas become accessible *(communication)*, people settle down *(settlement)* and then use the land, e.g. farming, i.e. *land use*.

So all aspects of Ordnance Survey Maps are inter-related, and any meaningful explanation of one aspect can only be answered in the context of the other four elements.

This 'system' provides a useful framework for answering questions on Ordnance Survey maps.

Sample Question 1 – Relief

Examine the influence of relief on the physical and human geography of the region shown on the map extract.

Because the question is on *relief*, we examine relief in relation to the four other factors.

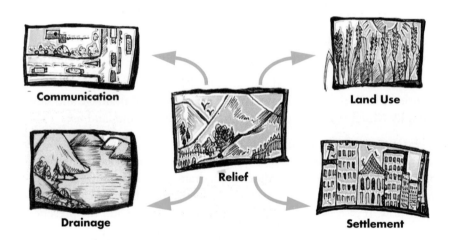

Fig 22.18

The five factors are inter-related, relief influences drainage – being the physical aspect, communication, settlement and land use being the human elements.

Sample Question 2 – Communication

Examine with reference to evidence from the map, the ways in which the development of the road network can be seen to be related to other elements of the landscape.

Because roads = communication, we approach the question by examining roads in relation to the other four elements.

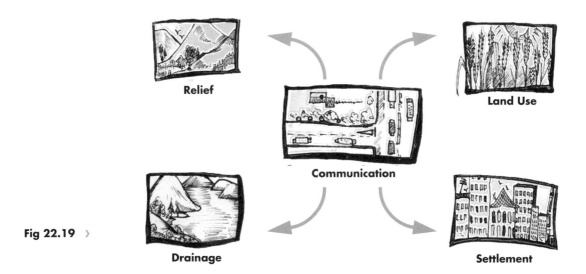

Fig 22.19

Communication (roads) are influenced by the relief, drainage, settlement and land use of the region.

Sample Question 3 – Settlement

Why has the town of New Ross developed at this location?

New Ross = Settlement, so

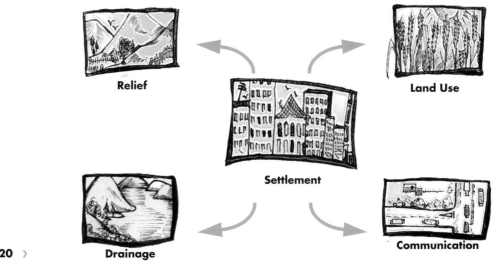

Fig 22.20

The factors which influence the development of New Ross then can be analysed under the headings: Relief; Drainage; Communication and Land Use.

We will return to this method of map analysis later, when we have a better understanding of each individual element of both physical and human geography.

Ordnance Survey Maps and Aerial Photographs

1. Relief

When we are describing the physical features on a map at Leaving Certificate level, we should make use of the technical vocabulary learned in physical geography. The more common features to be seen on Ordnance Survey maps are listed below with definitions adopted from a Glossary of Geographical Terms *(D. Stamp)*.

Col (See Figure 22.14): A depression in a range of mountains or hills.

Escarpment (See Figure 22.11(d)) : A landform with sides sloping in opposite directions consisting of a steep scarp slope and a more gentle dip or back slope. The highest point is called a *crest*.

Floodplain: A plain bordering a river formed of deposits of sediments carried down by a river.

Hill: An elevated landform generally under 600 metres.

Mountain: An area of land over 600 metres.

River basin: The area drained by a river and its tributaries.

Saddle (See Figure 22.13(a)): Similar to a depression but generally larger and higher.

Watershed: The line or area separating rivers flowing in different directions.

Valley: A long depression between stretches of high ground. It may be a river valley with V-shaped contours or glaciated with U-shaped contours as in Figure 22.21.

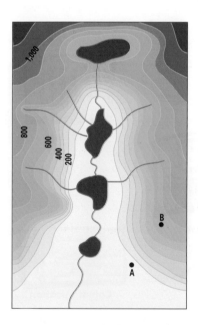 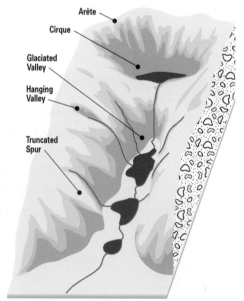

< **Fig 22.21**

Questions based on the Relief of an Ordnance Survey Map:

Sample Question 1

With the aid of a sketch map describe the main physiographic (relief) units on the map extract.

Drawing a Sketch Map
1. Always draw the frame the same shape as that of the map.
2. A sketch map is a freehand drawing – so never trace.
3. Mark and label only what is required.
4. Be accurate.

Any Ordnance Survey map may have some or all of the following physiographic units.

(i) Coastal area.
(ii) Lowland – land area up to 200 metres.
(iii) Upland – land area between 200 – 600 metres.
(iv) Highland – land area over 600 metres.

N.B. When identifying units on a sketch map, always use the same number for a particular unit. So if the map has a coastline – let that be marked 1: Lowland = 2; Upland = 3; Highland = 4. If there are two separate units of lowland then label them as 2(a) and 2(b) – this gives a particular number a certain identity.

Describing Physiographic Units

1. How to Describe a Coast

Give as much of the following information as possible.

(i) Where is it on the map?
(ii) In what direction(s) does it run?
(iii) Is it an upland or lowland coast – or both?
(iv) Comment on the processes which are active – and the features of erosion and/or deposition present, e.g. cliff, beach etc.

2. How to Describe a Lowland Region

Give as much of the following information as possible:

(i) Where is it positioned on the map? (North or South etc.).
(ii) How much of the map does it cover? – i.e. its approximate area, e.g. half or one-third of the map area etc.
(iii) What is its approximate height?
(iv) What agents have shaped the region? – river or glacial deposition or both. Refer to features using grid references.

3. How to describe Upland and Highland Regions

(i) Where is it on the map?
(ii) How much of the map does it cover i.e. its approximate area?
(iii) What is the general height of the region?
(iv) What agents of denudation have shaped the region?
Refer to rivers – and features of river erosion, and glaciation (if appropriate) – and features of glacial erosion. Be sure to refer to specific examples – use grid references.

Ordnance Survey Maps and Aerial Photographs

Sample Question 2

Examine some of the geomorphic agents which have shaped the region shown on the map extract.

OR

Examine some of the agents of denudation which have shaped the region shown on the map extract.

The essential difference here between these two questions is in the meaning of geomorphic and denudation. (See Chapter 1).

> Geomorphic refers to both internal and external forces.
> Denudation refers to external agents only.

Generally speaking, it is sufficient to refer to any three agents of landscape change. For each you should name the actual agents, the processes involved and the resultant landforms referring to actual examples by grid reference.

The actual detail however has to be decided by the marks allocated for that particular question.

Sample Question 3

An example of a more difficult question on relief is as follows:

The region shown on the map extract displays an obvious structural trend. Identify this trend and explain three ways in which it has influenced the physical and human development of the landscape.

'Structural Trend' refers to the direction in which the mountains are running, e.g. east to west or north east to south west. The influence of this trend can be examined by using the framework, discussed earlier.

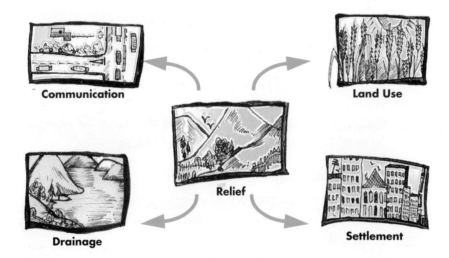

◁ **Fig 22.22**

Since the structural trend can be interpreted as the direction of the relief (mountains) we can use its influence on *Drainage* – as being a physical influence and any two of the three from *Communication, Settlement* and *Land Use* as being human influences.

The actual layout of our answer then will be:

1. Drainage – { Statement / Development
2. Communication – { Statement / Development
3. Settlement – { Statement / Development

This question can be attempted when the other physical and human elements have been analysed.

2. DRAINAGE

Drainage forms the second of the two elements of physical geography which are shown on Ordnance Survey Maps. Drainage refers to all water features, e.g. rivers, lakes and the sea – shown in blue on the OS maps. It is important to remember that, while the length of rivers are drawn to scale, the width of rivers are not. To denote width, a single blue line represents a stream, but a river is shown by a double blue line with blue shading between the lines.

At this stage it is essential to be familiar with the importance of rivers as an agent of landscape change (see Chapter 4, Rivers and River Basins).

> **N.B.**
>
> The Department of Education recently issued a number of sample questions based on the 1:50,000 OS (Discovery Series) which included the following sample question:
>
> 'The Basin of the River Dargle shows distinct upper and lower course characteristics'.
>
> Examine this statement using a sketch map to illustrate your answer.
>
> So a knowledge of river processes and features is essential.

How to Describe a River (Basin)

To describe a river basin you should give as much of the following information as possible:

(i) Name the river.
(ii) Identify its source.
(iii) Identify where it enters the sea or lake (its mouth).
(iv) State its direction of flow.
(v) Say how many tributaries (many or few) and whether they flow into the left or right bank (or both).
(vi) Identify the pattern of drainage.
(vii) State whether the river is eroding or depositing or both.
(viii) Comment on the processes involved, identifying some of the features formed as a result (use grid references).

Ordnance Survey Maps and Aerial Photographs

	YOUTH	MATURITY	OLD AGE
Plan of rivers with contours			
Long profile			
Cross section			
	Fast-flowing, waterfalls, rapids, interlocking spurs. Used for H.E.P. and resevoirs.	River cliffs, valley widening, water meadows, more tributaries, river terraces.	Gentle gradient, meanders and ox-bow lakes, wide flood plain, embankments.

‹ **Fig 22.23**

The Origin of Lakes on a Map

Lakes in upland regions often owe their origin to glacial erosion, e.g. cirques, ribbon lakes, etc. Lakes in lowland regions can be more difficult to explain.

They may again be due to glaciation, e.g. kettlehole lakes or solution lakes. Lakes near coasts are often the result of marine deposition, e.g. lagoons.

Drainage and Human Geography

Because human geography is particularly associated with lowland regions it is important to be able to recognise poorly drained regions (marshes) on a map.

Signs of poor drainage in lowland regions include:

- An absence of settlement.
- The presence of forests/woodlands (lowland is normally used for farming).
- The absence of roads.
- Roads avoiding the floodplains of rivers and keeping to the edge (bluffs).

Patterns of Drainage

How to name, describe and explain patterns of drainage:

Name: You should name the pattern you have identified from the map extract, giving a four-figure grid reference.

Description: You must give as much information from the map as possible, i.e. name of rivers, direction of flow, number of tributaries, etc.

Explanation: You must now explain why the rivers and tributaries flow to form the particular pattern which you have identified.

Explaining River Patterns

While the type of rock in any region obviously helps to explain the actual rate of erosion in that area, the essential reason why streams and rivers flow in a particular direction is that they simply follow the slope of the land in that area.

An explanation of a pattern of drainage then is obtained by describing the shape and slope of the underlying relief.

Radial Pattern

This develops in an upland area where the relief is circular or oval-shaped. As a result the surface water (rivers or streams) will flow off in different directions.

Dendritic Pattern

Apart from being associated with a homogenous bedrock, the essential reason for the development of this pattern is the slope of the underlying topography. We normally associate this pattern then with the upper-middle course of river valleys. As a result the tributaries which join the main river have to negotiate two slopes, i.e. the slope of the main river and that of the valley sides. The route taken then is a compromise between both and so the tributaries approach the main river at acute angles.

Trellised Pattern

While this pattern is often associated with varying bedrock, it is generally well developed in the middle-lower course of the river. The tributaries here flow into the main river which essentially has little or no slope of its own. As a result they approach the river at right angles.

Deranged Pattern

Deranged drainage is one which lacks any definite direction of flow. Rivers often tend to criss-cross each other exhibiting a chaotic appearance. It develops in a lowland area (lacking any particular direction of slope) where deposition of glacial material is often widespread.

Sample Question

Name, describe and explain three patterns of drainage.

Sample Answer

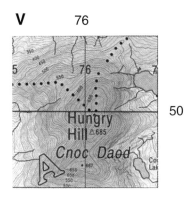

Fig 22.24

Ordnance Survey Maps and Aerial Photographs

Radial Pattern

Name: Radial – V7650

Description: From the hydrographic centre (watershed) at Hungry Hill at an altitude of 685 metres a number of streams and rivers flow in different directions, e.g. north, east, south-west, and north-west.

Explanation: The pattern is radial because the upland topography at Hungry Hill is essentially circular in shape. As a result surface water (rivers and streams) flows off in different directions.

Dendritic Pattern

Name: Dendritic – V8153

Description: The Clashduff river flows in a southerly direction and is joined by many tributaries from both the left and right banks which approach the river at acute angles.

Explanation: Apart from reflecting a homogenous bedrock the pattern is dendritic because the Clashduff river is in its upper-middle course. As a result the tributaries which flow down the valley sides have two slopes to negotiate, i.e. the slope of the main river and that of the valley sides. As a result they approach the river at acute angles.

‹ **Fig 22.25**

Trellised Pattern

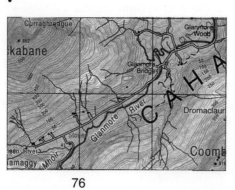

‹ **Fig 22.26**

Name: Trellised – V7652

Description: The Glanmore river flows in a north-east direction and is joined by a number of tributaries, both from the left and right banks which approach the river at right angles.

Explanation: The pattern is trellised because apart from reflecting a varied bedrock the Glanmore River is in its middle-lower course evident by the absence of gradient and the presence of meanders. As a result the tributaries have only one slope to contend with, i.e. the valley sides and so approach the river at right angles.

Exercise

Examine the map of Clifden on page 383.

Describe and explain two patterns which you can identify in the drainage system. Draw a sketch map to support your answer.

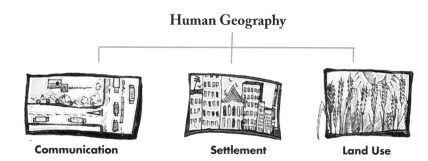

Fig 22.27
Human Geographical Factors on a Map

3. Communication

Ordnance Survey maps show the evolution of transport over time. There are many types of communication shown on Ordnance Survey Maps, e.g. rivers, canals, railways, roads, etc.

Rivers

Rivers were used from earliest times as a method of transport from the coast into the interior. The extent to which a river on an OS map still remains an important form of communication cannot always be discovered from a map. The presence of a port on a river does not in itself prove that a particular river was or is used as a method of transport into the interior. We should look for other features – along the river such as locks, piers for unloading or a high density of buildings near the river side.

Canals

Canals were developed in the eighteenth century to carry the bulk goods of the Industrial Revolution. They were costly to construct and movement was very slow. The land needed to be quite level and a river had to be close by to supply water.

Railways

By the 1830s railways were replacing canals as the main form of transport for bulk goods. They were easier and cheaper to construct than canals and were less tied by relief and drainage. Apart from their ability to carry bulk, their main advantage was speed.

Roads

The development of motorised transport with its flexibility and speed over short distances (see Chapter 19) has seen roads replace railways as the dominant form of transport in the twentieth century.

Ordnance Survey Maps and Aerial Photographs

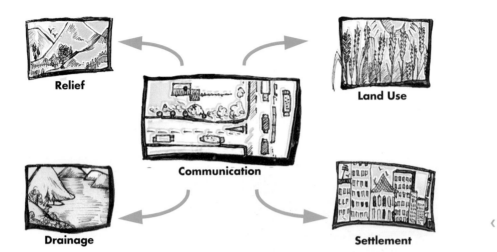

< Fig 22.28

We therefore have four influences on communication, i.e. Relief; Drainage; Settlement and Land Use. It is sufficient to comment on any three.

Before examining the various physical and human influences on communication in general and roads in particular, it is important to remember that:

- The word 'influence' has both positive and negative connotations.
- Main roads will always follow straight line routeways where possible, as it reduces distance and therefore reduces the cost of construction and travel time.
- Routes avoid high altitudes where possible (over 200 metres) – due to adverse weather conditions (i.e. fog, ice, etc.).
- Routes avoid steep slopes where possible: 1:25 for roads, 1:40 for railways. Where steep slopes have to be crossed, roads often ascend at a very oblique angle or in zig zags.

Influences on Roads

Relief

The relief or topography then can have both positive and negative influences on the road network.

Positive – Lowland areas: You should select a main road which travels in a relatively straight line. This suggests that the relief in terms of both altitude and gradient is positive.

Negative – Upland areas: You should then select a road which is formed to change direction. You should now identify what topographical features are avoided by these detours in order to maintain the most even gradient and keep to a low altitude.

Map Interpretation

Drainage

Fig 22.29

As with relief, drainage can have both positive and negative effects on roads.

Positive: Rivers carve valleys which often provide useful corridors in upland regions through which roads can pass.

Rivers in their lower courses have peneplained the landscape and so provide low level routeways with gentle gradients.

Negative: Rivers have to be bridged which is usually expensive.

Floodplains usually have to be avoided to ensure a firm site and freedom from flooding.

Lakes and marshy ground have to be avoided.

Railways are even more sensitive than roads to the physical geography of a region.

A loaded train cannot negotiate a gradient of less than 1:40. As a result, railways will often make long detours in the form of large loops to avoid steep slopes. Cuttings, embankments, bridges and tunnels are common constructions on railway lines for similar reasons.

> Remember that relief and drainage may help (positive) or hinder (negative) communication but they always influence it.

It is not enough then to describe in what direction a route travels – why it takes a particular direction must be explained.

Any description of routes, unrelated to the physical geography is meaningless.

Settlement

While the physical geography of a region may determine the actual direction taken by various routes, we must always remember that communication serves settlement. It follows then that the greater the concentration of settlement the greater the number and importance (type) of routeways. As a result three patterns can be identified:

Absence of Settlement: Areas which are devoid of settlement (due to either a negative relief or drainage – or both) will have few routes. Those that do exist will generally be linking other regions on the map and will negotiate the most favourable routeways (in terms of both relief and drainage) in the area.

Dispersed Settlement: Dispersed settlement often indicates farming (see land use) so that access to roads is essential. However minor routes are sufficient to serve farm needs provided they are in relatively close proximity to major routes.

Nucleated Settlement: The importance of a settlement can often be gauged by the number and type of routes which service it. You should comment on the possible evolution of transport over time, i.e. how the different forms of transport which serves the town have changed (evolved), i.e. river, canal, rail and road.

Sea Ports

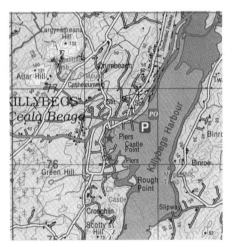

◁ **Fig 22.30**
Killybegs, Co. Donegal

All ports, whether commercial ports or fishing ports need basic, physical advantages if they are to develop. Deep, sheltered harbours (estuaries) are important as a protection against storms. Silt-free passages are important to allow vessels easy access. Level ground is also important for the development of facilities such as warehouses, processing plants, etc. Good road and rail routes are important if the port requires access to the surrounding hinterland.

Airports

The physical geography of a region is particularly important in the siting of airports. Low altitude will help to reduce the risk of adverse weather (ice, snow, fog, etc.) while a gentle gradient is important for runways. Airports need accessibility and the importance of an airport can, in part, be gauged by the number and type of roads which serve it. The undesirable noise and danger of modern aircraft means that airports are usually located some distance from the cities which they serve (see S6304 on the Waterford Map on page 379).

4. SETTLEMENT

In trying to explain the presence or absence of settlement it is essential to remember that as a species human beings are essentially lowland creatures. Consequently, we are particularly sensitive to the physical geography of a region, i.e. avoiding upland-mountainous topography, preferring lowland areas which are well drained.

To comprehensively explain the presence or absence of settlement in any region, it is useful to refer to both the physical and human geography of the map.

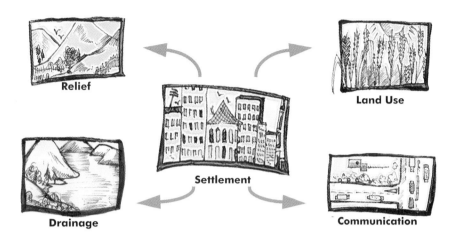

Fig 22.31

The factors which influence the presence or absence of settlement are both physical (relief and drainage) and human (communication and land use).

As a general rule, the regions which have the highest density of people are those which are more physically attractive.

At this point it is convenient to divide settlement into two categories:

(i) Rural Settlement (ii) Urban Settlement.

Rural Settlement: Physical Influences

Relief

Two aspects of relief are important when explaining the absence or presence of settlement.

Altitude: Settlement is rarely found above 200 metres due to a deterioration in climatic conditions. Above this altitude temperature is often too low, precipitation too high and winds too strong, especially in winter, to make habitation attractive.

Gradient: Settlement avoids steep slopes. Apart altogether from accessibility, steep slopes are difficult to build on and costly to develop. They also offer poor potential for development (farming, etc.).

Drainage

Settlement then is attracted to lowlying topography with gently sloping or level ground. It is essential however that this land is well drained, so the floodplains of rivers, while offering other obvious attractions have to be respected.

Rural Settlement: Human Influences

Communication

The physical geography of a region provides the natural passages on the landscape (e.g. cols, saddles, valleys, coastal plains). People exploit these routeways with rail and road links to make a region more accessible.

Ordnance Survey Maps and Aerial Photographs

Land Use

The degree to which the land can be used and its resources exploited has a major influence on the density of population in any region.

Low-lying regions with gentle gradients which are well-drained often support some type of agricultural land use, evident by the presence of dispersed settlement indicated by scattered black dots.

Patterns of Rural Settlement

A study of patterns of rural settlement is useful at this stage as a method of describing and explaining the distribution of rural settlement on a map.

Absence of A Pattern

This applies to a region which has a complete absence of settlement as is indicated by the absence of black dots.

Explanation: Settlement is absent here because the relief is negative.

Elevated areas like this (over 200 metres) are colder and wetter than lowland regions.

The slopes also are too steep, evident by the closeness of the contours. Altitude and gradient combine to make soils thin and unproductive and hence unattractive for agriculture. Apart from its negative relief, the region is also inaccessible as indicated by the absence of routeways.

Upland-mountainous relief is not the only reason for the absence of settlement. Low-lying regions which are marshy or prone to flooding are also avoided. River floodplains in the lower courses are prone to flooding in the wet season (winter) and during times of heavy rainfall.

Dispersed Settlement

Description: This applies to an area where settlement, indicated by black dots not clustered and houses are separated from each other by large tracts of land.

This pattern emerged with the decline of the feudal system. Farms were enclosed and farmers built their own dwelling house.

‹ Fig 22.32

Explanation: The pattern in Figure 22.32 is dispersed – as each dot probably represents a farm house. Farming is suggested because of the attractive relief – low-lying with a gentle gradient. Also the region is well drained and is easily accessible.

N.B. The actual density of this dispersed pattern may vary from one area to the next. While a high density of farms may indicate a high fertility of soil and intensive farming, it may also reflect the social history in which farming evolved, e.g. the small farms in the west of Ireland as opposed to the rather larger farms in the east of the country.

Sample Question:

Name, describe and explain two patterns of rural settlement.

Fig 22.33

Linear Pattern

Description: N.B. When describing a linear pattern of settlement you should give as much information from the map as possible.

- Name the type of road, e.g. regional or third class, etc.
- Is settlement on one side or both sides?
- Are the houses close together or far apart?

Settlement in Figure 22.33 is linear, along the road N16. It is located on both sides of the road and the houses are relatively close together.

Explanation: The pattern is linear because the houses enjoy access to the road and public transport. The houses also avail of services such as public lighting, water and sewage facilities. Some dots may represent commercial outlets such as shops and filling stations and so benefit from the passing traffic.

Urban Settlement

Towns and cities are generally referred to as urban settlement. When discussing any urban settlement it is important to distinguish between its site, situation and function.

Site

This is the actual building site, i.e. the actual land on which the settlement stands. It normally determines the shape or form of the original settlement, e.g. a market centre would tend to produce a nucleated shape.

Situation

This refers to the position of the town in relation to the surrounding area. The surrounding landscape might offer natural routeways so that the town might develop as a route centre.

Function

This refers to the services provided by the town for its people and those in the surrounding area.

It is interesting to examine some of the following reasons which are often used to explain the growth of a particular settlement.

Dry Point Site

This refers to settlements which were founded around wells and springs where a pure and regular water supply was available. They are sometimes referred to as spring-line settlements.

Defensive Settlement

This refers to sites which possessed the natural advantages for defence.

Confluence Towns

Where two or more rivers meet and the valley routes converge, a confluence town develops.

Market Towns

These are centrally situated in the areas they serve. This is their main advantage and they become centres of trade. Over time roads and railways converge on them, emphasising their nodal position.

N.B: While a particular town may owe the initial settlement to one particular advantage, e.g. dry point site or route centre, it is never in itself sufficient to explain the development of a town, i.e. its evolution from the initial settlement to the town we see on the map today.

Such an explanation needs reference to a broader set of factors.

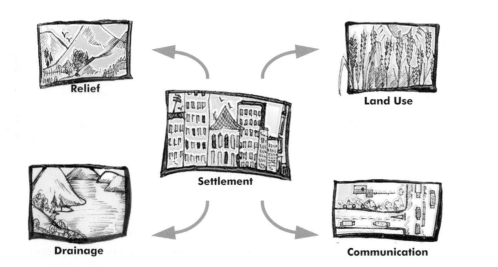

< Fig 22.34

Reasons for the Development of a Town

In answering examination questions it is generally sufficient to refer to any three of the above four factors.

Relief – Lowland:

Site: You should refer here to the altitude and the gradient of the site and the advantages which this offered to attract settlement.

Situation: You should refer to the surrounding land area in relation to its altitude and gradient and the advantages which this offers.

Drainage – Water

Site: You should comment on the availability and necessity of a ready water supply and how it contributed to the growth of the settlement, i.e. dry point site or wet point site.

Situation: You should mention the drainage of the surrounding area and how this contributed to the development of the town, e.g. agriculture (well-drained lowland) hence market town. Comment on any defensive attractions which the site possessed, e.g. island nature, etc.

Communication

Site: Draw a sketch map here of the actual site and the various methods of transport which made the site accessible, e.g. river, canal, rail, road.

Situation: You should comment on the surrounding landscape and explain how natural routeways might have contributed to the nodality of the town.

An alternative method of analysis could be:

(1) Site   (2) Situation   (3) Function

1. Site Statement

You should refer to the relief, drainage and accessibility of the site. It may have been a level site or been at the convergence of routeways. A dry point site with access to water. How the site helps to explain the shape (form) of the initial settlement.

2. Situation Statement

Again you should refer to the relief, drainage and communication of the surrounding area and how these factors contribute to the growth of the settlement.

3. Function Statement

This refers to the services which the town provides, e.g. tourism, markets, transport, commercial etc. The detail available in the new OS maps (Discovery Series) allows us to see many functions which were not shown in previous series.

How To Describe and Explain the Distribution of Settlement

Carefully study map evidence of settlement noting the major physiographic (relief) units on the map.

Now select three regions.

    (i) Where there is a notable absence of settlement.

    (ii) Where settlement is scattered or dispersed in isolated farms or small hamlets, i.e. rural settlement.

    (iii) Where there is a concentration of settlement, i.e. large villages and towns.

Now explain the absence or presence (and density) of settlement in each region.

5. LAND USE

Land use simply refers to the various ways the land on an Ordnance Survey map is (actual), or could be (potential) used by people.

1. Farming

Farming is suggested by the presence of black dots in a rural setting. Unlike aerial photographs where we can often see evidence of farming, on an OS map its presence can only be inferred by an examination of the physical and human geography of the region.

Physical Geography

Upland areas tend to be colder, wetter and more exposed to strong winds, especially in winter. Low lying areas with gentle slopes which are well drained tend to have more fertile soils and so are more attractive to agricultural land use. It is also worth remembering that arable farming is more sensitive than pastoral farming to the physical environment (see Chapter 13).

Human Geography

Successful commercial farming is equally dependant on human inputs. Accessibility to main roads is vital today in view of machinery and bulk milk carriers which need access to farms.

Proximity to urban settlement still remains important. Traditionally, the local town acted as a market centre. Today with modern farming techniques, farmers are not only suppliers but are also major consumers. The local town then may supply much of the farm needs, such as seeds, fertilisers, fencing, mechanical parts and a veterinary service.

2. Forestry – Woodland

The presence of forests and woodlands is shown by the use of symbols as shown in the legend. You should become familiar with different symbols which are used to show coniferous, deciduous and mixed woodlands.

The limitations on agricultural development imposed by the CAP (see Chapter 13) together with the economic benefits associated with forestry are resulting in a rapid programme of afforestation and reforestation throughout the country (see Chapter 15).

Distribution of Forests and Woodlands – Upland Area

Forestry is normally associated with marginal land, i.e. areas which due to either topographical or climatic extremes are beyond the economic, if not the absolute, limits of agricultural production. The following factors influence the distribution of forestry in upland areas.

Climate: As altitude increases, temperatures decrease while precipitation increases. Coniferous trees such as Norway Spruce and Scots Pine are planted here as they can endure the harsh climatic conditions (see Chapter 9). Aspect is important with the trees preferring the south-facing slopes with their higher temperatures and protection from cold northerly winds.

Soil Type: High altitudes, steep slopes combined with heavy precipitation means that leaching is common (see Chapter 10). Coniferous trees can adapt to these soil deficiencies and their roots also help to prevent erosion.

Forests in Lowland Areas

The presence of forests on lowland regions generally reflects poorly drained areas where soils have obvious deficiencies. These areas also lie beyond the commercial limits of agricultural production and so are planted with coniferous trees.

Absence of Woodland and Forests

Upland Areas: Topographical and climatic extremes combine to account for the absence of forests in these areas often lying above 600 metres. Trees need temperatures of 6°C and over to grow. These areas are often too cold and windy. A steep gradient also usually results in the removal of regolith down slope, thus preventing the development of soils.

Lowland Areas: Most lowlands which are well drained are devoted to agriculture. Farming, either arable (crops) or pastoral (animals) gives a greater return per hectare than forestry which is essentially a long-term investment. Forests in lowland areas are usually confined to areas which are poorly drained and subject to flooding.

Sample Question

With reference to both upland and lowland regions explain the distribution of woods and forests in this region.

Sample Answer

Select three regions on the map. Two regions with woodlands – forest cover and one region without. Now explain the presence and absence of forest cover in the regions selected.

3. Tourism

Sample Question

The area shown on the map extract has many attractions for tourists. Describe the main tourist attractions with reference to specific locations on the map extract.

Attractions

(i) Relief   (ii) Drainage   (iii) Settlement

Sample Answer

Relief: You should refer to upland topography – for mountain climbing, hill walking, etc.

Lowland topography might be used for pony-trekking, cycling, etc. Beaches would have a particular attraction for families with young children.

If the upland areas have woods and forest cover you might suggest nature trails etc.

Drainage: Rivers, lakes and the sea offer obvious attractions for angling, canoeing, cruising and water sports such as swimming and wind surfing, etc. Also bird sanctuaries, scenic beauty, etc.

Settlement: Towns on a map attract tourists for shopping and souvenir hunting. Also tourists enjoy looking around towns for places of historical interest. Towns also provide services, e.g. entertainment (bars, cinemas, etc.) and golf courses. Outside the towns, antiquities are clearly marked on the map and are themselves major tourist attractions.

N.B. Always support your answer with reference to the map, i.e. give grid references.

Ordnance Survey Maps and Aerial Photographs

Waterford

ORDINARY LEVEL

1. (a) On a sketch map of the region mark and name the following:
   - The N25
   - An area of woodland
   - A railway
   - The River Suir
   - A beach
   - Post office

   (b) There is evidence that both rural and urban settlement have taken place in the areas shown on the map over a long period of time. Discuss this statement, referring to any three types of settlement. Use map evidence to support your answer.

   (c) Identify, using map evidence only, any four services which are available in the built up area of Waterford.

2. Name, describe and explain any two patterns of rural settlement.

3. Describe the main attractions which Tramore possesses as a holiday centre.

4. Using map evidence explain three reasons why the town of Waterford developed at this location.

5. Using detailed map evidence to support your answer explain **two** of the reasons for the location of an industrial estate on the west of the city S5710.

6. Using detailed map evidence explain the advantages which Little Island (S6411) had for the development of a golf course.

HIGHER LEVEL

1. 'The map shows evidence of the development of human settlement and of land use over a long period'. Discuss this statement with reference to the map.

2. 'The Waterford region is well served by various means of transport'. With the aid of a sketch map examine this statement referring to map evidence.

3. Explain three reasons why Waterford developed at this location.

4. This area has many attractions for tourists. Describe the main tourist attractions with reference to specific locations on the map extract.

5. Describe and explain some of the variations in patterns of settlement which can be seen in this region.

6. Using grid references, locate three areas of different land use. For each area, name and explain two advantages which that area has for that particular land use.

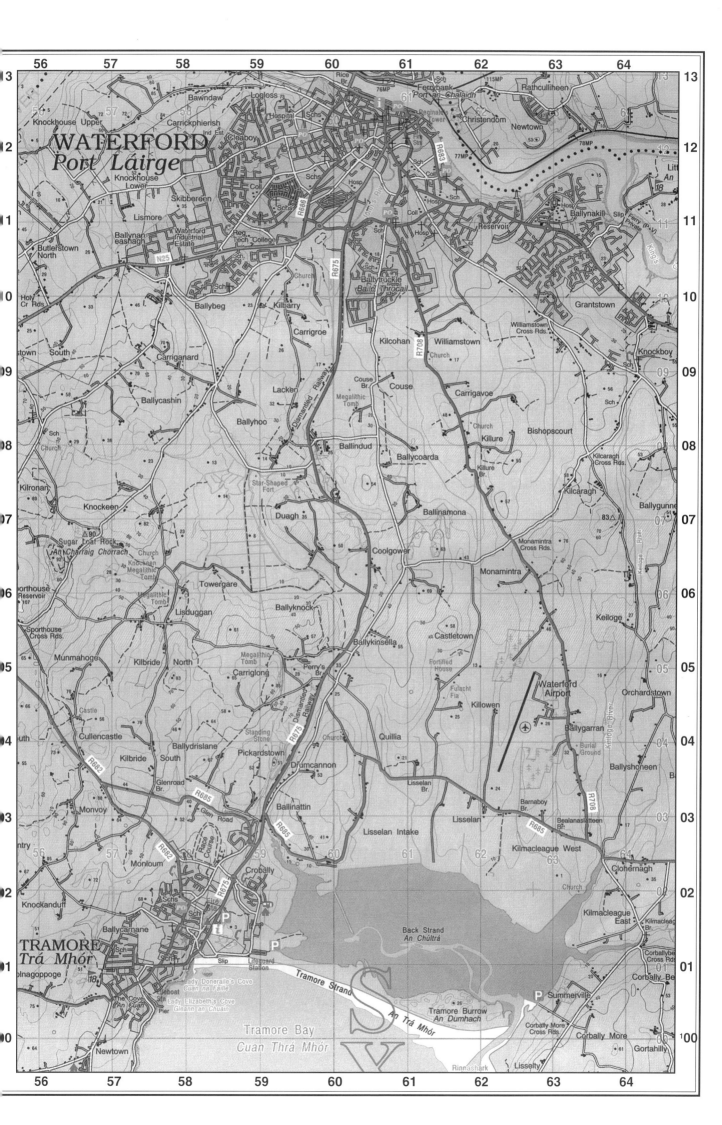

Ordnance Survey Maps and Aerial Photographs

Cork

Ordinary Level

1. On a sketch map of the region mark and name the following:
   - The N28
   - A golf course
   - A railway
   - An area of woodland
   - Haulbowline Island
   - The Owenboy River

2. Name and explain, using map evidence, why the town of Carrigaline developed at this location.

3. Name, describe and explain two patterns of rural settlement.

4. Using detailed map evidence explain two reasons why the island of Haulbowline was selected as a naval base.

5. Examine any two advantages which the site of W7862 has for the location of a pharmaceutical plant.

Higher Level

1. Explain three reasons why the town of Carrigaline developed at this location.

2. Name, describe and explain two patterns of rural settlement on the map extract.

3. 'Relief and drainage have had a considerable influence on the settlement patterns in this region'.
   Comment on the validity of the above statement.

4. Examine the suitability of Ringaskiddy (W7764) for the development of an industrial estate.

5. (i) Draw a sketch map (not a tracing) to illustrate the road network of the area shown on the extract.
   (ii) Explain with reference to the map, three ways in which the development of the road network can be seen to be related to other elements on the landscape.

6. (i) Identify three areas of different land use on the map extract.
   (ii) Explain two reasons for the suitability of each area for that particular land use.

7. There is a major pharmaceutical plant located in Curraghbinny (W7862). Comment on the suitability of the site for the location of that industry.

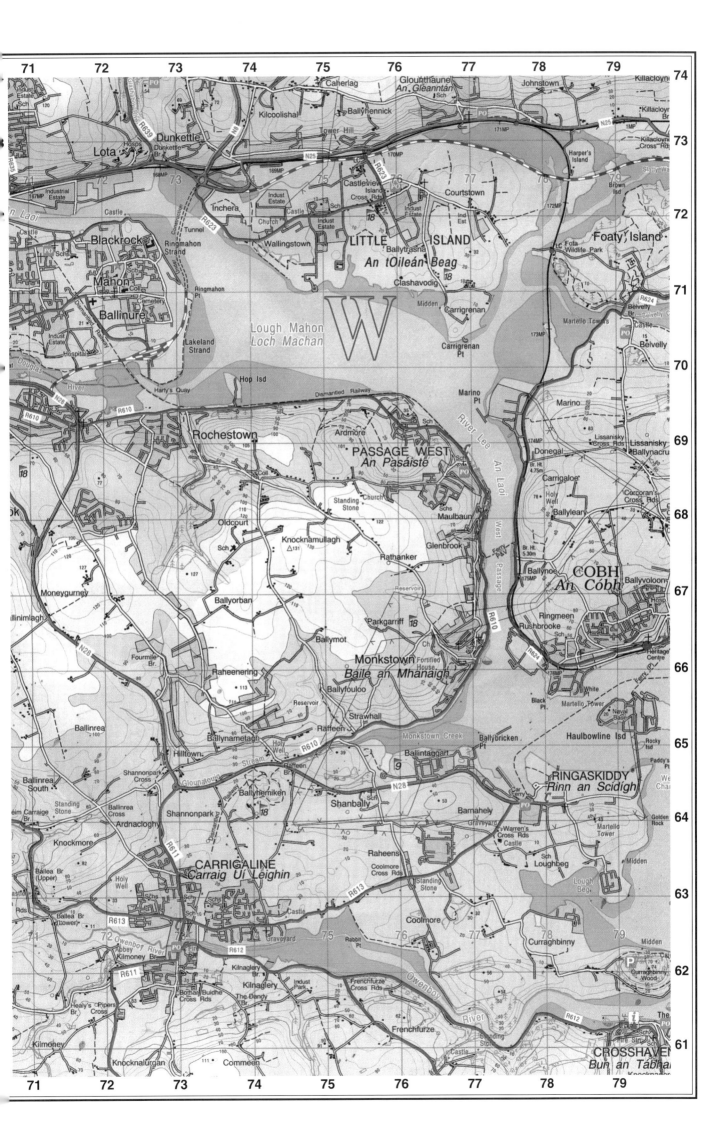

Ordnance Survey Maps and Aerial Photographs

Clifden

ORDINARY LEVEL

1. On a sketch map of the region mark and name:
   - The N59
   - A camping site
   - A major river
   - An area of ancient settlement
   - An area of woodland/forest

2. Name, describe and explain two patterns of drainage on the map extract.

3. Using map evidence, explain three reasons why the town of Clifden developed at this location.

4. The region possesses many attractions for tourists.
   Describe the main tourist attractions with reference to specific locations on the map.

5. Describe and explain two patterns of rural settlement on this map extract.

6. Suggest two reasons for the presence of the forested areas on the map.

HIGHER LEVEL

1. Name, describe and explain two patterns of drainage on the map extract.

2. Using map evidence explain three reasons why the town of Clifden developed at this location.

3. Describe and explain the distribution of forests in this region. Use a sketch map (not a tracing) to illustrate your answer.

4. Suppose you were given the task of choosing the best location for a new holiday complex somewhere in the region.
   Calculate a six-figure reference for the location you would select. Using map evidence explain two reasons for your choice.

5. Name, describe and explain two patterns of rural settlement on this map extract.

6. 'Rivers have played a major role in shaping the region shown on the map extract'.
   Examine the above statement with reference to specific locations on the map extract.

Leaving Certificate Geography

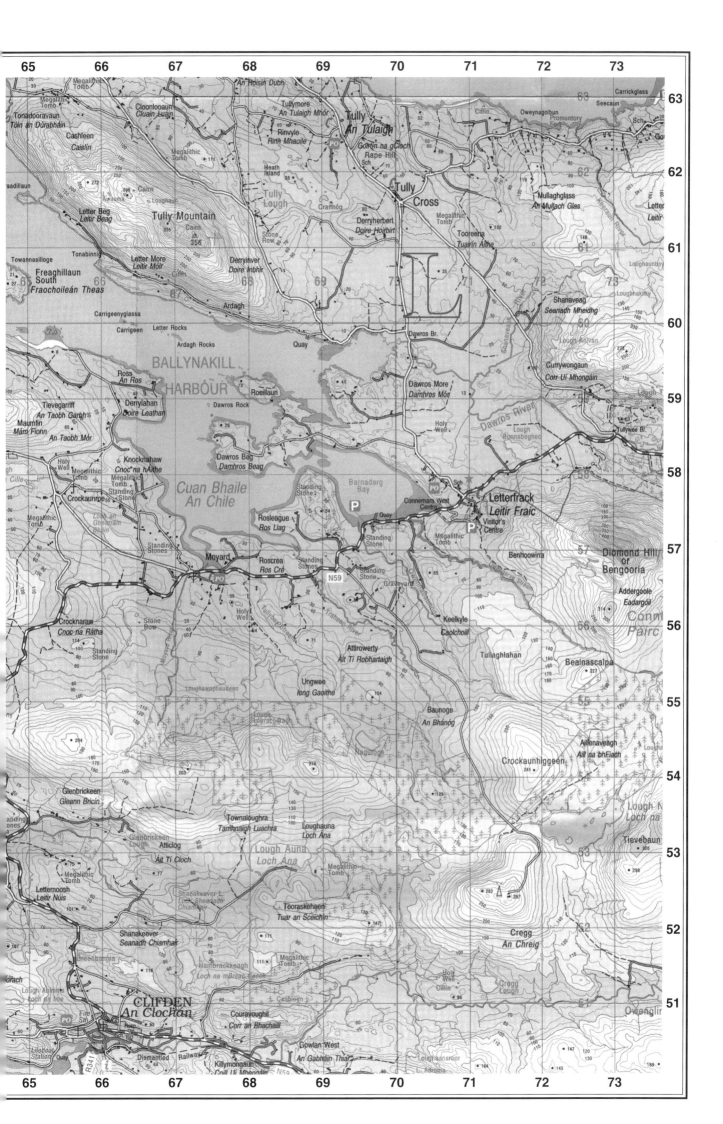

Aerial Photos

TYPES OF AERIAL PHOTOGRAPHS

Vertical Photographs

A vertical aerial photograph is taken with the camera pointing vertically downwards. Scale is therefore uniform. Vertical aerial photographs are used as a basis for modern OS maps.

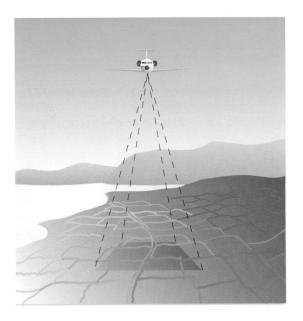

‹ Fig 22.35

Oblique Photographs

Oblique aerial photographs are taken with the camera pointing at an angle to the ground. A high oblique photograph, as in Figure 22.36, shows part of the horizon. A low oblique photograph does not take in any part of the horizon.

N.B. The terms 'high' and 'low' have nothing to do with the height of the aeroplane.

‹ Fig 22.36

LOCATING AREAS ON A PHOTOGRAPH

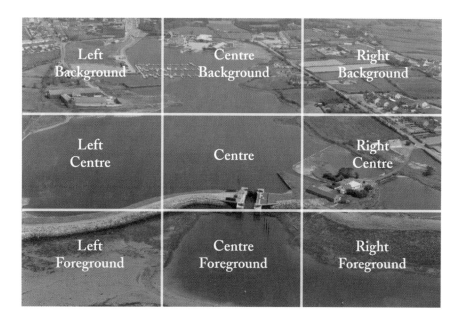

Fig 22.37

DIRECTION ON AERIAL PHOTOGRAPHS

If the direction is not stated on the photograph it is possible to calculate the direction in which the camera was pointing if the photograph was taken from the area shown on the Ordnance Survey map.

Steps

1. Draw a line on the photograph from the foreground to the background.
2. Identify two or three prominent objects on or near this line.
3. Locate these objects on the map and join them together.
4. Now read off the direction on this line which is similar to the direction of the camera.

TIME OF YEAR

It is often possible to work out the time of year when a photograph was taken. For example:

Spring
1. Ploughed fields.
2. Early signs of foliage on trees.

Summer
1. Deciduous trees in full foliage.
2. Crops growing in fields. Animals grazing.
3. Short shadows, indicating high altitude of the sun.
4. Crowded beaches.
5. Very clear visibility to the background of photograph.

Autumn

1. Various shades of colour on foliage.
2. Crops cut in fields, often in stacks or bales.

Winter

1. Deciduous trees without leaves.
2. Fields without crops or animals.

SCALE IN PHOTOGRAPHS

The scale is uniform on vertical photographs. On oblique photographs objects nearer the cameras (in the foreground) appear larger than objects of similar size in the background. It is possible to measure the actual distance if the OS map covers the same area by noting the distance between features, and identifying the same features on the map.

The history of a settlement can be studied through its buildings. From an aerial photograph, you should be able to do the following:

   (i) Name the historical period.
   (ii) Identify the building which suggested that period.
   (iii) Give its exact location on the photograph.
   (iv) Link or associate the building to the growth of the town.

HISTORICAL PERIODS

1. Middle Ages

You should look for features such as castles, round towers, walled towns, etc. These features would suggest military defence – essential in the Middle Ages.

2. Eighteenth and Nineteenth Centuries

Evidence of this period includes canals and railways. Canals appeared in the late eighteenth century and early nineteenth century. Railways were built from the 1830s onwards. They were constructed in response to the Industrial Revolution to carry bulk goods over long distances.

Neo-Gothic style churches too could suggest the nineteenth century as many of these were built after 1829 when Catholic Emancipation was granted. Many appeared after the 1850s when prosperity slowly returned after the Great Famine of 1845.

3. Twentieth Century

Evidence of the twentieth century includes features such as housing estates, regular road patterns, houses with garages, shopping complexes with car parking facilities.

Development in the twentieth century is marked by two important features:

(i) Necessity for Planning

Planning permission for development is essential in the twentieth century. This is evident in the photograph by the regular shapes of housing estates and roads.

ii) Motorised Transport

This is a twentieth century development and is evident in photographs by the provisions made for motor transport (garages, car parks, etc.).

DESCRIBING THE FUNCTIONS OF A SETTLEMENT

To answer this question you should:

1. Name the function (see Chapter 12) e.g. religious.
2. Prove it by naming the building which provides the function e.g. church.
3. Locate the building on the photograph, e.g. left background.
4. Comment on the suitability of the location of that building to provide that function, e.g.
   Is it centrally located?
   Is it easily accessible?
   You should try and identify both the advantages and disadvantages of its location.

Uses of Aerial Photographs

PRIMARY ACTIVITIES

Aerial photographs can be used to identify various land uses.

Agriculture: Various types of farming, e.g. arable (crops) or pastoral (animals).

Forestry: Both the types and extent of forest cover can be identified.

SECONDARY ACTIVITY

It is possible to identify the amount of land which is assigned to factories. Aerial photographs are also useful to identify potential areas for industrial development. Industries themselves such as the E.S.B. often make use of aerial photographs to estimate the volume of coal remaining at any given time.

TERTIARY ACTIVITY

Aerial photographs are particularly useful in the service sector. They appear on T.V. screens daily as meteorologists use them in weather forecasting. They are used by botanists to measure forest destruction by acid rain. They are also used by planning authorities to identify the most desirable routeways for road construction.

Ordnance Survey Maps and Aerial Photographs

Dunmore East

ORDINARY LEVEL

1. Draw a sketch map of the area shown and mark and name the road network and three zones of different land use.
2. Using evidence from the photograph in support of your answer suggest two likely functions of this settlement.
3. Describe any four major land uses which are evident in the aerial photograph.
4. Name and explain any three advantages that this area possesses for the development of a fishing port.
5. Examine briefly how human activities have changed the marine processes operating in this area.

HIGHER LEVEL

1. On a sketch map (not a tracing) illustrate the street pattern and three areas of different land use.
2. With reference to one land form in each case, explain how the processes of erosion and deposition operate in coastal areas such as this.
3. Explain why and how human activities have helped to change the natural physical processes operating in this area.
4. Examine some of the advantages which the town of Dunmore East possesses for the development of a fishing port.

Kinsale Map and Photograph

ORDINARY LEVEL

N.B. In part (ii) you are also required to use the Ordnance Survey map.

1. (i) Draw a sketch map of the aerial photograph of Kin
   and on it show and name:

   - The River
   - The Streets
   - A Clump of Trees
   - A Residential Are

Ordnance Survey Maps and Aerial Photographs

    (ii) Using evidence both from the aerial photograph and the ordnance survey map, describe any three reasons why you think the town of Kinsale grew up at this location.

2. Using evidence from the Aerial Photograph and the Ordnance Survey map suggest and explain any three likely present day functions of this settlement.
3. 'There is evidence from the photograph that this settlement has a long history'.
Refer to two areas in the photograph to support the above statement.

HIGHER LEVEL

1. Suggest and explain three likely present day functions of the settlement. You may refer to both the map and photograph to support your answer.
2. Suggest and explain how the history of this settlement might be studied through its buildings. Refer to both the map and photograph.
3. Draw a sketch map (not a tracing) and on it mark and name the street patterns and four zones of different land use.
4. Using evidence from the photograph identify two ways in which the street pattern is not suited to the present day functions of this settlement.

CHAPTER 23 — Areas of Recent Political Conflict

Fig 23.1 >

Conflict is an unfortunate element of the human condition. From Northern Ireland to east Asia, groups of people are engaged in varying degrees of violence, from relatively low-level terrorist campaigns to full-scale civil war. A knowledge and understanding of these conflicts is an important part of the study of geography. Because such situations are constantly changing, this chapter only attempts to provide brief notes on selected trouble spots. Students should use these notes as a starting point for research or project work and try to keep up to date with events. The amount of space given over to each conflict in no way reflects on its seriousness or complexity.

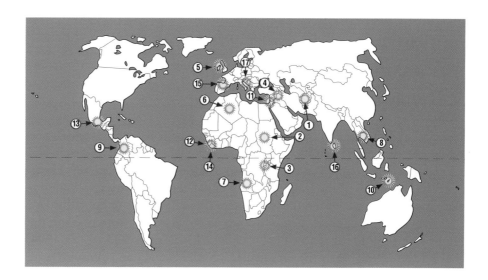

Fig 23.2 >

Map showing areas of conflict. The numbers correspond to the notes in this chapter

Leaving Certificate Geography

391

Areas of Recent Political Conflict

1. Afghanistan

Afghanistan remained neutral in the Cold War, between the USA and the Soviet Union, which followed World War II. In 1973, army leaders overthrew the government and made the country a republic.

In 1978, the government was taken over by a socialist group who turned to the Soviet Union for aid. Many people rebelled against the pro-Communist government and in late 1979, Soviet troops invaded the country. During the 1980s, Soviet troops fought unsuccessfully to put down the rebel Muslim forces, who were called the Mujaheddin ('holy warriors'). At the height of the war there were 120,000 Soviet troops in the country. By 1988, the Soviet army eventually accepted defeat and on 15 February 1989 the last Soviet troops left the country. Finally in 1992, the Mujaheddin forces entered the capital Kabul and set up an Islamic government. But the conflict between rival groups within the Mujaheddin continues.

2. Sudan

In 1952, Sudan became an independent state (from Britain and France). After Independence the black Africans in the South, who were either Christian or Communist, feared domination by the Muslims in the North. They strongly objected when the Muslims declared that Arabic was the only official language. In 1964, civil war broke out and continued until 1972 when the South was given regional self-government, though executive power remained vested in the military government in Khartoum.

In 1983, the government established Islamic law throughout the country. This led to further riots when the Sudan People's Liberation Army (SPLA) in the South launched attacks on government installations. Despite attempts to restore order, the fighting continued into the 1990s. In March 1991, the North reimposed the Shariah (the sacred law of Islam) as the basis of the legal system. Although the Southern States were to be excluded from this, the Southern Sudanese felt that acceptance of this division would reduce them to the status of second-class citizenship. The SPLA want a unified state with regional autonomy. They also want Sudan to distance itself from radical Arab States which are promoting Islamic fundamentalism.

3. Rwanda – Burundi – Zaire

The causes of the conflict in the Great Lakes region – Rwanda, Burundi and Zaire – are extremely complex: the legacy of colonialism; land ownership; and a long standing dominance of the Tutsis ('cattle owners') over the Hutus ('farmers'). With the same language and religion and intermarriage commonplace the difference between the Hutus and Tutsis has more to do with social class than ethnic groupings. Concerns over land ownership, farm size and the Hutus lack of political power finally exploded into mass killings. In 1994, civil war broke out when the presidents of both Rwanda and Burundi were killed when their aircraft was hit by a rocket as it approached Kigali. After the massacres of 1994 (up to 800,000 were slaughtered) and the Tutsi victory which followed, over one million Rwandans fled to refugee camps around Goma in Eastern Zaire. In 1995, Zaire decided to expel the refugees back to Rwanda. By November 1996, the refugees in Zaire (Congo) were streaming back into Rwanda. The situation however remained volatile and there remains a high risk of further conflict.

4. The Kurds

The Kurds are predominantly a mountainous people living on high plateaus and in valleys among peaks; the most famous of which is Mt Ararat, supposedly the landing place of Noah's ark. The region which they inhabit is Kurdistan an ill-defined territory lying between Turkey, CIS, Iran, Iraq and Syria. With 20-30 million people they are the world's largest ethnic group without a state. The region, often described as southern Kurdistan, lies in Iraq covering 17 per cent of that country's land area. In 1970, the Kurds were granted regional autonomy and a share in government. The rise of the Ba'ath Party however and in particular Saddam Hussein led to a programme of Arabisation. Towards the end of the Iraq-Iran war 1980-88, Saddam led a systematic attack to annihilate the Kurds in Northern Iraq – to establish an unpopulated safety zone along the frontiers with Turkey and Iran. Kurdist activists claim that 186,000 people were killed and 4,000 villages destroyed. During this *Anfal* campaign hundreds of thousands of people were rounded up. Many were deported to southern Iraq and simply disappeared. Chemical weapons which have the advantage of killing people but not damaging buildings were widely used. After the Gulf War the notion of 'safe haven' was born and Allied Forces imposed a 'no-fly zone' on Iraqi planes above the 36th parallel. Thus a sort of mini-state has been created. At present US war planes patrol the region from their base in Turkey. This US-Turkish agreement has to be revised every six months. Without it, Saddam's army stands ready to re-conquer the region.

The Turkish government, however, also wages a bitter war against Kurdish rebels in the south-east of the country and there have been thousands of human rights abuses documented.

Other Areas of Conflict

5. Northern Ireland

The division of the island of Ireland into two states, the Republic of Ireland and Northern Ireland, in 1922 led to the continuation of one of the world's oldest conflicts. Attempts are ongoing to find a solution to the problem on our own doorstep. The difficulty is to find an accommodation that satisfies the divergent aspirations of two communities: a unionist, Protestant majority, who wish to remain in the United Kingdom; and a sizeable nationalist, Catholic minority, who wish to join the Republic of Ireland.

Areas of Recent Political Conflict

6. Algeria

The fundamentalist Islamic Salvation Front won a general election in 1991. The refusal of the army to accept the result led to a military takeover. There has been a serious civil war ever since.

7. Angola

A war of independence that began in the 1960s turned into civil war after the Portuguese colonial power left in 1975. An uneasy, UN-backed peace deal provides hope.

8. Cambodia

A civil war raged from 1975–1993, when a peace deal was signed. The Khmer Rouge are, however, still active.

9. Columbia

There is ongoing conflict between government forces and the private armies of the drug barons.

10. East Timor

Portugal left East Timor in 1975 and it was annexed by Indonesia. Two hundred thousand people were massacred by the occupying Indonesian forces. There is ongoing resistance to the Indonesian occupation.

11. Israel/Palestine

There has been conflict between Israel and the Palestinians since the foundation of Israel in 1948. Since the granting of limited autonomy to the Palestinians, the area has been relatively peaceful, but the situation remains volatile. The conflict has occasionally spilt over into the Lebanon, Jordan, Syria and Egypt.

12. Liberia

A coup in 1989 led to civil war. The fighting escalated with factions forming their own armies.

13. Mexico

Separatist 'Zapatista' rebels in the Chiapas Province have been in conflict with the Mexican government for some years.

14. Sierra Leone

Civil war has been going on for some years with foreign mercenaries fighting on the government side.

15. Spain

The Basque separatist terrorist group ETA are engaged in a violent campaign against the Spanish government.

16. Sri Lanka

Civil war has raged since 1983 between the Hindu Tamils and Buddhist Sinhalese.

17. The Former Yugoslavia

As Yugoslavia began to disintegrate along with the other communist countries and at the end of the 1980s, old rivalries began to emerge. Yugoslavia was made up of several nations including, Serbs, Croats and Bosnians. Territorial disputes and nationalist aspirations led to a bloody civil war until a tentative peace was signed at Dayton, Ohio. The situation remains volatile.

Index

A

Abrasion 59, 93
Aeolian Deposition 95
Aeolian Erosion 94
Aerial Photographs 348
Afforestation 138
Afghanistan 392
Agriculture 181
Agriculture, Importance of 182
Agriculture, Origins of 183
Agriculture, Physical Environment 184
Agriculture, The Future 200
Air Mass 101, 102
Air Pollution 257
Algeria 394
Altitudes 353
Angola 394
Anticline 12
Anticyclones 104
Aquaculture 208
Aquifer 30
Artesian Wells 32
Asthenosphere 3
Atmosphere 1
Attrition 42, 77, 93

B

Bajadas 97
Barchan Dunes 96
Barometric Pressure 107
Batholiths 10
Bays 79
Block Mountains 13
Bolson 97
Burren 37, 72
Burundi 392

C

Calcutta 321
Calderas 9
Cambodia 394
Carbonation 23, 25
Caves 35
Chemical Action 41
Cirque 60

Climate 112, 114, 121, 126
Climatic Zones 25
Clints 33
Clouds 106
Coal 17, 234
Coastal Erosion 76, 77
Coastal Landforms 75
Coastal Management 88
Coastlines 77, 84, 85, 87
Cold Climate 126
Cold Front 103
Colonialism 312
Columbia 394
Column 35
Commercial Extensive Agriculture 200
Commercial Intensive Agriculture 198
Compressed Air 77
Cones, Composite 9
Continental Drift 3, 4, 71
Contour Lines 354
Core 2
Corrasion 41, 77
Crust 3
Cryoturbation 69

D

Dam Building 55
Deflation Hollows 94
Deltas 49
Deltas, Arcuate 49
Deltas, Bird's Foot 50
Deltas, Estuarine 50
Denudation 22
Denudation 2
Depression 103
Desert Landform 91
Desert Landscapes 92, 117
Desertification 99
Developing Economies 308, 318
Developing World, Problems in Cities 322
Discharge 55
Drainage, Dendritic, 53
Drainage, Deranged 53
Drainage, Patterns of 52
Drainage, Radial 53
Drainage, Trellised 53
Drumlins 65
Dry Valleys 34
Dunes 82
Dykes 10

E

Earth flows 27
Earthquakes 2, 5, 6, 7, 11
East Timor 394
End Moraine 65
Equatorial Climate 25, 114, 115
Equatorial Forests 115
Erratic 63
Erosion 36, 40, 75, 76, 87, 93, 94, 137
Esker 67
Estuary 49
Eustatic Movement 84
Evaporation 91
Extensive Subsistence Agriculture 197

F

Fast Mass Movement 27
Faulting 2, 5, 11, 13, 20
Fiord, *also* fjord 68, 86
Fish Farming 209
Fishing 203
Fishing, Areas 210
Fishing, Changes in 204
Fishing, Employment 203
Fishing, Importance of 203
Fishing, Methods 204
Fishing, SeinePurses 205
Flood Plain 47
Fluvio-Glacial Landforms 66
Fold mountains 5, 12
Folding 2, 5, 11, 12, 13, 20
Foreset Bed 49
Forest, Management and Conservation 225
Forest, Types of 220
Forestry 217
Forestry, Woodland 376
Forests, Distribution of 218
Frost 24

G

Geomorphology 2
Glacial Deposition 63, 64
Glacial Erosion 60
Glacial Spillways 67
Glaciation 46, 58, 72
Glaciation, Causes of 71
Glaciers Erosion 59, 60
Glaciers, Formation of 59
Glaciers, Movement of 59

Glaciers, Types of 59, 66
Graben, *see* Rift Valley
Granite 15
Grikes 33
Groundwater 30
Groynes 87

H

Hanging Valley 62
H.E.P. 55
Headlands 79
Headward Erosion 54
Horticulture 199
Horn 60
Horst 13
Hot Climates 114
Hot Deserts 25
Hum 36
Human Activity 72, 87
Human Societies 54
Hydration 23, 24
Hydraulic Action 41, 76
Hydrolysis 23, 24
Hydrosphere 2

I

Ice Age 58, 70, 71, 86, 146
Igneous Rocks 14, 15, 16
Inselberg 98
Intercultivation 196
Interlocking Spurs 45
Irish Landscape 99
Irrigation 190
Israel 394
Isostatic Movement 85

K

Kame 67
Karst Regions 32
Kettle hole 67
Kurds 392

L

Laccoliths 10
Landforms 58
Landslides 27

Index

Lateral Moraine 64
Laurentian Climate 125
Lava 8
Lava Dome 9
Levees 47, 55
Liberia 394
Limb 12
Limestone 16
Limestone Regions 29, 32
Limestone, Pavement 33
Lithosphere 2, 3, 4
Longitudinal Dunes 96
Longshore Drifting 80

M

Magma 8
Mantle 2
Manufacturing, Definition of 248
Manufacturing, Industry 247
Manufacturing, Transport in 252
Map Reading 348
Map Interpretation 357
Marble 18
Marine Deposition 75, 80
Marine Erosion 75, 76
Marine Polution 207
Mass Movement 22, 26, 28
Mass Wasting 26
Meanders 46
Meanders, Incised 52
Measuring Distance 350
Medial Moraine 64
Mediterranean Climate 121
Melting Ice 68
Mercali Scale 6
Metamorphic rock 14, 17
Mexico 394
Migration 149
Migration, Demographic Effects 151
Migration, International 149
Migration, Socio-Economic Effects 152
Migration, Voluntary 150
Minerals 230
Minerals, The Importance of Mineral Wealth 230
Mining, Surface 231
Mining, Underground 232
Monsoon Climate 118
Moraines, Lateral 64
Moraines, Medial 65
Moraines, Recessional 65
Moraines, Terminal 65

Mountains, Block 14
Mountains, Fold 11
Mud flows 27
Multicropping 196
Multinational Companies 327

N

Nomadic Herding 197
Non-Government Organisations 327
Northern Ireland 393
Nuclear Energy 243

O

Ordnance Survey Maps 348
Outwash Plain 66
Over-population 317
Ox Bow Lakes 48
Oxidation 23, 24

P

Pangaea 3
Pater Noster lakes 61
Patterned Groun 69
peneplanation 2
Periglaciation 68
Permafrost 68
Physiographic Units 361
Pingos 69
Plate Movements 5
Plate Tectonics 3, 4, 5, 20
Plateaux 46
Playa 98
Plucking 59
Poljes 34
Population 145,
Population and Food Supply 159
Population, Change 146
Population, Density 153
Population, Distribution of 153
Population, Future of 157
Population, Growth of 158
Population, Optimum 158
Population, Over-Population 158
Population, Problems 158
Population, Pyramids 156
Population, Structure 155
Population, Under-Population 159
Pot Holes 46

Leaving Certificate Geography

Index

Power 250
Precipitation 91
Problem Regions 332
Problem Regions, An Overview 340
Problem Regions, Congested Regions 336
Problem Regions, Origin of the Problems 340
Problem Regions, Solutions to 335
Problem Regions, Types of 333
Problem Regions, Urban Problems 338
Problem Regions, Urbanisation 338

R

Radio Carbon Dating 70
Rain Forest 117
Reafforestation 219
Recessional Moraine 65
Rejuvenation 51
Rice Farming 196
Richter Scale 6
Rift Valley 13
River Basin 39, 41
River Patterns 365
River Rejuvenation 51
River Capture 53
River Valleys 43, 44, 46, 54
Rivers 39
Rivers, Erosion 40, 44
Rivers, Origin 39
Rivers, Transportation 42
Rivers, Valley 43
Roches Moutonées 63
Rock Dating 18
Rock Pedestal 94
Rock, Types of 14
Rwanda 392

S

Saltation 42, 93
Sand Bars 83
Sand Dunes 82
Scree 24
Sea Arch 78
Sea Stack 78
Sea Walls 88
Sea-floor spreading 4
Sedimentary rock 14, 15
Seine Purses 205
Seismology 6
Settlement 369
SIAL 3

Sierra Leone 394
Sills 10
SIMA 3
Slopes 353
Slow Mass Movement 27
Soil 130
Soil Acidity 133
Soil Creep 26
Soil Humus 133
Soil Moisture 133
Soil, Azonal 137
Soil, Characteristics of 132
Soil, Classification of 133
Soil Erosion and Conservation 137
Soil, Intrazonal 136
Soil Profile 132
Soil, Structure 133
Soil, Zonal 134
Solar Cells 240
Solar Energy 238
Solifluction 26, 27
Solution 42
Spain 394
Spot Heights 354
Spring 31
Sri Lanka 395
Stalactites 35
Stalagmites 35
Structure 3
Structure of Earth 3
Subduction 5
Subsoil 132
Submergent Coast 85
Sudan 392
Surface Creep 94
Suspension 42, 93
Swallow Holes 34
Syncline 12
Synoptic Charts 101, 105

T

Taiga 127
Tarn 60
Tectonics 75
Temperate Climate 121
Temperature 107
Terminal Moraine 65
Third World, Capital 324
Third World, Causes of 317
Third World, Cities of 177, 320
Third World, Ireland's Aid to 325
Third World, Markets 327

Leaving Certificate Geography

Index

Third World, Poverty 309
Tombolo 84
Tourism 266, 377
Tourism in Ireland 284
Tourism, Effects of 274
Tourism, Future of 276
Tourism, Government Support of 268
Tourism, Growth of 267
Tourism, Reasons For Growth 267
Tourism, Reasons for Growth 277
Tourism, Study of the Mediterranean area 277
Tourist Attractions 269
Traction 42
Trade 289, 303
Trade, Free Trade and Trade Protection 305
Trade, Governments and International 304
Trade, Major Trade Alliances 304
Transhumance 189
Transport 289
Transport, Air 299
Transport, Costs 292
Transport, Factors Influencing the Development of 290
Transport, Ocean 301
Transport, Pipelines 298
Transport, Rail 296
Transport, Road 293
Transport, Water 301
Tropical Climate 116
Tropical Forest 220
Truncated Spurs 61
Tsunami 7
Tundra Climate 128

U

Urban Areas 166
Urban Areas, Manufacturing Industry in 174
Urban Areas, Residential 175
Urban Functions, Central Place Theory 169
Urban Functions, Hierarchy 171
Urban Problems 176
Urbanisation 164
Uvulas 34

V

Valley, U-shaped 61
Valley, V-Shaped 44
Valleys, Hanging 62
Varves 70

Vegetation 112, 115, 128
Volcanic Landforms 8
Volcano, Geysers 11
Volcanoes 2, 5, 7, 8, 9, 11, 20
Vulcanicity 7

W

Wadis 97
Warm Front 103
Water in Deserts 97
Water Pollution 257
Water, Cycle 29
Water, Table 30
Water, Underground 29
Waterfalls 45
Wave-Cut Platforms 77
Weather 101
Weather Charts 101
Weather Forecasting 107
Weathering 22
Weathering, Types of 23
Wegener 3
Wind 91, 106, 240
Wind Erosion 93
Wind Gap 54
Wind Power 242
Wind Transportation 93

X

Xerophyte 120

Y

Yardangs 95
Yugloslavia, Former 395

Z

Zaire 392
Zeugens 95

Leaving Certificate Geography